Topics in Current Genetics

4

Series Editor: *Stefan Hohmann*

Springer

Berlin
Heidelberg
New York
Hong Kong
London
Milan
Paris
Tokyo

Heribert Hirt • Kazuo Shinozaki (Eds.)

Plant Responses to Abiotic Stress

With 31 Figures, 1 in Color; and 5 Tables

Springer

Professor Dr. HERIBERT HIRT
Institute for Microbiology
and Genetics
Biocenter
Dr. Bohrgasse 9
1030 Vienna
Austria

Dr. KAZUO SHINOZAKI
Plant Functional Genomics
Research Group
RIKEN Genomic Sciences Center &
Laboratory of Plant Molecular Biology
RIKEN Tsukuba Institute
3-1-1 Koyadai, Tsukuba
Ibaraki 304-0074
Japan

The cover illustration depicts pseudohyphal filaments of the ascomycete *Saccharomyces cerevisiae* that enable this organism to forage for nutrients. Pseudohyphal filaments were induced here in a wild-type haploid MATa Σ1278b strain by an unknown readily diffusible factor provided by growth in confrontation with an isogenic petite yeast strain in a sealed petri dish for two weeks and photographed at 100X magnification (provided by Xuewen Pan and Joseph Heitman).

ISSN 1610-2096
ISBN 978-3-540-20037-6 Springer-Verlag Berlin Heidelberg New York

Cataloging-in-Publication Data applied for

Bibliographic information published by Die Deutsche Bibliothek
Die Deutsche Bibliothek lists this publication in the Deutsche Nationalbibliografie; detailed bibliographic data is available in the Internet at <http:/dnb.ddb.de>

Springer-Verlag Berlin Heidelberg New York
a member of BertelsmannSpringer Science+Business Media GmbH
http://www.springer.de

Typesetting: Camera ready by editors
Data-conversion: PTP-Berlin, Stefan Sossna e.K.
Cover Design: Design & Production, Heidelberg
39/3150-WI - 5 4 3 2 1 0 - Printed on acid-free paper

Table of contents

List of contributors

Bartels, Dorothea
Faculty of Earth and Life Sciences, Department of Ecology and Physiology of Plants, Vrije Universiteit, de Boelelaan 1085, 1081 HV Amsterdam, The Netherlands

Chinnusamy, Viswanathan
Department of Plant Sciences, University of Arizona, Tucson, AZ 85721, USA

Christmann, Alexander
Lehrstuhl für Botanik, Technische Universität München, Am Hochanger 4, 85354 Freising, Germany

Desikan, Radhika
Centre for Research in Plant Science, Faculty of Applied Sciences, University of the West of England, Bristol, Coldharbour Lane, Bristol BS16 1QY, UK

Fujita, Miki
Plant Mutation Exploration Team, Plant Functional Genomics Research Group, RIKEN Genomic Sciences Center (GSC), RIKEN Yokohama Institute, 1-7-22 Suehiro-cho, Tsurumi-ku, Yokohama 230-0045, Japan

Grill, Erwin
Lehrstuhl für Botanik, Technische Universität München, Am Hochanger 4, 85354 Freising, Germany

Hancock, John T.
Centre for Research in Plant Science, Faculty of Applied Sciences, University of the West of England, Bristol, Coldharbour Lane, Bristol BS16 1QY, UK

Heino, Pekka
Department of Biosciences, Division of Genetics, University of Helsinki, Box 56, 00014 University of Helsinki, Finland

Hirt, Heribert
Gregor-Mendel-Institute for Molecular Plant Sciences and Max-Perutz Laboratories of the University of Vienna, Vienna BioCenter, Dr. Bohrgasse 9, 1030 Vienna, Austria

Kamei, Ayako
Laboratory of Plant Molecular Biology, RIKEN Tsukuba Institute, 3-1-1 Koyadai, Tsukuba 305-0074, Japan

Krishna, Priti
Department of Biology, University of Western Ontario, London ON, Canada N6A 5B7

Meinhard, Michael
Lehrstuhl für Botanik, Technische Universität München, Am Hochanger 4, 85354 Freising, Germany

Mikami, Koji
Department of Regulation Biology, National Institute for Basic Biology, Okazaki 444-8585, Japan

Murata, Norio
Department of Regulation Biology, National Institute for Basic Biology, Okazaki 444-8585, Japan

Neill, Steven J.
Centre for Research in Plant Science, Faculty of Applied Sciences, University of the West of England, Bristol, Coldharbour Lane, Bristol BS16 1QY, UK

Oono, Youko
Laboratory of Plant Molecular Biology, RIKEN Tsukuba Institute, 3-1-1 Koyadai, Tsukuba 305-0074, Japan and
Master's Program in Biosystem Studies, University of Tsukuba, Tennoudai, Tsukuba, Ibaraki, 305-0074, Japan

Palva, E. Tapio
Department of Biosciences, Division of Genetics, University of Helsinki, Box 56, 00014 University of Helsinki, Finland

Polle, Andrea
Forstbotanisches Institut, Georg-August Universität, Büsgenweg 2, 37077 Göttingen, Germany

Sakurai, Tetsuya
Plant Mutation Exploration Team, Plant Functional Genomics Research Group, RIKEN Genomic Sciences Center (GSC), RIKEN Yokohama Institute, 1-7-22 Suehiro-cho, Tsurumi-ku, Yokohama 230-0045, Japan

Satou, Masakazu
Plant Mutation Exploration Team, Plant Functional Genomics Research Group, RIKEN Genomic Sciences Center (GSC), RIKEN Yokohama Institute, 1-7-22 Suehiro-cho, Tsurumi-ku, Yokohama 230-0045, Japan

Schützendübel, Andres
Department of Microbiology Ecology, University Lund, Ecology Building, SE-22362 Lund, Sweden

Seki, Motoaki
Plant Mutation Exploration Team, Plant Functional Genomics Research Group, RIKEN Genomic Sciences Center (GSC), RIKEN Yokohama Institute, 1-7-22 Suehiro-cho, Tsurumi-ku, Yokohama 230-0045, Japan and
Laboratory of Plant Molecular Biology, RIKEN Tsukuba Institute, 3-1-1 Koyadai, Tsukuba 305-0074, Japan

Shinozaki, Kazuo
Plant Mutation Exploration Team, Plant Functional Genomics Research Group, RIKEN Genomic Sciences Center (GSC), RIKEN Yokohama Institute, 1-7-22 Suehiro-cho, Tsurumi-ku, Yokohama 230-0045, Japan and
Laboratory of Plant Molecular Biology, RIKEN Tsukuba Institute, 3-1-1 Koyadai, Tsukuba 305-0074, Japan

Souer, Erik
Faculty of Earth and Life Sciences, Department of Ecology and Physiology of Plants, Vrije Universiteit, de Boelelaan 1085, 1081 HV Amsterdam, The Netherlands

Suzuki, Iwane
Department of Regulation Biology, National Institute for Basic Biology, Okazaki 444-8585, Japan

Ulm, Roman
Institute of Biology II/ Botany, University of Freiburg, Schänzlestrasse 1, D-79104 Freiburg, Germany

Yamaguchi-Shinozaki, Kazuko
Biological Resources Division, Japan International Research Center for Agricultural Sciences (JIRCAS), Ministry of Agriculture, Forestry, and Fisheries, 2-1 Ohwashi, Tsukuba 305-0074, Japan

Zhu, Jian-Kang
Department of Plant Sciences, University of Arizona, Tucson, AZ 85721, USA

Introduction

Heribert Hirt

Everywhere in the world, environmental stresses represent the most limiting factors for agricultural productivity. Besides plant-specific endogenous traits, a large proportion of the annual crop yield is lost to pathogens (biotic stress) or the detrimental effects of abiotic stress conditions including extremes in temperatures, drought, or salinity. In many cases, both biotic and abiotic factors contribute to the severity of disease and yield losses.

All wild type plants have been selected on the basis of competition and their performance under certain environmental conditions. In addition, the existing crop varieties underwent man-made selection for traits such as yield, size, taste, etc. However, all plant life is presently challenged by rapid environmental changes. As amply discussed in the news, greenhouse gases in the form of CO_2 or methane have a tremendous impact on global environmental conditions, resulting in changes of extreme temperatures and weather patterns in many areas of the world. In contrast to animals, plants are sessile organisms and cannot escape changes in ambient conditions. Greenhouse gases also influence the stratospheric ozone layer causing much higher UV radiation levels to reach the ground. Besides resulting in increased rates of skin cancer in humans, UV radiation also induces mutations in plants and poses a direct danger to plant species and agricultural performance. Another area of concern is the intense use of chemical fertilizers and artificial irrigation in agriculture. In many areas of the world, these practices have increased the salinity of the soils to such an extent that the land cannot support growth of any agriculturally important plant any more. Under these conditions, it is no wonder that abiotic stress resistance belongs to the most wanted traits of future crop plants. In summary, the factors discussed above, together with the growing transformation of agriculturally useful land into houses, roads, and industrialized areas, are one of the biggest challenges for future mankind with respect to a functioning agriculture and the conservation of the existing genetic diversity of plant species.

Besides tackling these problems at the political level, science has an important role in elucidating the limits and mechanisms of plant stress adaptation. In this regard, the development of new techniques in molecular biology and genetics has opened up novel possibilities in understanding plant physiology and development. It comes as no surprise that the last two decades have seen major advances in the fields of pathogen defence as well as adaptation to abiotic stresses. The advances in the field of abiotic stress responses provided the impetus for compiling up-to-date reviews on cold stress, heat stress, salinity, drought, heavy metals, oxidative stress, and radiation. In addition, reviews are included on the latest plant transcriptome studies and the use of the genetic model organism *Synechocystis* for investigating abiotic stress.

Topics in Current Genetics, Vol. 4
H. Hirt, K. Shinozaki (Eds.) Plant Responses To Abiotic Stress

Water stress

Water is a central molecule in all physiological processes in plants, comprising between 80 and 95% of the biomass of non-woody plants. If the water status of a plant is insufficient, the plants experience water deficit, also described as drought. Water deficit is not only caused by lack of water but also by environmental stresses like low temperature or salinity, thus it is not surprising that they share many molecular compounds (see separate chapters by Chinnusamy and Zhu on salt stress and Heino and Palva on cold stress). All of these different stresses negatively impact on plant productivity, representing an intensive research area for improving plant performances.

Plants have developed various mechanisms to adapt their growth to limited water conditions. The review by Bartels and Souer focuses on the molecular genetic aspects, which allow plants to respond and adapt to water deficit. These reactions are dependent on the severity and duration of the water deficit but also on the developmental stage and morphological/anatomical parameters of the plants. All species from bacteria to eukarya possess sensors, transducers, and regulators that allow them to respond and adapt to changes in water availability. The cellular response machinery includes solute transporters like aquaporins, transcriptional activators, enzymes encoding compatible solutes, reactive oxygen scavengers, as well as protective proteins. A variety of organisms have evolved highly effective mechanisms to colonize ecological niches with limited water availability. Two principle strategies can be realized to defend dehydration damage: either the synthesis of protective molecules during the dehydration phase to prevent damage and a repair-based mechanism during rehydration to neutralize the damage.

Dehydration tolerance is a trait that exists in seeds of most higher plants but only in some species, such as the resurrection plant *Craterostigma plantagineum,* at the whole plant level. In their review, Bartels and Souer discuss the molecular responses to dehydration and desiccation, from sensing and signalling to the regulation of gene expression, mainly focusing on the models *A. thaliana* and *C. plantagineum.*

Salt stress

Soil salinity is a major abiotic stress that adversely affects crop productivity and quality. Saline soil is characterized by toxic levels of chlorides and sulphates of sodium. The problem of soil salinity is increasing due to irrigation, improper drainage, seawater in coastal areas, and salt accumulation in arid and semi-arid regions. Sodium is an essential micronutrient for some of the plants, but most crop plants are natrophobic. Salinity is detrimental to plant growth as it causes nutritional constraints by decreasing uptake of phosphorus, potassium, nitrate and calcium, ion cytotoxicity and osmotic stress. Under salinity, ions like Na+ and Cl- penetrate the hydration shells of proteins and interfere with the function of these proteins. Ionic toxicity, osmotic stress, and nutritional defects under salinity lead to metabolic imbalances and oxidative stress. Plant salt tolerance mechanisms can be grouped into cellular homeostasis (including ion homeostasis and osmotic adjustment), stress damage control (repair and detoxification), and growth regulation. Considerable efforts have been invested to unravel plant salt tolerance mechanisms. The success of breeding programs with the ultimate goal of improving crop productivity is limited by the lack of a clear understanding of the molecular basis of salt tolerance. Recent advances in the genetic analysis of *Arabidopsis* mutants defective in salt tolerance, and molecular cloning of these loci, have given some insight into salt stress signalling and plant salt tolerance. In their review, Chinnusamy and Zhu discuss these developments as well as the molecular and genetic evidence concerning the perception of salinity stress by plants, cellular signal transduction, and effectors of salt stress tolerance.

Low temperature stress

One of the most severe environmental challenges to plants is low temperature. Different plant species vary widely in their ability to tolerate low temperature stress. Chilling-sensitive tropical species can be irreparably damaged even at temperatures significantly higher than the freezing temperature of the tissues. Injuries are caused by impairment of metabolic processes, by alterations in membrane properties, changes in structure of proteins and interactions between macromolecules as well as inhibition of enzymatic reactions. Chilling tolerant but freezing sensitive plants are able to survive temperatures slightly below zero, but are severely damaged upon ice formation in the tissues. On the other hand, frost tolerant plants are able to survive variable levels of freezing temperatures, the actual degree of tolerance being dependent on the species, developmental stage, and duration of the stress.

Exposure of plants to subzero temperatures results in extracellular ice formation, efflux of water, and cellular dehydration. Therefore, freezing tolerance is strongly correlated with tolerance to dehydration (caused by e.g. drought or high salinity). Freeze-induced dehydration can cause various perturbations in the membrane structures, including membrane fusions and phase transitions. Although freeze-induced cellular dehydration is a central cause of freezing damage, additional factors contribute to freezing injury. Growing ice crystals can cause mechanical damage to cells and tissues and freezing temperatures per se or freeze-induced dehydration can cause denaturation of proteins and disruption of macromolecular complexes. A common denominator in several stresses, including low temperature is the production of reactive oxygen species (ROS), which can generate damage to different macromolecules in the cells. Low temperatures, especially in combination with high light can cause excessive production of ROS and hence tolerance to freezing also correlates with effective scavenging systems for ROS to cope with oxidative stress.

Temperate plants respond to low temperature by activating a cold acclimation program leading to enhanced tolerance to freezing temperatures. This acclimation process is accompanied by altered expression of a number of stress response genes controlling production of proteins and metabolites that protect cellular structures and functions from the adverse effects of freezing and freeze-induced cellular dehydration. The changes in cold responsive gene expression are controlled by a set of dedicated transcription factors responding to the low temperature stimulus. Heino and Palva review the complex signal network that is required for sensing and transduction of the low temperature signal to altered gene expression and discuss the interactions of the signal pathways involved.

ABA as abiotic stress signalling hormone

Abscisic acid (ABA) is considered a ‘stress hormone’ integrating environmental constraints linked to changes in water activity with metabolic and developmental programs. Plants respond to environmental challenges like drought and salt stress by changes in ABA availability, but ABA is also an endogenous signal required for proper development. After exceeding certain threshold levels ABA causes complete closure of stomata and massive alteration of gene expression. ABA signalling comprises various cellular events including turgor-regulation and differential gene expression. Accordingly, a turgor-regulatory pathway can be distinguished from a nuclear signalling cascade. Cross-talk between ABA and other pathways is involved in coordinating primary metabolism, cell growth, and division.

The review by Christmann, Grill, and Meinhard on ABA signalling emphasizes the emerging regulatory circuits of ABA hormone biosynthesis, ABA signalling, and ABA-specific gene expression. The role of ABA in different abiotic stress responses is covered by Bartels and Souer (drought), Chinnusamy and Zhu (salt), and Heino and Palva (cold stress).

Heat stress

Heat stress response is invoked in organisms as diverse as bacteria, fungi, plants, and animals by sudden increases in temperature, and is characterized by elevated synthesis of a set of proteins called heat shock proteins (hsps). Hsps comprise several evolutionarily conserved protein families. A common feature of the heat stress response is that an initial exposure to mild heat stress provides resistance against a subsequent usual lethal dose of heat stress. This phenomenon is referred to as 'acquired thermotolerance'. Since thermotolerant cells express high levels of hsps, these proteins have been associated with the development of thermotolerance. High temperature stress causes extensive denaturation and aggregation of cellular proteins, which, if unchecked, lead to cell death. Through their chaperoning activity, hsps help cells to cope with heat-induced damage to cellular proteins. During stress, hsps function primarily to prevent aggregation and promote proper refolding of denatured proteins, but because protein conformation is important right from the time a protein is synthesized, hsps play important roles under normal conditions as well.

In nature, temperature changes are likely to occur more rapidly than other stress-causing factors. Due to their inability to translocate, plants are subject to wide variations in temperature both diurnally and seasonally, and must therefore adapt to temperature stress quickly and efficiently. The heat stress response is characterized by inhibition of normal transcription and translation, higher expression of heat shock proteins (hsps) and induction of thermotolerance. If stress is too severe, signalling pathways leading to apoptotic cell death are also activated. As molecular chaperones, hsps provide protection to cells against the damaging effects of heat stress and enhance survival. The enhanced expression of hsps is regulated by heat shock transcription factors (HSFs). While knowledge about hsp expression and functions has been gained, our understanding of the regulatory mechanisms is still limited. In her review, Krishna discusses the recent progress made in understanding the molecular mechanism of the heat shock response in plants. Her review outlines our current knowledge of the functions of plant hsps and the regulation of HSFs, and offers a comparative view of heat stress responses in plants and other organisms.

Oxidative stress

For plants, as for all aerobic organisms, oxygen is a double-edged sword. It is absolutely required for normal growth and development, yet continuous exposure to oxygen can result in cellular damage and ultimately death. This is because molecular oxygen is continually reduced within cells to several forms of Reactive Oxygen Species (ROS), in particular the superoxide free radical anion (O_2 .-) and hydrogen peroxide (H_2O_2), that react with various cellular components to bring about acute or chronic damage sufficient to result in cellular death. In plant cells, ROS are generated in high amounts by both constitutive and inducible routes, but under normal situations, the redox balance of the cell is maintained via the constitutive action of a wide range of antioxidant mechanisms that have evolved to remove ROS.

Various environmental stresses and endogenous stimuli perturb this redox balance via increased ROS production or reduced antioxidant activity, such that oxidative stress ensues. In response to increased ROS, the expression of genes encoding antioxidant proteins is induced, as well as that of genes encoding proteins involved in a wider range of cellular rescue processes. In addition, it is increasingly clear that ROS also have signalling functions outside of oxidative stress. The review by Desikan, Hancock, and Neill gives an outline of the mechanisms that regulate redox balance in plant cells, discuss cellular responses to ROS and the potential signalling mechanisms involved, and highlight some of the developmental and physiological processes in which ROS may participate.

Heavy metal stress

Heavy metals are defined as metals with a density higher than 5 g cm^{-3}. From a biological perspective, this definition is not very useful because it comprises the majority of naturally occurring elements. However, only a limited number of these elements is soluble under physiological conditions and, thus, may become available for living cells. Among them are elements which serve plant metabolism as micronutrients or trace elements (Fe, Mo, Mn, Zn, Ni, Cu, V, Co, W, Cr) and which become toxic when present in excess, as well as others with no known biological functions and high phytotoxicity such as As, Hg, Ag, Sb, Cd, Pb, and U. The regulatory limits of heavy metals in the environment are defined by national legislation. Apart from confined natural habitats, there is growing concern about an increasing release of heavy metals into the environment. Sources of heavy metals include traffic, refuse dumps, and sewage sludge. Emissions of dust, aerosols, and ashes from metal processing industries lead to spreading of heavy metals into rural areas. In agricultural soils, heavy metal pollution is an increasing problem because of soil amendment with municipal sewage sludge and intense use of phosphate fertilisers, which contain Cd as a contaminant. The long biological lifetime and retention in soils favours heavy metal accumulation in the food web with potentially negative effects for human health. The bioavailability for heavy metals is plant specific and depends on the demand of specific metals as micronutrients and on the plant's ability to regulate actively metal mobilisation by exudation of organic acids or protons into the rhizosphere. In addition, soil properties influence the chemical mobility of metals, thereby regulating their release into the soil solution.

The ability of plants to extract metals from soil, plant internal metal allocation, and cellular detoxification mechanisms are research areas currently attracting increasing attention. The review by Polle and Schützendübel focuses on metals with contrasting action in plants cells, discussing the chemical properties of these metals with respect to their toxicity and summarising current knowledge how heavy metals interfere with cellular signalling and which signalling cascades lead to plant adaptation or injury.

Genotoxic stress

All organisms have the capacity to dynamically respond to environmental challenges as a result of the activation of complex signalling networks. One of the most extreme challenges is damage to the genetic information itself. The genomes of all living organisms are under continuous assault by environmental agents (e.g. UV irradiation and reactive chemicals) as well as by-products of endogenous metabolic processes (e.g. reactive oxygen species and erroneous DNA replication). As a result of the perception of the genotoxic stress, the cell cycle is halted to gain the time necessary for DNA repair, and genes required for repair and protection of other cellular components endangered by the genotoxic treatment are activated. Alternatively, particularly in multicellular eucaryotes, cells may respond by undergoing apoptosis, thereby eliminating damaged cells.

Research on genotoxic stress perception and signalling in mammalian cells is of particular importance due to its implications in human health and disease, including carcinogenesis. In plants, however, owing to the static nature of their cells anchored by cell walls, tumourous tissue cannot metastasise and plants do not die of cancer. On the other hand, the reproductive tissues of plants are derived from cells that went through many rounds of DNA replication producing the entire organism, before forming gametes. This feature makes plants particularly sensitive to the potential accumulation of mutations in the germline, which finally opens the way for the passage of somatic mutations to the next generation. In contrast to animals, plants are sessile organisms that depend on solar radiation as the vital source of biological energy and thus are continuously exposed to environmental mutagens, including UV-B radiation, and tolerance to this abiotic stress factor is critical for plant fitness. Repair of DNA damage is essential for the maintenance of genomic integrity and substantial information is available on DNA repair processes in plants. In contrast, knowledge on perception and signalling of DNA-damaging threats in plants is rather limited and genetic support for proteins involved in genotoxic signalling in *Arabidopsis* is only emerging. Importantly, as deduced from the mammalian system, these might include signalling components engaged by both "nuclear" and "non-nuclear" targets of genotoxic agents. The review by Ulm focuses on recent advances in the identification of genetically defined components in genotoxic stress signalling in plants.

Stress transcriptome analysis

Upon sensing of stress, cells respond and adapt to a given stress in order to survive. In almost all cases, the stress responses are based primarily on the expression of specific stress-induced genes, followed by specific biochemical and physiological reactions. Several genes that respond to drought, high-salinity or cold stress have been studied at the transcriptional level. Recently, gene expression profiling using cDNA microarrays or gene chips identified many hundred genes that are regulated by these abiotic stresses. The products of the stress-inducible genes can be classified into two groups: those that directly protect against environmental stresses and those that regulate gene expression and signal transduction in the stress response. The first group includes proteins that likely function by protecting cells from dehydration, such as the enzymes required for biosynthesis of various osmoprotectants, late-embryogenesis-abundant (LEA) proteins, antifreeze proteins, chaperones, and detoxification enzymes. The second group of gene products includes transcription factors, protein kinases, and enzymes involved in phosphoinositide metabolism. Stress-inducible genes have been used to improve the stress tolerance of plants by gene transfer and to analyze the functions of stress-inducible genes. The review by Seki et al. reports on the recent progress on microarray gene expression studies in response to abiotic stresses discussing these findings with respect to current and future strategies of improving stress tolerance of crop plants.

Stress sensors in the model organism *Synechocystis*

Cells perceive a particular stress and react to it by expressing specific sets of stress-inducible genes, the gene products of which appear to play important roles in the acclimation to the stress. In photosynthetic organisms, such as cyanobacteria and plants, and in the simple eukaryote yeast, but not in animals, various two-component systems contribute to the perception and transduction of environmental signals. Two-component systems are minimally built up of a histidine kinase that perceives the stress and a response regulator that transduces the stress signal. In bacteria, a two-component system is all that is needed for signal transduction and gene expression is usually mediated by the response regulator acting as transcription factor. In yeast and plants, additional components are involved in the signalling adding the potential of fine-tuning and cross-talk with other pathways.

What is yeast for animals is *Synechocystis* for higher plants. Being an easily tractable genetic model organism that can be readily mutated and transformed, homologous recombination is the routine in Synechocystis and not the exception as in plants. *Synechocystis* only has a small genome (3.7 MBp) that is fully sequenced and has not more than 12 % non-coding regions. In addition, the development of full genome based microchips allows genome-wide gene expression analysis. Based on the screening of mutant libraries, sensors for various abiotic stresses, nutrients, and metals were identified in *Synechocystis*. Mikami, Suzuki, and Murata review the current status of these findings and their implications for our understanding of stress signalling in plants.

1 Molecular responses of higher plants to dehydration

Dorothea Bartels and Erik Souer

Abstract

A massive amount of data has been accumulating on the molecular responses of plants to water deficit. In plants, dehydration activates a protective response to prevent or repair ensuing damage to cells. The plant hormone, abscisic acid plays a central role in this process. The genetic model *Arabidopsis thaliana* tolerates a low level of dehydration. Analysis of the abscisic acid signalling pathway and of pathways induced by dehydration in *A. thaliana* have made a major contribution to our knowledge of the molecular responses of plants to dehydration. Desiccation tolerance is a trait found in the seeds of most higher plants, but also at the level of the whole plant in some species, such as the resurrection plant *Craterostigma plantagineum*. Here, we discuss the molecular responses to dehydration and desiccation, from the sensing and signalling of water deficit to the regulation of gene expression, focusing mainly on the model systems *A. thaliana* and *C. plantagineum*.

1.1 Introduction

The availability of water determines the distribution of plants and their productivity. Water is central to all physiological processes in plants; at the cellular level, it is the major medium for transporting metabolites and nutrients. Water accounts for between 80 and 95% of the biomass of leaves and roots in non-woody plants. The water status of a plant is described by measuring water potential and relative water content. If the water status is unbalanced due to an insufficiency of water, the plant experiences water deficit and subsequently suffers from water stress, often referred to as drought. The expression 'drought' derives from the agricultural context. Here, we will use 'water deficit' or 'dehydration' to mean an inadequate water supply that has an immediate effect on cellular metabolism and negatively influences growth and development.

The movement of water molecules is determined by the water potential gradient across the plasma membrane, which in turn is influenced by the concentrations of solute molecules inside and outside the plant cell. Fluctuations in the availability of extracellular water cause transmembrane water and solute fluxes that perturb cellular structures, alter the composition of the cytoplasm, and modulate cell function. Water deficit is caused not only by a simple lack of water, but also by envi-

Topics in Current Genetics, Vol. 4
H. Hirt, K. Shinozaki (Eds.) Plant Responses To Abiotic Stress

ronmental stresses like low temperature or salinity; thus it is not surprising that responses to these various stresses involve many shared molecular components. These different stresses have an enormous negative impact on plant productivity, and intensive research is underway with a view to defining strategies for improving plant performance under stress. The effects of salt and cold stress on plants are covered in separate chapters of this volume (see chapters 6 and 9)

Plants have developed many mechanisms to adapt their growth to the availability of water. Their responses to water limitation at the whole plant level have been described in detail recently (Black and Pritchard 2002), and these will not be dealt with in this review. Here the focus is on molecular genetic aspects of the reactions that allow plants to respond and adapt to water deficit. These are dependent on the severity and duration of the water deficit, and also on the developmental stage and morphological/anatomical parameters of the plants. In general, rapid emergency responses and slow adaptive responses can be distinguished.

Cells across all species from bacteria to eukarya possess sensors, transducers, and regulators that allow them to respond and adapt to changes in water availability. The cellular response machinery includes solute transporters like aquaporins, transcriptional activators, enzymes that synthesize compatible solutes, scavengers of reactive oxygen species, and protective proteins. A large number of publications report on the synthesis and accumulation of such molecules, but down-regulated processes have been comparatively neglected.

A variety of organisms has evolved highly effective mechanisms that allow them to colonize ecological niches characterized by limited water availability. Two main strategies can be used to restrict dehydration damage: synthesis of protective molecules during the dehydration phase to prevent damage, or activation of repair mechanisms during rehydration to neutralize the damage incurred. Within the plant kingdom, the first strategy seems to be preferred by higher plants; the repair strategy has been reported only for bryophytes (Phillips et al. 2002). This leads to interesting evolutionary questions, namely whether the same tolerance mechanisms have evolved in different groups of organisms or whether different strategies have been invented.

1.2 Plant species and experimental systems used in molecular studies

Molecular reactions to water stress in higher plants have been studied mainly in the genetic model system *Arabidopsis thaliana*, in desiccation-tolerant resurrection plants, and in some crop plants including trees. Studies on *Arabidopsis* have been very informative with respect to the identification of general components in the water stress signalling network. This knowledge has been obtained principally from the analysis of *Arabidopsis* mutants that show defects in water balance (Kirch et al. 2002; Leung and Giraudat 1998). Different ecotypes of *Arabidopsis* grow in different habitats and exhibit various degrees of tolerance to water stress

(Meyre et al. 2001). Nevertheless, *Arabidopsis* can only tolerate moderate water loss, and the tissues collapse irreversibly under extreme dehydration.

In addition to studies on whole plants, research using guard cells, mainly from *Arabidopsis*, has advanced our understanding of water stress-related signalling molecules. Besides being responsible for the uptake of CO_2 for photosynthesis, guard cells control water loss via transpiration to the atmosphere. In the context of understanding how water balance is controlled, guard cells have been instrumental in the identification of specific calcium signatures, second messengers that regulate calcium levels, phosphorylation signals, specific ion channels and transporters, and the dissection of the abscisic acid (ABA)-induced closure of stomata, which is mediated by a reduction in the turgor pressure of guard cells. There are several excellent recent reviews, which discuss these aspects of guard cell function (e.g. see Assmann and Wang 2001; Schroeder et al. 2001). It remains to be seen how responses at the guard cell level are integrated with responses in the whole plant.

In contrast to *Arabidopsis*, seeds and a small group of vascular plants, termed resurrection plants, can tolerate extreme water loss and endure in this desiccated, dormant state until sufficient water is available for further growth (Black and Pritchard 2002). These systems are being exploited with a view to understanding the molecular basis of the phenomenon. Resurrection plants express desiccation tolerance in all tissues, including callus. Most molecular studies on resurrection plants have been done with *Craterostigma plantagineum* (Bartels and Salamini 2001). The ability to induce desiccation tolerance in callus tissue from *C. plantagineum* by treatment with ABA allows one to study its basis in undifferentiated cells (Bartels et al. 1990). This is an important advantage over seed systems, where it is difficult to separate the acquisition of desiccation tolerance from other processes involved in seed development.

In this review, we will focus on molecular studies of desiccation tolerance carried out on resurrection plants and seeds, and we will attempt to place these data within the context of the knowledge derived from *Arabidopsis*. Many of the regulatory genes involved appear to belong to closely related gene families. The assignment of functions to the different members of these families is probably only possible via mutant analysis, an approach that is largely restricted to *Arabidopsis*. For this reason, we will also comment on aspects of this work.

1.3 Abscisic acid (ABA)

The importance of ABA in multiple stress responses, including dehydration, is indisputable. Dehydration in plants leads to an increase in levels of ABA, which in turn induces the expression of multiple genes involved in defence against the effects of water deficit. ABA triggers stoma closure, thus reducing water loss via transpiration. In *Arabidopsis*, dehydration causes prompt ABA synthesis, which is detectable within 2 hours and reaches a maximum after 10 hours (Kiyosue et al. 1994). The increase in ABA content is relatively slow, and thus ABA-induced

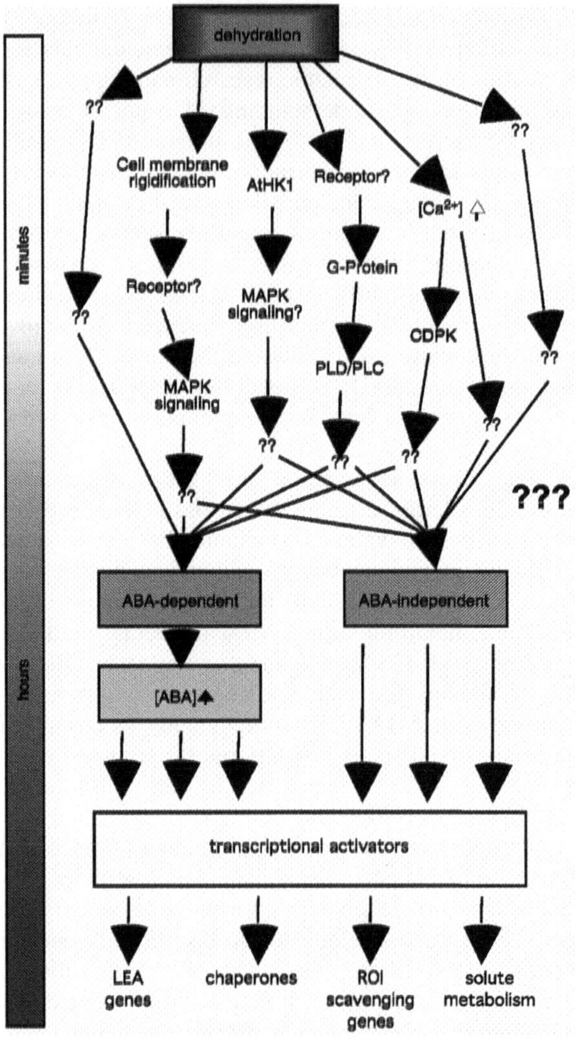

Fig. 1. Complexity of molecular responses to dehydration. Upon water deficit, plant cells activate a number of pathways to regulate down-stream defence mechanisms. The earliest events can be detected within minutes. Potentially, some pathways still need to revealed. Some of the early events, which remain to be elucidated, lead to the biosynthesis of the plant hormone abscisic acid (ABA). Signals coming from either the ABA-dependent or the ABA-independent pathway activate transcriptional activators that subsequently induce the expression of genes that encode a variety of enzymes and proteins that are required to survive dehydration. These include genes encoding molecular chaperones, reactive oxygen intermediate (ROI) scavenging enzymes, sucrose metabolism enzymes, and a variety of late embryogenesis abundant (LEA) protein that are supposed to have protective functions. As some of the pathways also influence each other (not shown), the response is more complex than illustrated here.

genes may be correlated with adaptation mechanisms (Fig. 1). The ABA biosynthetic pathway is a side-branch of the carotenoid pathway, and many enzymes of the ABA biosynthetic pathway are upregulated by dehydration (Seo and Koshiba 2002). Most genes involved in responses to dehydration are also induced by ABA. Therefore, the treatment of plants with exogenous ABA has been used to mimic dehydration responses. Screens for mutants affected in seed germination or plants that tolerate mild dehydration have led to the identification of many ABA mutants, including ABA biosynthesis mutants, ABA-hypersensitive mutants and ABA-insensitive mutants. Cloning of the corresponding genes identified a number of ABA signalling compounds (Finkelstein et al. 2002; Leung and Giraudat 1998). A challenge for the future will be to link all of these components functionally, thereby ultimately revealing the complete ABA signalling network. It has to be emphasized that ABA is not the only small molecule involved in water deficit signalling, since several ABA-independent pathways have also been identified (Frank et al. 2000; Shinozaki and Yamaguchi-Shinozaki 1997). The role of ABA is discussed in detail in chapter 2.

1.4 The perception of water stress

How is dehydration sensed, and how is this perception translated into a molecular signal? Time course experiments in several plants have shown that water deficit is sensed very rapidly -- long before symptoms such as wilting become manifest, and before the relative water content decreases significantly. Transcripts and proteins indicative of a dehydration response are detectable within 60 minutes after the onset of dehydration in the resurrection plant *C. plantagineum* and in *A. thaliana* (Bartels et al. 1990; Urao et al. 1994; Nakashima et al. 1997). The question of how shifts in water availability are sensed in plants is completely open; the nature of the physical signal and the mode of its translation into a biochemical signal are unknown.

In classical signalling pathways, environmental stimuli are sensed by receptor molecules. In the case of water deficit in plants, the nature of the biochemical receptor/ligand interaction, if there is any, is not yet known. Some information on the sensing of osmotic stress is available from bacteria, and for some eukarya including yeast. There are extremely well adapted species among these organisms. These possess sensors, transducers, and regulators that allow them to attenuate the cellular consequences of water deficit. However, even in these organisms, it is still not completely clear how osmotic stress is sensed (Hohmann 2002).

1.4.1 Histidine kinases

A well studied group of sensor molecules which are undoubtedly involved in the initial response to osmotic stress are protein histidine kinases, which form part of so-called two-component systems that were first identified in bacteria (Wurgler-

Murphy and Saito 1997). These kinases sense environmental changes, which trigger autophosphorylation of a histidine residue, and subsequently the phosphate is transmitted to an aspartic residue in the receiver. One such histidine kinase, Sln1, has been identified as an osmosensor in yeast (Maeda et al. 1994). In plants, histidine kinases function as receptors for the plant hormones ethylene and cytokinin (Chang and Stewart 1998; Inoue et al. 2001). In addition, the histidine kinase AtHK1 has been shown to be involved in the response to dehydration in *Arabidopsis* (Urao et al. 1999). AtHK1 shares significant structural homology with the Sln1 osmosensor from yeast. Indeed, *AtHK1* is able to complement the yeast *sln1* mutant, allowing it to grow in high-salt medium. Moreover, *Arabidopsis* AtHK1 is able to interact with and activate the yeast mitogen-activated-like protein (MAP) kinase pathway downstream of Sln1. This implies that a similar cascade might exist in *Arabidopsis*. However, at the moment, the function of *AtHK1* in plants is still unclear.

1.4.2 The role of kinases and phosphatases in the response to water deficit

Besides receptor-ligand and protein-DNA interactions, protein modification represents another potential level of control. Phosphorylation is an effective and rapid mechanism of post-translational modification, which alters the activities of DNA-binding factors and a plethora of intermediate molecules. Understanding the specificity of these reactions is difficult, because eukaryotes have very large numbers of genes that encode phosphorylating and dephosphorylating enzymes. It has been estimated that the *Arabidopsis* genome codes for around 120 protein phosphatases and 90 MAP kinases (Kerk et al. 2002).

1.4.2.1 MAP kinase signalling

There are many indications that a very early step in transducing the water deficit signal in plant cells involves MAPK pathways. For instance, a change in the fluidity of the plasma membrane might induce a conformational change in a receptor that activates a downstream kinase cascade. Most information on MAPK signalling in plants is derived from work on pathogen defence reactions and on the response to cold and salt stress (Zhang and Klessig 2001). The same MAP kinase pathways might be used to signal dehydration, cold and salt stress, with the specificity of the response being determined by the duration of the activation state and formation of complexes with other proteins. In alfalfa, the stress-activated MAP kinase SAMK (also referred to as MKK4) was found to be activated by touch (Bögre et al. 1996) and by cold and dehydration (Jonak et al. 1996). Interestingly, SAMK is not activated by ABA, indicating that SAMK either acts in an ABA independent pathway or upstream of ABA. Recently, it was shown that a reduction in the fluidity of the membrane is the trigger that leads to activation of SAMK (Sangwan et al. 2002). This indicates that one way in which a plant cell senses dehydration is through a change in the fluidity of the cell membrane. In *A. thaliana*, the levels of the MAP kinases ATMPK4 and ATMPK6 and their mRNAs remain unaltered upon stress, but the activities of these enzymes are rapidly increased by a variety of stresses including dehydration (Ichimura et al. 2000). Yeast two-hybrid screening and complementation of yeast mutants led to the identification of a MAPK pathway that involves ATMPK4, comprising a MAPKKK (AtMEKK1) and a MAPKK (ATMKK2) (Ichimura et al. 1998). Therefore, this MAPK signalling pathway seems to play an important role in the molecular response to dehydration in plants. Characterization of multiple MAPK proteins should eventually reveal how they are activated and how signals are transmitted to the specific targets that are ultimately responsible for protection against dehydration damage.

1.4.2.2 Phosphatases

One theme that is emerging from mutant analyses is that, besides kinases, phosphatases are essential modifiers in regulatory networks. Some kinase signals appear to act very early in the temporal hierarchy of signals, but this is not so evident for the action of phosphatases. The involvement of phosphatases in dehydration-stress signal transduction has been established using the ABA-insensitive *Arabidopsis* mutants, *abi1*and *abi2*. These mutants display pleiotropic phenotypes affecting seed dormancy, stoma regulation, and signal transduction during water stress (Koornneef 1984; Merlot and Giraudat 1997). Both mutant genes encode homologous, type 2C, Ser/Thr protein phosphatases with identical amino acid substitutions at equivalent positions, which result in reduced phosphatase activity and a dominant-negative phenotype (Bertauche et al. 1996; Leung et al. 1997). The phenotype of intragenic null suppressor alleles of *abi1-1* and *abi2-1*, which exhibit higher seed dormancy and enhanced ABA-dependent sensitivity to inhibition of germination and stoma closure, led to the conclusion that *ABI1* and *ABI2* act as negative regulators in the ABA signal transduction pathway (Gosti et al. 1999; Merlot et al. 2001). Although the similarity between *ABI1* and *ABI2* suggests that they may act in overlapping pathways, careful physiological analysis revealed that *ABI1* and *ABI2* do not show complete functional equivalence (Murata et al. 2001). A search for potential targets of the phosphatases using a yeast two-hybrid approach has identified two possible candidates. A member of the homeodomain leucine-zipper transcription factor family (ATHB6, see below) was shown to interact with the catalytic site of ABI1 (Himmelbach et al. 2002). Furthermore, a protein kinase interacts with ABI2 and, to a lesser extent, with ABI1 (Guo et al. 2002). Double mutant analysis of *abi-1* and *abi-2* with a protein kinase mutant and a calcium binding protein mutant suggests that *ABI1* and *ABI2* act in conjunction with a calcium and a H_2O_2 signal (Guo et al. 2002). This for the first time provides genetic evidence for a link between phosphatases and second messenger molecules in the transcriptional control of genes relevant for osmotic stress responses. The role of phosphatases in signalling pathways is also supported by the observation that a phosphatase2C from alfalfa negatively regulates a MAP kinase (Meskiene et al. 1998) and that a MAP kinase phosphatase plays a role in the response to genotoxic stress (Ulm et al. 2001). Comparison of the data on MAP kinases and phosphatases suggests the following unifying hypothesis: MAP kinases and possibly other kinases rapidly relay signals which redirect cellular metabolism toward the synthesis of compounds that attenuate the effects of dehydration; phosphatases, on the other hand, repress this stress response. It is now possible through a strategic genomic approach to address the function of the many other phosphatases encoded in the *Arabidopsis* genome. This should reveal whether and how other phosphatases are involved in stress signalling.

1.4.2 Calcium signalling

Studies on animal cells first established that oscillations in cytosolic calcium concentrations are an important intermediate step in the activation of specific signalling cascades, which then determine downstream physiological responses. In plants, transient increases in cytosolic [Ca^{2+}] have been reported in response to a diverse range of abiotic and biotic stimuli (Kiegle et al. 2000; Evans et al. 2001), but the specificity of the physiological responses is not understood. It has been suggested that duration, magnitude, and cellular location of the changes in [Ca^{2+}] determine specificity (McAinsh and Hetherington 1998; Evans et al. 2001).

A change in membrane fluidity might alter the activity of Ca^{2+} channels, leading to a change in [Ca^{2+}] in the cytosol. An increase in cytosolic [Ca^{2+}] as a result of influx from extracellular sources and/or extrusion from the vacuole has been recognized as one of the early responses to dehydration (Sanders et al. 1999). With respect to water balance the change in cytosolic [Ca^{2+}] has been well studied in guard cells (Schroeder et al. 2001). There, repetitive Ca^{2+} transients play a role in both stoma opening and closure. The diverse and opposing effects of [Ca^{2+}] are puzzling. Presumably, the effect of the Ca^{2+} wave depends on the Ca^{2+} channel, the cellular location, and the dynamics of the change in [Ca^{2+}] and the availability of downstream signalling pathways at the time the change in [Ca^{2+}] occurs (Sanders et al. 1999). Recent data on the closure of stomata showed that a defined range of calcium oscillations determines stoma movements (Allen et al. 2001).

An interesting class of Ca^{2+} binding proteins are the Ca^{2+} dependent kinases (CDPKs) that combine a calmodulin-like calcium binding module with a kinase domain (Cheng et al. 2002). Some of these CDPKs have been shown to be inducible by dehydration (Urao et al. 1994; Patharkar and Cushman 2000). Two *Arabidopsis* CDPKs, CPK10 (*AtCDPK1*) and CPK11 (*AtCDPK2*), are induced within 10 minutes upon dehydration stress (Urao et al. 1994). CPK10 (*AtCDPK1*) is capable of transactivating a stress-induced promoter (Sheen 1996). The activity of CPK10 is stimulated by 14-3-3 proteins, but this is only apparent in the presence of Ca^{2+} (Camoni et al. 1998). Therefore, Ca^{2+} binding by CPK10 (*AtCDPK1*) seems to precede the formation of the 14-3-3 protein complex. Ectopic expression of a rice CDPK gene, *OsCDPK7*, increases stress tolerance in rice (Saijo et al. 2000). The identification of the target(s) of the dehydration-induced CDPKs promises to be an important breakthrough in the understanding of dehydration signal transduction.

1.4.3 Heterotrimeric G-proteins

Another group of signalling molecules that have an established role in signal transduction in animal cells are the G-proteins. Evidence for the involvement of G-proteins in the dehydration signalling cascade in plant cells is beginning to emerge. The heterotrimeric G-proteins consist of three subunits, Gα, Gβ, and Gγ. The G-protein interacts with a ligand-bound receptor domain in the cytoplasmic membrane, which leads to exchange of a bound GDP nucleotide for a GTP on the α-subunit. The trimeric G-protein then dissociates into the α-monomeric subunit and the βγ dimer, and these two components activate downstream molecules such as phospholipase C or phospholipase D. Active G-proteins and the activation of phospholipases have been observed in plants, but associated receptors have not been identified (Jones 2002a). In contrast to the situation in animals - where 41 different subunits are known - plants only have four G-protein subunits, a single Gα, a single Gβ, and two Gγ.

Evidence for functional G-proteins in plants is mainly derived from pharmacological approaches. G-protein mimicking compounds and antagonists have been administered to plant systems, and subsequently putative downstream reactions were measured. The agents most widely used for this purpose have been mastoparan and its more potent analogue MAS7. Mastoparan is a 14-amino acid polypeptide isolated from wasp venom, which stimulates the replacement of bound GDP by GTP, corresponding to the formation of active G-protein subunits. The effects of mastoparan have implicated heterotrimeric G-proteins in, among other processes, the regulation of stoma opening and water deficit-induced phospholipase D activity (Frank et al. 2000, Wang et al. 2001). In guard cells, ABA signalling has been shown to require a G protein. In *Arabidopsis* plants carrying a mutant allele of the Gα gene *GPA1*, ABA can no longer inhibit the closure of stomata (Wang et al. 2001). The presence of only two different G protein complexes and the apparent lack of an associated receptor for G-proteins in plants is puzzling. In plants, apparently, the specificity of signal transduction is likely to be conferred not by G proteins, but by other signalling compounds present in the cell.

1.4.4 Phospholipid signalling

The activation of phospholipid signalling in response to a variety of stresses, including dehydration, is now well documented in plants (Munnik 2001). Phospholipase activity generates phosphatidic acid, which acts as a second messenger to activate downstream targets. The phosphatidic acid signal is generated by two pathways involving phospholipase C (PLC) and phospholipase D (PLD).

PLD cleaves phospholipids into a polar head group and phosphatidic acid. Analysis of the *A. thaliana* genome has identified twelve *PLD* genes, which can be grouped into five classes: α, β, γ, δ, and ζ (Qin and Wang 2002). Two PLD genes, *PLD1* and *PLD2*, have been isolated from *C. plantagineum* (Frank et al. 2000). In both *A. thaliana* and *C. plantagineum*, water deficit leads to a response at two levels, PLD enzyme activity and *PLD* transcription. The enzymatic activa-

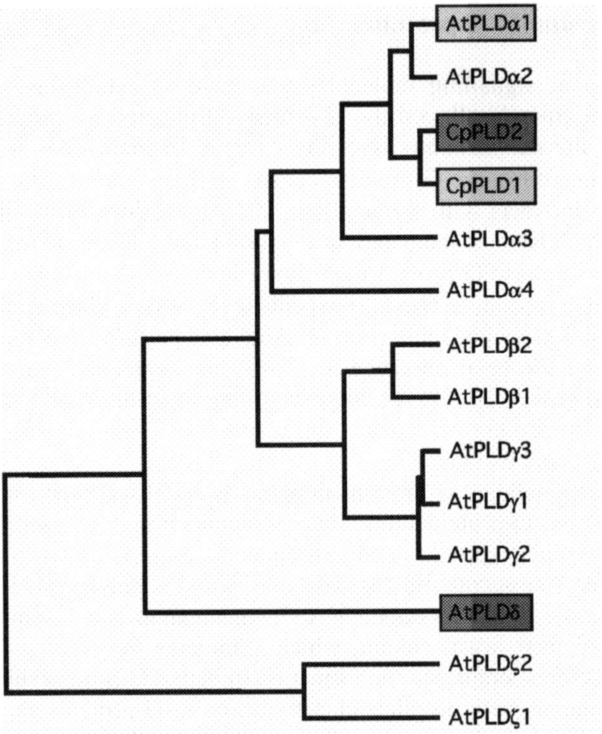

Fig. 2. Phylogenetic tree illustrating the relationship between all *A. thaliana* and two *C. plantagineum phospholipase D* (*PLD*) genes. Both *C. platagineum* genes group with the *A. thaliana PLDα* genes, although the expression pattern of *CpPLD2* matches that of *AtPLDδ*. Matching expression patterns are illustrated by different grey boxes.

tion occurs within minutes of the dehydration stimulus. In A. thaliana (Katagiri et al. 2001) and *C. plantagineum* PLD enzyme activity is not activated by ABA, suggesting that ABA signalling operates via another pathway or enters the phospholipid signalling pathway further downstream. At the transcriptional level *PLDδ* from *A. thaliana* and *PLD2* from *C. plantagineum* are induced by dehydration some hours after the onset of water stress, while *PLDα* from *A. thaliana* and *PLD1* from *C. plantagineum* are constitutively expressed. This suggests that *PLD2* is the *C. plantagineum* orthologue of *PLDδ*, although phylogenetic analysis shows that both *C. plantagineum* genes fall within the *PLDα* group (Fig. 2). In *A. thaliana* plants expressing antisense *PLDδ* transcripts, total PLD activity was significantly reduced during the first hours of dehydration stress, which supports a role for *PLDδ* in the early signal transduction cascade (Sang et al. 2001).

The second phospholipid signalling pathway that is implicated in the response to dehydration is the phospholipase C (PLC) pathway. PLC converts phosphatidylinositol 4,5-biphosphate into inositol 1,4,5-triphosphate and diacylglycerol (DAG). Inositol 1,4,5-triphosphate causes the release of calcium into the cytosol, while DAG is converted into the second messenger phosphatidic acid by DAG kinase. PLC seems to account for most of the phophatidic acid generated upon exposure to osmotic stress (Munnik et al. 2000). Like PLD, PLC can be activated at both the enzymatic level and the transcriptional level by stress. The *Arabidopsis AtPLC1* gene is activated by a variety of stresses, including dehydration (Hirayama et al. 1995). A constitutively expressed PLC gene, *AtPLC2*, has also been found (Hirayama et al. 1997). PLC is activated rapidly upon dehydration (Takahashi et al. 2001). PLC also seems to activate an ABA-independent pathway, as PLC inhibitors block the expression of dehydration-induced, but not of ABA-induced, target genes. PLC genes have been identified in a number of plants, including dehydration-induced PLC genes in potato (Kopka et al. 1998).

Some of the molecular targets of phosphatidic acid, the second messenger produced by phopholipases, are now being revealed in plants. For instance, phosphatidic acid increases the kinase activity of a Ca^{2+} activated, calcium-dependent protein kinase (Farmer and Choi 1999) and activation of a wound-induced MAP kinase is also dependent on phosphatidic acid (Lee et al. 2001). It can be expected that other components of the downstream signalling pathway and the mechanism(s) by which phosphatidic acid acts will be identified in the near future.

1.5 Transcriptional control

One of the effects of the dehydration signal transduction cascade is the activation of transcription factors, each of which activates a set of target genes, including those required for the synthesis of protective molecules (Fig. 1). A number of transcription factors that are activated by dehydration have been isolated (Table 1). Most of these factors were identified because they are differentially expressed in untreated versus stress-treated tissue, or by virtue of their ability to bind to promoters of dehydration-induced genes. The dehydration response seems to involve members of several groups of plant transcription factors. At present, it is not known whether closely related transcription factors from one family have overlapping functions -- and thus show a certain degree of redundancy -- or whether each family member has a distinct function. The functional analysis of transcription factors belonging to large families is particularly difficult, and the assignment of target genes to specific factors may not be possible even with targeted mutations. In this part of the review, we will focus on the model systems *A. thaliana* and *C. plantagineum*. We first describe promoter elements that are important for the dehydration-induced expression of genes, and then discuss the corresponding transcription factors and DNA binding proteins.

Dehydration triggers high-level expression of many genes; of which the most prominent are the so-called *late embryogenesis abundant* (*lea*) genes (see 1.6.2). Promoters of various *lea* genes were initially analyzed to define sequences, which are of general importance for dehydration-induced gene expression. The most widely distributed and best investigated elements are the ABA response elements (ABREs) and the dehydration response elements (DREs). However, it is becoming

Table 1. Transcription factors controlling the dehydration response in *C. plantagineum* and *A. thaliana*

Trx factor[a]	Species	Gene	Binding site	Transcription induced by drought/ABA[b]	Reference
bZIP	*C. plantagineum*	?	-	-	-
	A. thaliana	AtbZIP35 (ABF1)	ACGTGGC	Not by dehydration ABA	Choi et al. 2000
		AtbZIP36 (ABF2/AREB1)	GGACACGTG GCG	(Not by) Dehydration[c] ABA	Choi et al. 2000 Uno et al. 2000
		AtbZIP37 (ABF3)	ACGTGGC	Not by dehydration ABA	Choi et al. 2000
		AtbZIP38 (ABF4/AREB2)	GGACACGTG GCG	Dehydration ABA	Choi et al. 2000 Uno et al. 2000
		AtbZIP39 (ABI5)	?	ABA Dehydration not known	Finkelstein et al. 2000
		AtbZIP40 (GBF4)	CACGTG	?	Menkens et al. 1994
		AtbZIP66 (AREB3)	ACGTG	?	Uno et al. 2000
ERF (AP2)	*C. plantagineum*	?	-	-	-
	A. thaliana	CBF1 (DREB1B)	CCGAC	Not by dehydration Not by ABA	Stockinger et al. 1997 Liu et al. 1998
		CBF2 (DREB1C)	?	Not by dehydration Not by ABA	Medina et al. 1999 Liu et al. 1998
		CBF3 (DREB1A)	TACCGACAT	Not by dehydration Not by ABA	Medina et al. 1999 Liu et al. 1998
		CBF4	?	Dehydration ABA	Haake et al. 2002
		DREB2A	TACCGACAT	Dehydration ABA	Liu et al. 1998
		DREB2B	?	Dehydration ABA	Liu et al. 1998
		ABI4	?	Not by dehydration Not by ABA	Finkelstein et al. 1998 Söderman et al. 2000

Table 1. Continued

Trx factor[a]	Species	Gene	Binding site	Transcription induced by drought/ABA[b]	Reference
SAP	*C. plantagineum*	CpR18	pr3.75[d]	Dehydration ABA	Hilbricht et al. 2002
	A. thaliana	At5g66840	?	?	Hilbricht et al. 2002
Myb	*C. plantagineum*	Cpm7	?	Dehydration Not by ABA	Iturriaga et al. 1996
	A. thaliana	Atmyb2	CTAACCA	Dehydration ABA	Urao et al. 1993
bHLH	*C. plantagineum*	?	-	-	-
	A. thaliana	AtMyc2 (rd22BP1)	CACATG	Dehydration ABA	Abe et al. 1997
HD-Zip	*C. plantagineum*	CpHB-1	CAAT(A/T)ATTG	Dehydration Not by ABA	Frank et al. 2000
		CpHB-2	?	Dehydration ABA	Frank et al. 2000
		CpHB-3	CAAT(A/T)ATTG	Downregulated by dehydration and ABA	Deng et al. 2002
		CpHB-4	CAAT(A/T)ATTG	Downregulated by dehydration	Deng et al. 2002
		CpHB-5	CAAT(A/T)ATTG	Downregulated by dehydration	Deng et al. 2002
		CpHB-6	CAAT(A/T)ATTG	Dehydration ABA	Deng et al. 2002
		CpHB-7	CAAT(A/T)ATTG	Dehydration ABA	Deng et al. 2002
	A. thaliana	AtHB-6	?	Dehydration ABA	Soderman et al. 1999
		AtHB-7	?	Dehydration ABA	Soderman et al. 1996
		AtHB-12	?	Dehydration ABA	Lee and Chun 1998

[a] transcription factor or DNA binding factor family
[b] other abiotic stresses omitted
[c] conflicting results in literature
[d] a promoter element required for the ABA induced expression of the *lea* type gene *CDeT27-45* (GCAAGCCCAAATTTCACAGCCCGATTAACCG)

increasingly clear that these elements alone are not always sufficient to determine stress-activated transcription of genes, and additional motifs have to be considered (see e.g., Nelson et al. 1994). An important new line of research will be to investigate how the actively transcribed genes are organised in chromatin and how they are made accessible for transcription during stress.

1.5.1 The ABA responsive element

A promoter element that is important for dehydration- and ABA-responsive expression was first identified in the *lea* gene (see X.6.2) *rab16A* of rice (Mundy et al. 1990). This so-called ABA responsive element (ABRE) has the core sequence PyACGTGGC and represents a subgroup of so-called G-box elements (Williams et al.1992). A bZIP protein, EmBP-1, was found to bind to the ABRE of the wheat gene *Em*, another representative of the *lea* genes (Guiltinan et al. 1990). Several bZIP transcription factors, which have recently been classified as group A bZIPs, have been shown to be involved in ABA signalling in *Arabidopsis* as well (Jakoby et al. 2002). This set of proteins has been referred to as ABA responsive element binding factors (ABF) or ABA responsive element binding (AREB) proteins (Menkens and Cashmore 1994; Choi et al. 2000; Finkelstein and Lynch 2000; Uno et al. 2000). The nomenclature used for stress-induced bZIP genes is somewhat confusing, as different names have been used for the same genes (see Table 1). Four members of group A of the bZIP family, AtbZIP35-38, were identified as binding to ABRE promoter elements in a yeast one-hybrid screen (Choi et al. 2000). AtbZIP36 and AtbZIP38 can activate an ABA-inducible promoter in a transient protoplast assay (Uno et al. 2000). Ectopic expression of *AtbZIP37* and *AtbZIP38* resulted in hypersensitivity to ABA and induced multiple ABA-associated phenotypes, including increased resistance to dehydration (Kang et al. 2002). These data suggest a central role for these bZIP proteins in mediating ABA signalling. How ABA is linked to the transcriptional activation of bZIP genes is, however, currently unknown.

Besides the ABREs, so-called coupling elements (CEs) have been identified as *cis*-regulatory promoter elements involved in ABA-mediated gene expression (Shen et al. 1996). The core sequence of the CE1 element is CACCGC. These elements occur in several genes from *Arabidopsis*, barley, maize, and other plants. They are activated by ABA, particularly during seed maturation. Interestingly, CE1 elements also occur in sugar response genes (Huijser et al. 2000), which indicates cross-talk between ABA and sugar sensing. Recently, Niu et al. (2002) showed that the maize homologue of the *Arabidopsis* AP2-domain protein ABI4 binds to the CE1 element in a number of ABA-related genes. This may imply cooperation between bZIP-type factors and AP2-domain proteins in ABA-mediated responses to dehydration.

1.5.2 The dehydration-responsive element

Analysis of the promoter regions of other dehydration-induced genes led to the discovery of the dehydration-responsive element (DRE) in *Arabidopsis*; the same motif, here referred to as the C-repeat (CRT), was also identified in promoters of genes that respond to cold stress (Baker et al. 1994; Yamaguchi-Shinozaki and Shinozaki 1994). The sequence was subsequently found in other cold and dehydration inducible promoters, a finding which indicates a general role for this element. Yeast one-hybrid screens using this CRT/DRE element led to the identification of CRT/DRE binding factors, CBF and DREB, respectively (Table 1) (Stockinger et al. 1997; Liu et al. 1998). Characteristic for these factors is an AP2 domain, a 60-amino acid DNA-binding motif originally identified in the product of the floral organ identity gene *APETALA2* (Jofuku et al. 1994). Other members of this AP2-domain subfamily are involved in ethylene and jasmonate signalling, and in pathogen defence reactions, and are now often referred to as ethylene responsive element-binding-factors or ERF proteins (Singh et al. 2002). *CBF1*, *CBF2*, and *CBF3* are transcriptionally induced by cold, but not by ABA or dehydration (Medina et al. 1999). *CBF4*, on the other hand, is induced by dehydration and ABA treatment. Ectopic expression of *CBF4* confers increased tolerance to cold and dehydration in *Arabidopsis* (Haake et al. 2002). Similarly, in tomato, over-expression of *CBF1* leads to enhanced dehydration tolerance (Hsieh et al. 2002).

The gene *DREB1A* and its homologues *DREB1B* and *DREB1C* are identical to the *CBF1*, *CBF2*, and *CBF3* genes (see Table 1), and are transcriptionally induced by cold. *DREB2A* and *DREB2B* encode new ERF factors, and they are induced by dehydration (Liu et al. 1998). Although over-expression of *DREB1A* seems to activate downstream target genes and confer stress tolerance, the transformants showed growth retardation. The negative effect on plant growth can be avoided by using a stress inducible promoter for overexpression (Kasuga et al. 1999). When the cold inducible *DREB1A* gene was overexpressed, the transformants showed increased tolerance to dehydration and salinity in addition to cold tolerance, supporting the idea that the pathways that confer protection against the different stresses overlap.

1.5.3 The SAP domain

In *C. plantagineum*, detailed mapping of the promoter of one of the dehydration-induced *lea*-like genes, *CDeT27-45*, revealed a novel element that is important for ABA-induced expression (Michel et al. 1993; Nelson et al. 1994). Using yeast one-hybrid interaction cloning, a gene for a SAP (SAF-A/B, Acinus, PIAS) domain protein, *CpR18*, was isolated, and its product was found to bind to this promoter in vitro (Hilbricht et al. 2002). CpR18 is localized in the nucleus and, importantly, it is able to activate the *CDeT27-45* promoter in tobacco protoplasts. The SAP domain is a DNA binding domain that is able to bind to scaffold or matrix attachment regions (Aravind and Koonin 2000). Unlike the *CDeT27-45* promoter, these regions are relatively AT-rich. Therefore, it seems likely that the C-terminal zinc finger in CpR18, and not the SAP domain, recognises and binds the *CDeT27-45* promoter. The SAP domain might be necessary to make the *CDeT27-45* promoter accessible to other transcription factors that directly regulate transcription of the *CDeT27-45* gene (Fig. 3). Interestingly, immediately adjacent to the SAP binding domain in the *CDeT27-45* promoter are four potential bZIP binding sites that are similar to ABA responsive elements. Therefore, binding of *CpR18* might facilitate the binding of bZIP proteins and thereby induce transcription of the *CDeT27-45* gene in response to dehydration.

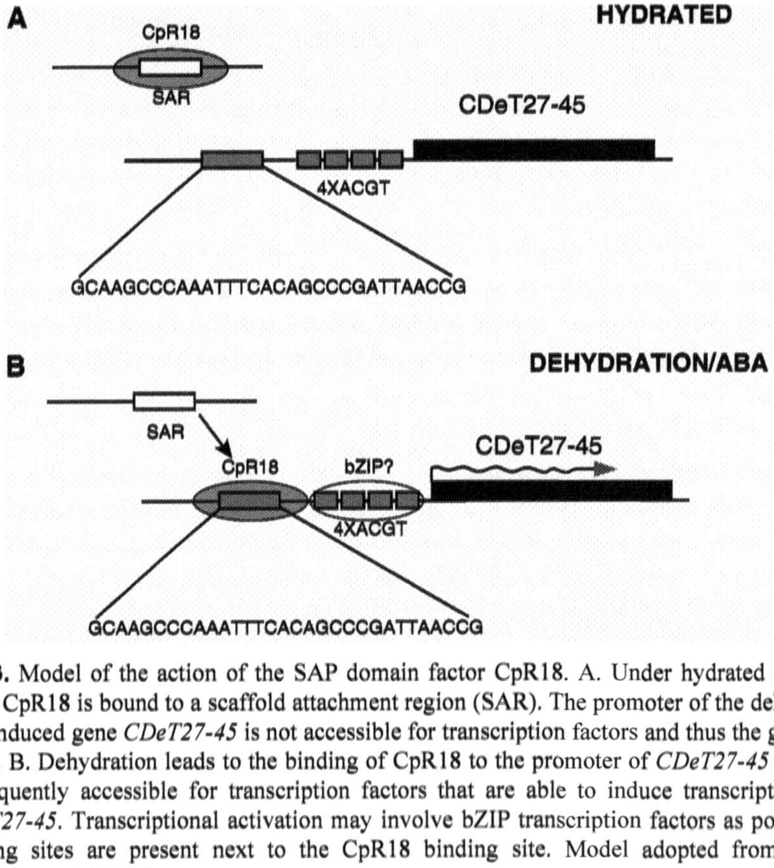

Fig. 3. Model of the action of the SAP domain factor CpR18. A. Under hydrated conditions, CpR18 is bound to a scaffold attachment region (SAR). The promoter of the dehydration-induced gene *CDeT27-45* is not accessible for transcription factors and thus the gene is silent. B. Dehydration leads to the binding of CpR18 to the promoter of *CDeT27-45* that is subsequently accessible for transcription factors that are able to induce transcription of *CDeT27-45*. Transcriptional activation may involve bZIP transcription factors as potential binding sites are present next to the CpR18 binding site. Model adopted from after Hilbricht et al. (2002).

1.5.4 Myb and helix-loop-helix domains

Various Myb-type transcription factors have been isolated which are induced by dehydration (Urao et al. 1993; Iturriaga et al. 1996; Abe et al. 1997). The *AtMyb2* gene of *Arabidopsis* is induced by dehydration, salt, and ABA (Urao et al. 1993). AtMYB2 was found to bind to the promoter of the dehydration-responsive gene *rd22* (Abe et al. 1997). In addition, a dehydration- and ABA-inducible helix-loop-helix type transcription factor, AtMYC2, was found to bind the *rd22* promoter. Over-expression of both genes led to hypersensitivity to ABA and to increased expression of *rd22* (Abe et al. 2003). This indicates that both *AtMYB2* and *AtMYC2* play a major role in the control of gene expression in response to water deficit. Three Myb transcription factors have been identified in *C. plantagineum*, which share high sequence homology with AtMYB2 from *A. thaliana* (Iturriaga et al. 1996). The gene for one of these, *cpm7*, is induced in roots after dehydration treatment. One of the *lea*-like genes, *pcC11-24*, contains a Myb binding site in its promoter and may thus be a target for CPM7 (Velasco et al. 1998).

1.5.5 Homeodomain proteins

Various homeodomain-leucine zipper proteins (HD-Zip) are induced by a variety of stress conditions, including dehydration (Söderman et al. 1996; Lee and Chun 1998; Söderman et al. 1999; Frank et al. 1998; Deng et al. 2002). In *Arabidopsis*, the dehydration induced transcription of both *ATHB-6* and *ATHB-7* is completely dependent on ABA, as no mRNA can be detected in the ABA biosynthetic mutant *aba-3* (Söderman et al. 1996; Söderman et al. 1999). In *C. plantagineum*, genes for two dehydration-induced HD-ZIP proteins, *CpHB-1* and *CpHB-2*, have been identified. These proteins are capable of forming both homo- and heterodimers in a yeast two-hybrid assay. While *CPHB-2* is induced by ABA, *CPHB-1* is not, distinguishing two pathways that are induced by dehydration, one via ABA and one independently of ABA. The yeast two-hybrid interaction suggests that the ABA-independent and ABA-dependent pathways can converge through these HD-Zip proteins. The analysis of a number of other HD-Zip proteins from *C. plantagineum* identified genes that are transcriptionally induced by dehydration, while others are downregulated by dehydration (Deng et al. 2002). As some HD-Zip proteins seem to function as repressors (Meijer et al. 1997; Steindler et al. 1999), an intriguing possibility is that the latter encode repressors of dehydration-induced genes.

1.5.6 An RNA as a signalling molecule?

Callus tissue of *Craterostigma* is normally not resistant to dehydration unless it is pre-treated with ABA (Bartels et al. 1990). An activation tagging experiment that screened for mutants that are capable of bypassing this ABA dependence identified a gene (*cdt-1*) with unusual characteristics (Furini et al. 1997). Only two very small open reading frames are present in the *cdt-1* RNA. Whether these small open reading frames are of biological relevance or the *cdt-1* transcript itself acts as the active molecule remains to be established. The latter possibility is intriguing in light of the recent breakthroughs that have identified many potential functions for non-coding RNAs, including the numerous small RNA species that act as important regulators via RNA interference (Matzke et al. 2001; Jones 2002b; Storz 2002). Under conditions of extreme stress, these RNAs may have a similar function to retroelements and transposons, which have been implicated in stress-triggered alterations in gene expression (Weil and Wessler 1990). For example, it was recently reported that retrotransposons alter the expression of adjacent genes in wheat (Kashkush et al. 2003).

1.5.7 Positioning of signals in the network

In the preceding sections, we have described many different molecules, which are presumed to be involved in dehydration stress signalling. The evidence for their involvement has been derived mainly from the observation that the molecules accumulate or are modified upon dehydration. This approach identifies single compounds without determining their interacting partners or their positions in the signalling network. Results are now emerging from studies on double mutants, which allow us to determine the hierarchy of some signals. Thus, analysis of double mutants deficient in the phosphatases ABI1 and ABI2 revealed a link between these phosphatases and calcium as well as reactive oxygen signals (Guo et al. 2002). Recently, a protein kinase (OST1) was identified which is an essential positive regulatory element in ABA-mediated stoma opening in response to dehydration (Mustilli et al. 2002). OST1 was shown to act downstream of ABA perception and upstream of a reactive oxygen signal (ROS).

1.6 Dehydration-activated proteins

The last step in the dehydration signalling cascade is the activation of genes responsible for the synthesis of compounds that serve to protect cellular structures against the deleterious effects of dehydration (Fig.1). Plants that are capable of surviving under dry conditions have adopted a variety of different strategies. We will discuss three mechanisms that seem important in enabling plants to withstand dehydration: the accumulation of solutes, scavenging of reactive oxygen species and synthesis of proteins with protective functions.

1.6.1 The accumulation of compatible solutes

In many species, dehydration leads to the accumulation of a variety of compatible solutes. Compatible solutes are soluble molecules of low molecular weight that are non-toxic and do not interfere with cellular metabolism. The chemical nature of the compatible solutes differs among plant species. They include betaines including glycine betaine, amino acids (especially proline), polyols and sugars such as mannitol, sorbitol, sucrose, or trehalose. These compounds help to maintain turgor during dehydration by increasing the number of particles in solution. Furthermore, they may modulate the fluidity of membranes and keep proteins hydrated, thus stabilizing their structure (Hoekstra et al. 2001). Ultimately the sugars replace the water molecules, converting the cytosol into a so-called glassy state. Using transgenics, increases in the levels of a number of solutes have been shown to increase tolerance to dehydration. Examples are glycine betaine (Huang et al. 2000), mannitol (Tarczynski et al. 1993), fructan (Pilon-Smits et al. 1995), D-ononitol (Sheveleva et al. 1997), and trehalose (Holmström et al. 1996). However, the level of newly synthesized solutes was in most cases not sufficient to enable the tolerance effect to be attributed to osmotic adjustment; therefore other mechanisms are under discussion: such solutes may themselves act as signalling molecules to induce protective pathways or function as scavengers of reactive oxygen species.

Dehydration and desiccation tolerance are associated with the presence of considerable quantities of non-reducing di- and oligosaccharides, as illustrated in desiccation-tolerant resurrection plants (Hoekstra et al. 2001; Phillips et al. 2002). In *C. plantagineum*, the unusual sugar 2-octulose is present in leaves under normal growth conditions. This sugar is converted into sucrose upon water loss, and 2-octulose accumulates again upon rehydration (Bianchi et al. 1991). The accumulation of sucrose upon dehydration seems to be correlated with the acquisition of desiccation tolerance, as it has been reported for a number of different species of resurrection plants. It remains to be established whether the presence of this unusual sugar is really required for desiccation tolerance. The molecular mechanisms that regulate the accumulation of sucrose are unknown. However, genes encoding enzymes of sucrose metabolism have been found to be upregulated by dehydration and it has been proposed that these enzymes are involved in the conversion of octulose to sucrose (Ingram and Bartels 1996).

1.6.2 Genes that encode proteins with protective functions

Many genes, which are abundantly expressed in response to dehydration, have been isolated from numerous species by differential screening approaches using dehydrated versus non-stressed plant tissues. In spite of the molecular characterization of so many genes, our knowledge of the biochemical functions of their products is remarkably limited. The resurrection plant *C. plantagineum* and seeds of various plants have been rich sources of genes that are expressed upon dehydration and are likely to be involved in acquired resistance to desiccation (Bartels et al. 1990). Seeds of many species are able to survive without water for a long time, which requires maintenance of viable embryo structures. A number of the genes isolated from resurrection plants share sequence homologies with genes that are expressed in maturing seeds, suggesting that the desiccation tolerance mechanisms that operate in these situations show some similarity. The *late embryogenesis abundant* or *lea* genes are a case in point. Their expression is correlated with dehydration (Galau et al. 1986). The corresponding transcripts accumulate to high levels both in dormant seeds and in vegetative tissues of desiccation-tolerant and -sensitive plants upon dehydration. Based on sequence characteristics, they can be assigned to different groups, although there is some disagreement in the literature with regard to their classification (for details see Bray 1993; Dure 1993; Close 1997). Most LEA proteins are very hydrophilic, which allows them to remain soluble after boiling, and determines their particular biochemical properties. LEA proteins are probably ubiquitous in higher plants. Recently a LEA protein has also been reported to accumulate in response to desiccation in the anhydrobiotic nematode *Aphelenchus avenae* (Browne et al. 2002). This finding corroborates the predicted protective role of LEA proteins. The genomes of some microorganisms also contain sequences that may encode LEA-like proteins, although they have not been shown to be expressed (Browne et al. 2002). Despite correlative evidence for a protective function of LEA proteins during water deficit, direct biochemical evidence for this is still lacking. Transgenic approaches designed to demonstrate the protective role of LEA proteins by overexpressing them have yielded contradictory results. Thus, overexpression of the barley *lea* gene *HVA1* resulted in transgenic plants with increased tolerance (Xu et al. 1996). In contrast, overexpression of *C. plantagineum lea* genes did not lead to enhanced tolerance (Iturriaga et al. 1992). That LEA proteins may act synergistically with non-reducing sugars to form a glassy matrix and thus confer protection is an attractive hypothesis (Hoekstra et al. 2001, and references herein). This hypothesis is supported by the abundance of LEA proteins and of reducing sugars in desiccation-tolerant plant tissues and also in desiccation-tolerant nematodes (Browne et al. 2002; Phillips et al. 2002).

1.6.3 Reactive oxygen intermediates

One consequence of many stresses, including dehydration, is an increase in the concentration of reactive oxygen intermediates (ROI) (Mittler 2002). ROI cause irreversible damage to membranes, proteins, DNA, and RNA. However, a low concentration of ROI is vital for the plant cell, as some ROI molecules are themselves essential signalling components in stress defence, as was discovered recently and has been discussed in this chapter (Pei et al. 2000, Musitlli et al. 2002). When the concentration of ROI increases because of dehydration, prevention of concurrent damage is essential for survival. It is well known that ROI accumulation is largely controlled by intrinsic antioxidant systems that include enzymatic scavengers like superoxide dismutase, peroxidases, and catalases. Besides these general defence strategies, dehydration seems to trigger other enzymes, which are likely to be involved in ROI scavenging. In *C. plantagineum*, a gene encoding an aldehyde dehydrogenase (*Cp-ALDH*) is induced by dehydration and ABA (Kirch et al. 2001). The gene product is localised in plastids. Presumably, it functions in the detoxification of reactive aldehydes produced by peroxidation of lipids by ROI (Sunkar et al. 2003). An *Arabidopsis* homologue, *Ath-ALDH3*, was also identified which exhibits a similar expression pattern. Another similar example is the ABA- and dehydration-induced gene *MsALR*, which encodes an aldose/aldehyde reductase in alfalfa (Oberschall et al. 2000). Ectopic expression of *MsALR* in tobacco improves tolerance to dehydration. Thus, detoxification of aldehydes seems to be an important defence against dehydration-induced damage by ROI (Bartels 2001). Details of the effects of oxidative stress are described in chapter 5 of this volume.

1.7 Conclusions and outlook

The analysis of differential gene expression and, more recently, analysis of global gene expression patterns using macro- and microarray approaches have identified a broad spectrum of transcripts, whose expression is modified in response to dehydration (Fowler and Thomashow 2002; Kreps et al. 2002, Seki et al. 2002). These studies have provided a fairly comprehensive overview of the types of transcripts modulated by dehydration in plants. They have shown that at least several hundred genes are affected by dehydration and that their products can be assigned to many pathways. This makes it difficult to prioritize genes and define primary dehydration genes. No simple solution can be expected from a genetic approach for two reasons: dehydration tolerance is a polygenic trait, and the resurrection plants - the best characterized desiccation-tolerant species - do not have a diploid genome. For this reason, no simple mutational approach has been reported for the identification of genes involved in desiccation tolerance. Desiccation tolerance is a complex character which cannot be dissected genetically in the same way as monogenic resistances e.g. to some plant pathogens (Crute and Pink 1996). However, the analysis of quantitative trait loci may shed some light on the influence of the different genetic parameters.

Progress in understanding survival under dehydration conditions may be expected from an interdisciplinary approach involving cell biology and physical biochemistry. Such an approach must include studies on water structure-function relationships in the context of molecules like compatible solutes and LEA proteins. Analyses of water-solute-macromolecule-solid matrix interactions may offer new perspectives on the requirements for maintaining the integrity of cellular structures under otherwise lethal conditions.

We are far from being able to provide a comprehensive picture of the signalling network necessary for a coordinated cellular response to dehydration. Large projects designed to produce collections of specific knockout mutants will shed some light on the role of individual members of gene families encoding regulatory molecules. This approach is at present restricted to *Arabidopsis*. However, one must exercise caution in extrapolating the results obtained from *Arabidopsis*. There is some diversity with respect to gene function and gene identity, as pointed out for some examples in this review. In case of the *phospholipase D* genes, the genes with the closest sequence homology between *C. plantagineum* and *A. thaliana* do not respond in the same way to dehydration. Likewise, *Arabidopsis* does not have homologues of *cdt-1* genes, nor does it synthesize the same sugars as some resurrection plants. Despite the enormous progress derived from studies in *Arabidopsis*, it is necessary to keep in mind that there is diversity within the different plant species in their molecular response to dehydration. A comparison of different systems will also contribute to identifying essential genes, as illustrated by the similarities in desiccation responses between resurrection plants and certain nematodes.

Hardly any information is available on the post-transcriptional control level, a research area that can be addressed by proteomics approaches, which are just beginning. Despite the expenditure of a great deal of effort and the application of a variety of approaches, the identity of the molecules that sense dehydration remains a complete mystery, and it is difficult to see how this challenge can be resolved soon. These are research areas that should receive most attention in the near future.

Acknowledgements

The work of D.B. on this subject is supported by grants from the European commission, Deutsche Forschungsgemeinschaft (DFG) and Fonds der chemischen Industrie. We thank D. Hoonhout for secretarial assistance. We would like to apologize to those whose work was not cited due to space restrictions.

References

Abe H, Urao T, Ito T, Seki M, Shinozaki K, Yamaguchi-Shinozaki K (2003) *Arabidopsis* AtMYC2 (bHLH) and AtMYB2 (MYB) function as transcriptional activators in abscisic acid signaling. Plant Cell 15:63-78

Abe H, Yamaguchi-Shinozaki K, Urao T, Iwasaki T, Hosokawa D, Shinozaki K (1997) Role of *Arabidopsis* MYC and MYB homologs in drought-and abscisic acid-regulated gene expression. Plant Cell 9:1859-1868

Allen GJ, Chu SP, Harrington CL, Schumacher K, Hoffmann T, Tang YY, Grill E, Schroeder JI (2001) A defined range of guard cell calcium oscillation parameters encodes stomatal movements. Nature 411:1053-1057

Aravind L, Koonin EV (2000) SAP - a putative DNA-binding motif involved in chromosomal organization. Trends Biochem Sci 25:112-114

Assmann SM, Wang X-Q (2001) From milliseconds to millions of years: guard cells and environmental responses. Curr Opin Plant Biol 4:421-428

Baker SS, Wilhelm KS, Thomashow MF (1994) The 5'-region of *Arabidopsis thaliana cor15a* has cis-acting elements that confer cold-, drought- and ABA-regulated gene expression. Plant Mol Biol 24:701-713

Bartels D (2001) Targeting detoxification pathways: an efficient approach to obtain plants with multiple stress tolerance? Trends Plant Sci 6:284-286

Bartels D, Salamini F (2001) Desiccation tolerance in the resurrection plant *Craterostigma plantagineum* A contribution to the study of drought tolerance at the molecular level. Plant Physiol 127:1346-1353

Bartels D, Schneider K, Terstappen G, Piatkowski D, Salamini F (1990) Molecular cloning of abscisic acid-modulated genes which are induced during desiccation of the resurrection plant *Craterostigma plantagineum.* Planta 181:27-34

Bertauche N, Leung J, Giraudat J (1996) Protein phosphatase activity of abscisic acid insensitive 1 (ABI1) protein from *Arabidopsis thaliana.* Eur J Biochem 241:193-200

Bianchi G, Gamba A, Murelli C, Salamini F, Bartels D (1991) Novel carbohydrate metabolism in the resurrection plant *Craterostigma plantagineum.* Plant J 1:355-359

Black M, Pritchard HW (2002) Desiccation and survival in plants. Drying without dying CABI Publishing, New York

Bögre L, Ligterink W, Heberle-Bors E, Hirt H (1996) Mechanosensors in plants. Nature 383:489-490

Bray EA (1993) Molecular responses to water deficit. Plant Physiol 103:1035-1040

Browne J, Tunnacliffe A, Burnell A (2002) Plant desiccation gene found in a nematode. Nature 416:36

Camoni L, Harper JF, Palmgren MG (1998) 14-3-3 proteins activate a plant calcium-dependent protein kinase (CDPK). FEBS Lett 430:381-384

Chang C, Stewart RC (1998) The two-component system regulation of diverse signaling pathways in prokaryotes and eukaryotes .Plant Physiol 117:723-731

Cheng S-H, Willmann MR, Chen H-C, Sheen J (2002) Calcium signaling through protein kinases. The *Arabidopsis* calcium-dependent protein kinase gene family. Plant Physiol 129:469-485

Choi H, Hong J, Ha J, Kang J, Kim SY (2000) ABFs, a family of ABA-responsive element binding factors. J Biol Chem 275:1723-1730

Close TJ (1997) Dehydrins: A commonalty in the response of plants to dehydration and low temperature. Physiol Plantarum 100:291-296

Crute JR, Pink DAC (1996) Genetics and utilization of pathogen resistance in plants. Plant Cell 8:1747-1717

Deng X, Phillips J, Meijer AH, Salamini F, Bartels D (2002) Characterization of five novel dehydration-responsive homeodomain leucine zipper genes from the resurrection plant *Craterostigma plantagineum.* Plant Mol Biol 49:601-610

Dure L (1993) Structural motifs in LEA proteins In: Timothy J, Bray CAEA (eds) Plant responses to cellular dehydration during environmental stress. American Society of Plant Physiologists, pp 91-103

Evans NH, McAinsh MR, Hetherington AM (2001) Calcium oscillations in higher plants. Plant Biol 4:415-420

Farmer PK, Choi JH (1999) Calcium and phospholipid activation of a recombinant calcium-dependent protein kinase (DcCPK1) from carrot (Daucus carota L) Biochim Biophys Acta 1434:6-17

Finkelstein RR, Gampala SSL, Rock CD (2002) Abscisic acid signaling in seeds and seedlings. Plant Cell 14:S15-S45

Finkelstein RR, Lynch TJ (2000) The *Arabidopsis* abscisic acid response gene ABI5 encodes a basic leucine zipper transcription factor. Plant Cell 12:599-609

Fowler S, Thomashow MF (2002) *Arabidopsis* transcriptome profiling indicates that multiple regulatory pathways are activated during cold acclimation in addition to the CBF cold response pathway. Plant Cell 14:1675-1690

Frank W, Munnik T, Kerkmann K, Salamini F, Bartels D (2000) Water deficit triggers phospholipase D activity in the resurrection plant *Craterostigma plantagineum.* Plant Cell 12:111-124

Frank W, Phillips J, Salamini F, Bartels D (1998) Two dehydration-inducible transcripts from the resurrection plant *Craterostigma plantagineum* encode interacting homeodomain-leucine zipper proteins. Plant J 15:413-421

Furini A, Koncz C, Salamini F, Bartels D (1997) High level transcription of a member of a repeated gene family confers dehydration tolerance to callus tissue of *Craterostigma plantagineum.* EMBO J 16:3599-3608

Galau GW, Hughes DW, Dure L (1986) Abscisic acid induction of cloned cotton late embryogenesis abundant(LEA) messenger RNAs. Plant Mol Biol 7:155-170

Gosti F, Beaudoin N, Serizet C, Webb AA, Vartanian N, Giraudat J (1999) ABI1 protein phosphatase 2C is a negative regulator of abscisic acid signaling. Plant Cell 11:1897-1910

Guiltinan MJ, Marcotte WRJ, Quatrano RS (1990) A plant leucine zipper protein that recognizes an abscisic acid response element. Science 250:267-271

Guo Y, Xiong LM, Song C-P, Gong DM, Halfter U, Zhu JK (2002) A calcium sensor and its interacting protein kinase are global regulators of abscisic acid signaling in *Arabidopsis*. Dev Cell 3:233-244

Haake V, Cook D, Riechmann JL, Pineda O, Thomashow MF, Zhang JZ (2002) Transcription factor CBF4 is a regulator of drought adaptation in *Arabidopsis*. Plant Physiol 130:639-648

Hilbricht T, Salamini F, Bartels D (2002) CpR18, a novel SAP-domain plant transcription factor, binds to a promoter region necessary for ABA mediated expression of the *CDeT27-45* gene from the resurrection plant *Craterostigma plantagineum* Hochst. Plant J 31:293-303

Himmelbach A, Hoffmann T, Leube MP, Hohener B, Grill E (2002) Homeodomain protein ATHB6 is a target of the protein phosphatase ABI1 and regulates hormone responses in *Arabidopsis*. EMBO J 21:3029-3038

Hirayama T, Mitsukawa N, Shibata D, K S (1997) AtPLC2, a gene encoding phosphoinositide-specific phospholipase C, is constitutively expressed in vegetative and floral tissues in *Arabidopsis thaliana*. Plant Mol Biol 34:175-180

Hirayama T, Ohto C, Mizoguchi T, Shinozaki K (1995) A gene encoding a phosphatidylinositol-specific phospholipase C is induced by dehydration and salt stress in *Arabidopsis thaliana*. Proc Natl Acad Sci USA 92:3903-3907

Hoekstra FA, Golovina EA, Buitink J (2001) Mechanisms of plant desiccation tolerance. Trends Plant Sci 6:431-438

Hohmann S (2002) Osmotic stress signaling and osmoadaptation in yeasts. Microbiol Mol Biol Rev 66:300-372

Holmström KO, Mantyla E, Welin B, Mandal A, Palva ET (1996) Drought tolerance in tobacco. Nature 379

Hsieh T-H, Lee J-t, Charng Y-y, Chan M-T (2002) Tomato plants ectopically expressing *Arabidopsis* CBF1 show enhanced resistance to water deficit stress. Plant Physiol 130:618-626

Huang J, Hirji R, Adam L, Rozwadowski KL, Hammerlindl JK, Keller WA, Selvaraj G (2000) Genetic engineering of Glycinebetaine production toward enhancing stress tolerance in plants: Metabolic limitations. Plant Physiol 122:747-756

Huijser C, Kortstee A, Pego J, Weisbeek PJ, Wisman E, Smeekens S (2000) The *Arabidopsis* Sucrose uncoupled-6 gene is identical to Abscisic acid insensitive-4: Involvement of abscisic acid in sugar responses. Plant Journal 23:577-585

Ichimura K, Mizoguchi T, Irie K, Morris P, Giraudat J, Matsumoto K, Shinozaki K (1998) Isolation of ATMEKK1 (a MAP Kinase Kinase Kinase)-interacting proteins and analysis of a MAP kinase cascade in *Arabidopsis*. Biochem Biophys Res Comm 253:532-543

Ichimura K, Mizoguchi T, Yoshida R, Yuasa T, Shinozaki K (2000) Various abiotic stresses rapidly activate *Arabidopsis* MAP kinases ATMPK4 and ATMPK6. Plant J 24:655-665

Ingram J, Bartels D (1996) The molecular basis of dehydration tolerance in plants. Annu Rev Plant Physiol Plant Mol Biol 47:377-403

Inoue T, Higuchi M, Hashimoto Y, Seki M, Kobayashi M, Kato T, Tabata S, Shinozaki K, Kakimoto T (2001) Identification of CRE1 as a cytokinin receptor from *Arabidopsis*. Nature 409:1060-1063

Iturriaga G, Leyns L, Villegas A, Gharaibeh R, Salamini F, Bartels D (1996) A family of novel myb-related genes from the resurrection plant *Craterostigma plantagineum* are specifically expressed in callus and roots in response to ABA or desiccation. Plant Mol Biol 32:707-716

Iturriaga G, Schneider K, Salamini F, Bartels D (1992) Expression of desiccation-related proteins from the resurrection plant *Craterostigma plantagineum* in transgenic tobacco. Plant Mol Biol 20:555-558

Jakoby M, Weisshaar B, Droge-Laser W, Vicente-Carbajosa J, Tiedemann J, Kroj T, Parcy F (2002) bZIP transcription factors in *Arabidopsis*. Trends Plant Sci 7:106-111

Jofuku KD, Boer B, Montagu MV, Okamuro JK (1994) Control of *Arabidopsis* flower and seed development by the homeotic gene APETALA2. Plant Cell 6:1211-1225

Jonak C, Kiegerl S, Ligterink W, Barker PJ, Huskisson NS, Hirt H (1996) Stress signaling in plants: A mitogen-activated protein kinase pathway is activated by cold and drought. Prof Natl Acad Sci USA 93:11274-11279

Jones AM (2002a) G-protein-coupled signaling in *Arabidopsis*. Curr Opin in Plant Biol 5:402-407

Jones L (2002b) Revealing micro-RNAs in plants. Trends Plant Sci 7:473-475

Kang J-y, Choi H-i, Im M-y, Kim SY (2002) *Arabidopsis* basic leucine zipper proteins that mediate stress-responsive abscisic acid signaling. Plant Cell 14:343-357

Kashkush K, Feldman M, Levy AA (2003) Transcriptional activation of retrotransposons alters the expression of adjacent genes in wheat. Nature Genet 33:102-106

Kasuga M, Liu Q, Miura S, Yamaguchi-Shinozaki K, Shinozaki K (1999) Improving plant drought, salt, and freezing tolerance by gene transfer of a single stress-inducible transcription factor. Nat Biotechnol 17:287-291

Katagiri T, Takahashi S, Shinozaki K (2001) Involvement of a novel *Arabidopsis* phospholipase D, AtPLD∂, in dehydration-inducible accumulation of phosphatidic acid in stress signalling. Plant J 26:595-605

Kerk D, Bulgrien J, Smith DW, Barsam B, Veretnik S, Gribskov M (2002) The complement of protein phosphatase catalytic subunits encoded in the genome of *Arabidopsis*. Plant Physiol 129:908-925

Kiegle E, Moore CA, Haseloff J, Tester MA, Knight MR (2000) Cell-type-specific calcium responses to drought, salt and cold in the *Arabidopsis* root. Plant J 23:267-278

Kirch HH, Nair A, Bartels D (2001) Novel ABA- and dehydration-inducible aldehyde dehydrogenase genes isolated from the resurrection plant *Craterostigma plantagineum* and *Arabidopsis thaliana*. Plant J 28:555-567

Kiyosue T, Yamaguchi-Shinozaki K, Shinozaki K (1994) Cloning of cDNAs for genes that are early-responsive to dehydration stress (ERDs) in *Arabidopsis thaliana* L: identification of three ERDs as HSP cognate genes. Plant Mol Biol 25:791-798

Koornneef M (1984) The isolation and characterization of abscisic acid-insensitive mutants of *Arabidopsis thaliana*. Physiol Plant 61:377

Kopka J, Pical C, Gray JE, Muller-Rober B (1998) Molecular and enzymatic characterization of three phosphoinositide-specific phospholipase C isoforms from potato. Plant Physiol 116:239-250

Kreps JA, Wu Y, Chang H-S, Zhu T, Wang X, Harper JF (2002) Transcriptome changes for *Arabidopsis* in response to salt, osmotic, and cold stress. Plant Physiol 130:2129-2141

Lee S, Hirt H, Lee Y (2001) Phosphatidic acid activates a wound-activated MAPK in Glycine max. Plant J 26:479-486

Lee YH, Chun JY (1998) A new homeodomain-leucine zipper gene from *Arabidopsis thaliana* induced by water stress and abscisic acid treatment. Plant Mol Biol 37:377-384

Leung J, Giraudat J (1998) Abscisic acid signal transduction. Annu Rev Plant Physiol Plant Mol Biol 49:199-222

Leung J, Merlot S, Giraudat J (1997) The *Arabidopsis* ABSCISIC ACID-INSENSITIVE2 (ABI2) and ABI1 genes encode homologous protein phosphatases 2C involved in abscisic acid signal transduction. Plant Cell 9:759-771

Liu Q, Kasuga M, Sakuma Y, Abe H, Miura S, Yamaguchi-Shinozaki K, Shinozaki K (1998) Two transcription factors, DREB1 and DREB2, with an EREBP/AP2 DNA binding domain separate two cellular signal transduction pathways in drought- and low-temperature-responsive gene expression, respectively, in *Arabidopsis*. Plant Cell 10:1391-1406

Maeda T, Wurgler-Murphy S, Saito H (1994) A two-component system that regulates an osmosensing MAP kinase cascade in yeast. Nature 369:242-245

Matzke M, Matzke AJM, Kooter JM (2001) RNA: Guiding gene silencing. Science 293:1080-1083

McAinsh MR, Hetherington AM (1998) Encoding specificity in calcium signalling systems. Trends Plant Sci 3:32-36

Medina J, Bargues M, Terol J, Perez-Alonso M, Salinas J (1999) The *Arabidopsis* CBF gene family is composed of three genes encoding AP2 Domain-containing proteins whose expression is regulated by low temperature but not by abscisic acid or dehydration. Plant Physiol 119:463-470

Meijer AH, Scarpella E, van Dijk EL, Qin L, Taal AJC, Rueb S, Harrington SE, McCouch SR, Schilperoort RA, Hoge JHC (1997) Transcriptional repression by Oshox1, a novel homeodomain leucine zipper protein from rice. Plant J 11:263-276

Menkens A, Cashmore A (1994) Isolation and characterization of a fourth *Arabidopsis thaliana* G-Box- binding factor, which has similarities to Fos oncoprotein. Proc. Natl Acad Sci USA 91:2522-2526

Merlot S, Giraudat J (1997) Genetic analysis of abscisic acid signal transduction. Plant Physiol 114:751

Merlot S, Gosti F, Guerrier D, Vavasseur A, Giraudat J (2001) The ABI1 and ABI2 protein phosphatases 2C act in a negative feedback regulatory loop of the abscisic acid signalling pathway. Plant J 25:295-303

Meskiene I, Bogre L, Glaser W, Balog J, Brandstotter M, Zwerger K, Ammerer G, Hirt H (1998) MP2C, a plant protein phosphatase 2C, functions as a negative regulator of mitogen-activated protein kinase pathways in yeast and plants. Proc Natl Acad Sci USA 95:1938-1943

Meyre D, Leonardi A, Brisson G, Vartanian N (2001) Drought-adaptive mechanisms involved in the escape/tolerance strategies of *Arabidopsis* Landsberg *erecta* and Columbia ecotypes and their F1 reciprocal progeny. J Plant Physiol 158:1145-1152

Michel D, Salamini F, Bartels D, Dale P, Baga M, Szalay A (1993) Analysis of a desiccation and ABA-responsive promoter isolated from the resurrection plant *Craterostigma plantagineum.* Plant J 4:29-40

Mittler R (2002) Oxidative stress, antioxidants and stress tolerance. Trends Plant Sci 7:405-410

Mundy J, Yamaguchi-Shinozaki K, Chua N (1990) Nuclear proteins bind conserved elements in the abscisic acid-responsive promoter of a Rice Rab gene. Prof. Natl Acad Sci USA 87:1406-1410

Munnik T (2001) Phosphatidic acid: an emerging plant lipid second messenger. Trends Plant Sci 6:227-233

Munnik T, Meijer HJ, Ter Riet B, Hirt H, Frank W, Bartels D, Musgrave A (2000) Hyperosmotic stress stimulates phospholipase D activity and elevates the levels of phosphatidic acid and diacylglycerol pyrophosphate. Plant J 22:147-154

Murata Y, Pei ZM, Mori IC, Schroeder J (2001) Abscisic acid activation of plasma membrane Ca^{2+} channels in guard cells requires cytosolic NAD(P)H and is differentially disrupted upstream and downstream of reactive oxygen species production in abi1-1 and abi2-1 protein phosphatase 2C mutants. Plant Cell 13:2513-2523

Mustilli A-C, Merlot S, Vavasseur A, Fenzi F, Giraudat J (2002) *Arabidopsis* OST1 protein kinase mediates the regulation of stomatal aperture by abscisic acid and acts upstream of reactive oxygen species production. Plant Cell 14:3089-3099

Nakashima K, Kiyosue T, Yamaguchi-Shinozaki K, Shinozaki K (1997) A nuclear gene, *erd*1, encoding a chloroplast-targeted Clp protease regulatory subunit homolog is not only induced by water stress but also developmentally upregulated during senescence in *Arabidopsis thaliana.* Plant J 12:851-861

Nelson D, Salamini F, Bartels D (1994) Abscisic acid promotes novel DNA-binding activity to a desiccation- related promoter of *Craterostigma plantagineum.* Plant J 5:451-458

Niu X, Helentjaris T, Bate NJ (2002) Maize ABI4 binds coupling element1 in abscisic acid and sugar response genes. Plant Cell 14:2565-2575

Oberschall A, Deak M, Torok K, Sass L, Vass I, Kovacs I, Feher A, Dudits D, Horvath GV (2000) A novel aldose/aldehyde reductase protects transgenic plants against lipid peroxidation under chemical and drought stresses. Plant J 24:437-446

Patharkar OR, Cushman JC (2000) A stress-induced calcium-dependent protein kinase from *Mesembryanthum crystallinum* phosphorylates a two-component pseudo-response regulator. Plant J 24:679-691

Pei ZM, Murata Y, Benning G, Thomine S, Klusener B, Allen GJ, Grill E, Schroeder JI (2000) Calcium channels activated by hydrogen peroxide mediate abscisic acid signalling in guard cells. Nature 406:731-734

Phillips JR, Oliver MJ, Bartels D (2002) Molecular genetics of desiccation and tolerant systems In: Black M, Pritchard HW (eds) Desiccation and survival in plants: Drying without dying CAB International

Pilon-Smits E, Ebskamp M, Paul MJ, Jeuken M, Weisbeek PJ, Smeekens S (1995) Improved performance of transgenic fructan-accumulating tobacco under drought stress. Plant Physiol 107:125-130

Qin C, Wang X (2002) The *Arabidopsis* phospholipase D family Characterization of a calcium-independent and phosphatidylcholine-selective PLDζ1 with distinct regulatory domains. Plant Physiol 128:1057-1068

Saijo Y, Hata S, Kyozuka J, Shimamoto K, Izui K (2000) Over-expression of a single Ca^{2+}-dependent kinase confers both cold and salt/drought tolerance on rice plants. Plant J 23:319-327

Sanders D, Brownlee C, Harper JF (1999) Communicating with calcium. Plant Cell 11:691-706

Sang Y, Zheng S, Li W, Huang B, Wang X (2001) Regulation of plant water loss by manipulating the expression of phospholipase Dα. Plant J 28:135-144

Sangwan V, Örvar BL, Beyerly J, Hirt H, Dhindsa RS (2002) Opposite changes in membrane fluidity mimic cold and heat stress activation of distinct plant MAP kinase pathways. Plant J 31:629-638

Schroeder JI, Allen GJ, Hugouvieux V, Kwak JM, Waner D (2001) Guard cell signal transduction. Annu Rev Plant Physiol Plant Mol Biol 52:627-658

Seki M, Narusaka M, Ishida J, Nanjo T, Fujita M, Oono Y, Kamiya A, Nakajima M, Enju A, Sakurai T, Satou M, Akiyama K, Taji T, Yamaguchi-Shinozaki K, Carninci P, Kawai J, Hayashizaki Y, Shinozaki K (2002) Monitoring the expression profiles of 7000 *Arabidopsis* genes under drought, cold and high-salinity stresses using a full-length cDNA microarray. Plant J 31:279-292

Seo M, Koshiba T (2002) Complex regulation of ABA biosynthesis in plants. Trends Plant Sci 7:41-48

Sheen J (1996) Ca^{2+}-dependent protein kinases and stress signal transduction in plants. Science 274:1900-1902

Shen Q, Zhang P, Ho T-H (1996) Modular nature of abscisic acid (ABA) response complexes: Composite promoter units that are necessary and sufficient for ABA induction of gene expression in barley. Plant Cell 8:1107-1119

Sheveleva E, Chmara W, Bohnert HJ, Jensen RG (1997) Increased salt and drought tolerance by D-Ononitol production in transgenic *Nicotiana tabacum* L. Plant Physiol 115:1211-1219

Shinozaki K, Yamaguchi-Shinozaki K (1997) Gene expression and signal transduction in water-stress response. Plant Physiol 115:327-334

Singh KB, Foley RC, Onate-Sanchez L (2002) Transcription factors in plant defense and stress responses. Curr Opi Plant Biol 5:430-436

Söderman E, Hjellström M, Fahleson J, Engström P (1999) The HD-Zip gene ATHB6 in *Arabidopsis* is expressed in developing leaves, roots and carpels and up-regulated by water deficit conditions. Plant Mol Biol 40:1073-1083

Söderman E, Mattsson J, Engström P (1996) The *Arabidopsis* homeobox gene ATHB-7 is induced by water deficit and by abscisic acid. Plant J 10:375-381

Steindler C, Matteucci A, Sessa G, Weimar T, Ohgishi M, Aoyama T, Morelli G, Ruberti I (1999) Shade avoidance responses are mediated by the ATHB-2 HD-zip protein, a negative regulator of gene expression. Development 126:4235-4245

Stockinger EJ, Gilmour SJ, Thomashow MF (1997) *Arabidopsis* thaliana CBF1 encodes an AP2 domain-containing transcriptional activator that binds to the C-repeat/DRE, a cis-acting DNA regulatory element that stimulates transcription in response to low temperature and water deficit. Proc Natl Acad Sci USA 94:1035-1040

Storz G (2002) An Expanding Universe of Noncoding RNAs. Science 296:1260-1263

Sunkar R, Bartels D, Kirch, HH (2003) Overexpression of a stress-inducible aldehyde dehydrogenase gene from *Arabidopsis thaliana* in transgenic plants improves stress tolerance. Plant J, in press

Takahashi S, Katagiri T, Hirayama T, Yamaguchi-Shinozaki K, Shinozaki K (2001) Hyperosmotic stress induces a rapid and transient increase in inositol 1,4,5-trisphosphate independent of abscisic acid in *Arabidopsis* cell culture. Plant Cell Physiol 42:214-222

Tarczynski MC, Jensen RG, Bohnert HJ (1993) Stress protection of transgenic plants by production of the osmolyte mannitol. Science 259:508-510

Ulm R, Revenkova E, di Sansebastiano GP, Bechtold N, Paszkowski J (2001) Mitogen-activated protein kinase phosphatase is required for genotoxic stress relief in *Arabidopsis*. Genes Dev 15:699-709

Uno Y, Furihata T, Abe H, Yoshida R, Shinozaki K, Yamaguchi-Shinozaki K (2000) *Arabidopsis* basic leucine zipper transcription factors involved in an abscisic acid-dependent signal transduction pathway under drought and high-salinity conditions. Proc Natl Acad Sci USA 97:11632-11637

Urao T, Katagiri T, Mizoguchi T, Yamaguchi-Shinozaki K, Hayashida N, Shinozaki K (1994) Two genes that encode Ca^{2+}-dependent protein kinases are induced by drought and high-salt stresses in *Arabidopsis thaliana.* Mol Gen Genet 244:331-340

Urao T, Yakubov B, Satoh R, Yamaguchi-Shinozaki K, Seki M, Hirayama T, Shinozaki K (1999) A transmembrane hybrid-type histidine kinase in *Arabidopsis* functions as an osmosensor. Plant Cell 11:1743-1754

Urao T, Yamaguchi-Shinozaki K, Urao S, Shinozaki K (1993) An *Arabidopsis* myb homolog is induced by dehydration stress and its gene product binds to the conserved MYB recognition sequence. Plant Cell 5:1529-1539

Velasco R, Salamini F, Bartels D (1998) Gene structure and expression analysis of the drought- and abscisic acid-responsive CDeT11-24 gene family from the resurrection plant *Craterostigma plantagineum* Hochst. Planta 204:459-471

Wang XQ, Ullah H, Jones AM, Assmann SM (2001) G protein regulation of ion channels and abscisic acid signaling in *Arabidopsis* guard cells. Science 292:2070-2072

Weil CF, Wessler SR (1990) The effects of plant transposable element insertion on transcription initiation and RNA processing. Annu Rev Plant Physiol Plant Mol Biol 41:527-552

Williams ME, Foster R, Chua NH (1992) Sequences flanking the hexameric G-Box core CACGTG affect the specificity of protein binding. Plant Cell 4:485-496

Wurgler-Murphy SM, Saito H (1997) Two-component signal transducers and MAPK cascades. Trends Biochem Sci 22:172-176

Yamaguchi-Shinozaki K, Shinozaki K (1994) A novel cis-acting element in an *Arabidopsis* gene is involved in responsiveness to drought, low-temperature, or high-salt stress. Plant Cell 6:251-264

Xu DP, Duan XL, Wang BY, Hong BM, Ho THD, Wu R (1996) Expression of a late embryogenesis abundant protein gene, HVA1, from barley confers tolerance to water deficit and salt stress in transgenic rice. Plant Physiol 110:249-257

Zhang S, Klessig DF (2001) MAPK cascades in plant defense signaling. Trends Plant Sci 6:520-527

Abbreviations

ABA: Abscisic acid
ABI: Abscisic acid insensitive
ABRE: Abscisic acid responsive element
AtHK1: *Arabidopsis* thaliana histidine kinase 1
bZIP: Basic region/leucine zipper
CE: Coupling element
CDPK: Calcium dependent protein kinase
CRT: C-repeat
DRE: Dehydration responsive element
HD-ZIP: Homeodomain-leucine zipper
LEA: late embryogenesis abundant
MAPK: Mitogen activated-like protein kinase

PLC: Phospholipase C
PLD: Phospholipase D
ROI: Reactive oxygen intermediates

2 Abscisic acid signalling

Alexander Christmann, Erwin Grill and Michael Meinhard

Abstract

Signalling of abscisic acid (ABA) in plants is characterized by an amazing number of secondary messengers that are part of the pathway or modulate the specific hormonal responses by interference with other signal transduction chains. In guard cells, a fast turgor-regulatory pathway triggered by ABA can be distinguished from a slower signalling pathway to the nucleus. The former is characterized by changes in K^+ and anion channel activities mediated by ABA-induced Ca^{2+} oscillations and, subsequently, by vesicle trafficking due to alteration in cell size. The nuclear signalling pathway involves changes in the phosphorylation status of signalling components including transcriptional regulators thereby redirecting gene expression. Turgor- and nuclear-targeted steps of the ABA signalling cascade are to some extent shared by common components. Recent findings emphasize the importance of posttranscriptional regulation at the level of mRNA maturation and protein-turnover. In addition, the concept of reciprocal feed back loops of both pathways emerges.

2.1 Introduction

Abscisic acid (ABA) is considered a 'stress hormone' (Zeevaart and Creelman 1988). The phytohormone integrates environmental constraints linked to changes in water activity with plant's metabolic and developmental programs. Plants respond to environmental challenges like drought and salt stress by changes in ABA availability, either *via* re-distribution of the signal (Slovik et al. 1995, Wilkinson and Davies 1997) or increased biosynthesis (Zeevaart and Creelman 1988), and possibly by altering the sensitivity to the hormone signal. ABA is, however, not only a stress hormone but also an endogenous signal required for proper development. In the absence of environmental stress, a basal ABA level fine-tunes optimal growth of plants (Cheng et al. 2002a) possibly by reducing growth-inhibitory ethylene release (Sharp 2002). After exceeding certain threshold levels ABA precipitates the stress-related effects such as complete closure of stomata and massive alteration of gene expression (Hoth et al. 2002, Seki et al. 2002, Rock 2000).

ABA signalling comprises the cellular events initiated by ABA that result in the specific responses including turgor-regulation and differential gene expression. Accordingly, a turgor-regulatory pathway can be distinguished from a nuclear signalling cascade (Webb et al. 2001, Fig. 1). ABA signalling, however, is not iso-

Topics in Current Genetics, Vol. 4
H. Hirt, K. Shinozaki (Eds.) Plant Responses To Abiotic Stress

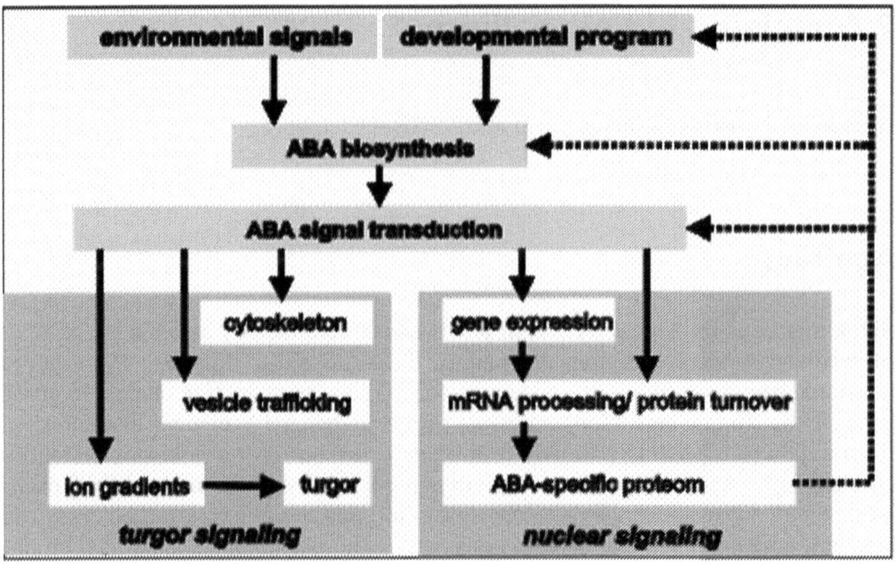

Fig. 1. Integration of ABA signal transduction into stress signalling and development. Stress initiates the release of the internal signal ABA. ABA activates a signalling cascade, which branches into several pathways including turgor-regulation and the nuclear pathway readdressing gene expression. The ABA-induced change in the proteom feeds back to ABA biosynthesis and ABA signalling as well as to ontogenesis of the plant.

lated but, in fact, is embedded in the transduction chains, which generate the signal in the first place, as well as in other signalling pathways that modulate and interfere. The extend of 'cross-talk' between ABA signal transduction and other cellular regulatory pathways can be envisaged as a central component of a clockwork interlinked to other essential circuits such as primary metabolism, cell growth and division. As a consequence, of this tight interaction it is difficult to discern a primary event from a secondary process.

Several recent reviews cover different aspects of ABA signalling such as control of germination (Finkelstein et al. 2002) or cover the role of ABA in guard cells in a broader context (Schroeder et al. 2001, Hetherington 2001, Luan 2002). Our contribution attempts to emphasize the emerging regulatory circuits of hormone biosynthesis, ABA signalling, and ABA-specific gene expression. Further chapters in this volume deal with the involvement of ABA in abiotic stress responses to drought (chapter 1), salt (chapter 9), and cold (chapter 6).

2.2 Systems used to study ABA signal transduction

A plethora of different experimental systems has emerged for studying ABA signal relay. In principle, any ABA-specific response provides a suitable basis, however, ABA-dependent control of seed germination (Finkelstein et al. 2002) and vegetative growth (Himmelbach et al. 1998) or stomatal closure of guard cells (Schroeder et al. 2001) are primarily studied. The use of hormone-specific reporter genes and transient expression systems has greatly facilitated the analysis of ABA-signalling in plantlets (Wu et al. 1997), protoplasts (Sheen 2001), and cell suspension cultures (Takahashi et al. 2001). Most of the work is performed with *Arabidopsis* providing an extensive collection of mutants in ABA biosynthesis or phytohormone-dependent responses (Finkelstein et al. 2002). However, there are unique systems to study specific actions of ABA such as the interaction of ABA and GA in barley aleurones (Shen et al. 2001) or the ABA-triggered developmental switch of the water fern *Marsilea* (Hsu et al. 2001).

An excellent system proved to be guard cells. Advantages of guard cells as a model system for signal transduction include the absence of plasmatic connections to neighbouring cells ideal for electrophysiological analysis and the sensitivity to ABA resulting in reversible stomatal closure. In addition, the target cells are readily accessible for manipulation and there are facile isolation procedures available (Kruse et al. 1989). However, regulation of stomatal closure is not only influenced by ABA and rather reflects the integration of internal and external stimuli for optimization of gas exchange (Grill and Ziegler 1998).

Different experimental systems are a prerequisite to test and generalize observations and conclusions drawn from specific analyses. ABA is just a signal and the very same signal can generate different outputs depending on the different 'wiring' or competence of the target cell or tissue. This feature is frequently observed in hormonal signalling. Guard cells from abaxial and adaxial surfaces of a leaf blade differ in the response-sensitivities of inward-rectifying K^+ channels (Wang et al. 1998). With the same token, different cell types can respond in an opposing manner to physiological ABA concentrations. Guard cells loose turgor via outward-rectifying K^+ channels in the presence of ABA (Blatt and Armstrong 1993) while cells of the mesophyll and root cortex cells keep or even increase turgor pressure (Sutton et al. 2000, Roberts and Snowman 2000).

2.3 ABA biosynthesis

ABA formation is the result of C_{40} carotenoid cleavage in plastids by a specific dioxygenase generating a C_{25} reaction product and the C_{15} compound xanthoxin (Fig. 2), which is subsequently converted in the cytosol to abscisic aldehyde and ultimately to ABA (Taylor et al. 2000). The plastidic origin of the carotenoid argues for a non-mevalonate derived biosynthesis of ABA. Carotenoid biosynthesis in chloroplasts occurs via the alternative pathway of activated isoprene formation starting from 1-deoxy-D-xylulose 5-phosphate (Arigoni et al. 1997, Lichtenthaler et al. 1997) and ABA biosynthesis *via* the non-mevalonate pathway has been demonstrated in tulip tree (Hirai et al. 2000). ABA biosynthesis is controlled by substrate availability and activity of the dioxygenase, the 9-*cis*-epoxycarotenoid dioxygenase (NCED), and by the reactions leading to the formation of ABA from xanthoxin.

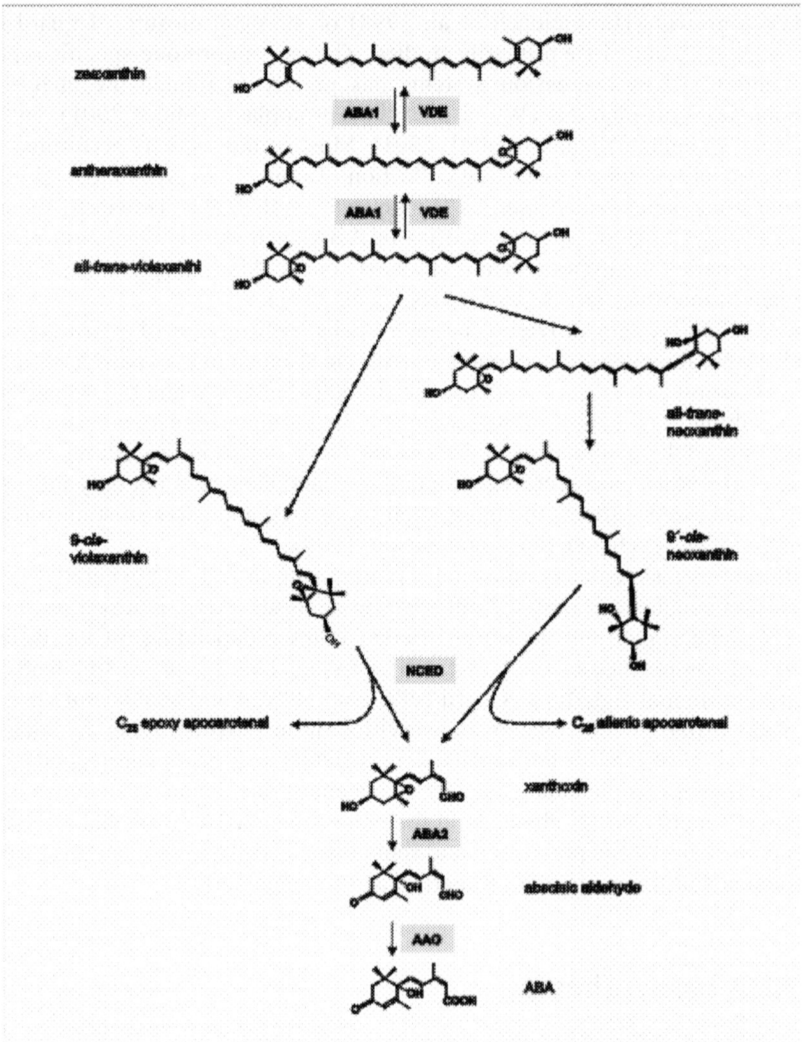

Fig. 2. Pathway of ABA biosynthesis in *Arabidopsis*. Zeaxanthin, antheraxanthin, and all-*trans*-violaxanthin are the carotenoids of the xanthophyll cycle. Interconversion of these compounds is accomplished *via* epoxidation or de-epoxidation reactions catalyzed by zeaxanthin epoxidase (ABA1) and violaxanthin de-epoxidase (VDE), respectively. All-*trans*-violaxanthin is converted either to 9-*cis*-violaxanthin or to 9-*cis*-neoxanthin. Both C_{40} carotenoids are cleaved by 9-*cis*-epoxycarotenoid dioxygenase (NCED) to the C_{15} aldehyde xanthoxin and a C_{25} compound. Xanthoxin is converted in a redox reaction catalyzed by the short-chain dehydrogenase/reductase ABA2 into abscisic aldehyde, which is finally oxidised to ABA by abscisic aldehyde oxidase (AAO). Proper function of AAO requires the sulfurylated form of the molybdenum cofactor formed by the specific sulfurase ABA3. The reaction sequences not entirely clarified yet are indicated by arrows with dotted lines.

2.3.1 Reactions generating substrates for NCED

The ABA biosynthesis pathway shares components of the xanthophyll cycle (Siefermann-Harms 1977) in which the carotenoids zeaxanthin, antheraxanthin, and all-*trans*-violaxanthin are interconvertible by the action of zeaxanthin epoxidase (ZEP) and violaxanthin de-epoxidase (VDE). ZEP-deficient mutants *aba1* of *Arabidopsis* and *aba2* of *Nicotiana plumbaginifolia* reveal ABA-deficiency and a wilty phenotype (Duckham et al. 1991, Marin et al. 1996). Likewise, ectopic expression of ZEP resulted in increased ABA levels (Frey et al. 1999). ZEP transcript levels were upregulated in roots not in leaves under drought conditions with enhanced ABA biosynthesis (Audran et al. 1998, Thompson et al. 2000a). All-*trans*-violaxanthin conversion into all-*trans*-neoxanthin is catalyzed by neoxanthin synthase (Bouvier et al. 2000).

2.3.2 NCED-catalyzed cleavage reaction

NCED catalyzes the formation of the first C_{15} intermediate, xanthoxin, by oxidative cleavage of C_{40} carotenoids. Deficiency in the NCED VP14 of maize results in precocious germination (Schwartz et al. 1997). *Arabidopsis* contains 5 NCEDs homologues (AtNCED2/3/5/6/9) to VP14 (Iuchi et al. 2001). Expression analysis of *AtNCED5*, *AtNCED6*, and *AtNCED9* imply a seed-specific developmental role in the regulation of ABA synthesis while *AtNCED2* and *AtNCED3* expression was associated with lateral root formation (Tan et al. 2003). Transcript levels of all five NCEDs, especially of *AtNCED3,* increased upon drought stress. NCED overexpression resulted in elevated ABA levels in tomato (Thompson et al. 2000b), tobacco (Qin and Zeevaart 2002), and *Arabidopsis* while in *Arabidopsis* AtNCED3 antisense lines revealed reduced ABA levels (Iuchi et al. 2001). The VP14 homologue of *Phaseolus vulgaris* is rapidly induced at the mRNA and protein level prior to ABA accumulation during drought stress in accordance with a key regulatory role of the dioxygenase in ABA biosynthesis (Qin and Zeevaart 1999). The carotenoid cleavage reaction probably occurs in plastids revealed by the presence of a plastidic transit signal in drought-induced precursor proteins of VP14 (Tan et al. 2001), AtNCED3 (Iuchi et al. 2001), and NCED1 of cowpea (Iuchi et al. 2000) that localized as a fusion protein in chloroplasts.

2.3.3 Formation of ABA from xanthoxin

The biosynthetic steps from xanthoxin (now also referred to as: 'xanthoxal') to ABA have been debated (Cowan 2000, Milborrow 2001), till ABA2, which catalyzes one of these steps, was cloned and further characterized (Cheng et al. 2002a, González-Guzmán et al. 2002). It was then proven that ABA2, a unique short-chain dehydrogenase/reductase in *Arabidopsis* converts xanthoxin to abscisic aldehyde in the cytosol. Abscisic aldehyde oxidase (AAO), present as four isoforms in *Arabidopsis* (Seo et al. 2000a), catalyzes the final step in ABA biosynthesis, the conversion of abscisic aldehyde to ABA. The enzymatic activity of ABA3 is required to provide the sulfurated form of a molybdenum cofactor that is necessary for AAO to generate ABA (Bittner et al. 2001, Xiong et al. 2001b).

2.3.4 Feedback regulation of ABA biosynthesis

Expression studies of ABA biosynthetic genes revealed feedback regulation of ABA synthesis by ABA. Increased ABA levels and drought or salt stress resulted in enhanced expression of zeaxanthin epoxidase (ABA1, Xiong et al. 2002) and of At-AAO3 and ABA3 (Xiong et al. 2001b, Xiong et al. 2002, Cheng et al. 2002a) while transcript abundance of At-AAO2 was reduced by ABA (Hoth et al. 2002). The changes are in line with an ABA-triggered mechanism to increase the capacity of the plant to generate ABA. NCED expression seems not, however, to be regulated by ABA and probably represents the rate-limiting step in ABA biosynthesis.

However, the effect of ABA on expression of ABA biosynthesis genes in most cases has not been confirmed at the protein level. Increase of At-AAO3 mRNA under water stress conditions was not accompanied by a parallel increase in protein abundance or activity (Seo et al. 2000b). In the ABA-insensitive mutant *abi1* but not in *abi2* (see 4.5.1) ABA induction of ABA1 and At-AAO3 was reduced indicating that feedback regulation of ABA synthesis was impaired (Xiong et al. 2002). However, ABA levels in response to drought in the ABA insensitive mutant *abi1* still seem to reach almost double the levels found in wild types (Lång et al. 1994), emphasizing the importance of posttranscriptional regulation of ABA turnover.

2.4. Signalling components

2.4.1 ABA- Receptor

The identification of an ABA-receptor is still at large despite a long history of promising reports to characterize ABA-binding proteins (Hornberg and Weiler 1984, Pedron et al. 1998, Zhang et al. 2001a, Zhang et al. 2002). So far, neither genetic nor biochemical approaches unequivocally identified candidates that fulfil the needs for a functional receptor protein that are specific as well as saturable binding of ABA and indispensable functionality in ABA signalling, at least for a subset of ABA-responses.

The existence of structurally different ABA-receptors has been implied from the observation that phaseic acid can substitute for ABA in some ABA-regulated responses (Walker-Simmons et al. 1997) and from a study of *Arabidopsis* mutants with an altered germination response in the presence of one ABA stereoisomer *vs.* the other (Nambara et al. 2002). It is not clear, however, whether the stereospecificity reflects discrimination of a postulated ABA-uptake system (Wilkinson and Davies 1997) or the hormone receptor itself.

An apoplastic perception of ABA is deducible from the capability of impermeable ABA-BSA conjugate to induce ABA dependent gene expression in *Arabidopsis* cell suspension cultures (Jeannette et al. 1999, Hallouin et al 2002), from the incapability of injected endogenous ABA to inhibit stomatal opening (Anderson et al. 1994), and from purification of an ABA-binding protein from *Vicia faba* plasma membranes (Zhang et al. 2002).

Both the paucity of data on a plasmamembrane-localised receptor and the failure to identify such a component in diverse genetic screens have fostered the idea of an intracellular perception site (Kushiro et al. 2003). The intracellular location of an ABA-receptor is in line with the induction of ABA-responses by cytoplasmic application of ABA (Allan et al. 1994, Schwartz et al. 1994) and the identification of ABA-binding sites in cytosolic protein fractions (Zhang et al. 2001a). The structural similarity of ABA with retinoic acid, a hormone signal generated by the breakdown of carotenoids in animals, stimulates the speculation that ABA could directly regulate transcription comparable to retinoic acid that activates its receptor, a transcription factor of the steroid receptor super family (Moriss-Kay and Ward 1999).

2.4.2 Intracellular messengers

In many phytohormone signalling processes, Ca^{2+} serves as an intracellular messenger including ABA-responses in which the Ca^{2+}-signal is triggered by secondary messengers like cyclic ADP ribose (cADPR), inositol 1,4,5 trisphosphate ($InsP_3$), or H_2O_2 (Schroeder et al. 2001). The elevation of cytosolic Ca^{2+} level (Ca^{2+}_{cyt}) during the ABA response of guard cells follows a distinct pattern of reiterated phases of increase and decrease, the so-called Ca^{2+}-oscillations that constitute a primary regulator of the output response (Allen et al. 2001). Analysis of ABA signalling events are frequently discerned in steps prior elevation of Ca^{2+}_{cyt} and post-Ca^{2+} signal transduction events. We might follow this line of categorizing but like to point out that a signal component downstream of Ca^{2+} can affect Ca^{2+}-release by regulatory circuits and thereby confound the analysis. For instance, a transcription factor acting downstream of Ca^{2+} may regulate the expression of Ca^{2+}-upstream components and, thus, alters in a feedback loop the cytosolic Ca^{2+} levels. These circuits are common to transduction pathways targeting gene expression.

In addition, it should be noted that the 'promiscuous' Ca^{2+}-signal of diverse signal pathways argues for hormone-specificity provided either by a characteristic Ca^{2+} signature, e.g. oscillation pattern (Klüsener et al. 2002) and/or parallel signalling events triggered by the hormone receptor interaction (Himmelbach et al. 1998).

While the central role of Ca^{2+} in the ABA response is little debated, there is a confounding plethora of other secondary signals involved in the transduction process. ABA induced changes in the cytosolic compartment might also involve alterations of the redox status (see 2.4.2.1) and elevation of pH (Irving et al. 1992, Blatt and Armstrong 1993, Wang et al. 2001). Recently, a number of lipid-derived secondary messengers have been identified to affect ABA-responses including myo-inositol hexakisphosphate ($InsP_6$, Lemtiri-Chlieh et al. 2000) and sphingosine-1-phosphate (Ng et al. 2001). In other instances, activation of phospholipases during ABA signalling is known but little is known about the identity of the cleavage products.

2.4.2.1 Redox signals

In the last years, a novel link between ABA perception and responses crystallized implicating hydrogen peroxide and nitric oxide (NO) as mediators or regulators of ABA signal transduction (Neill et al. 2002b). Both secondary messengers are associated with pathogen interaction (Klüsener et al. 2002) and are characterized by a short biological half-life and by their impact on the redox status of the cell.

Conclusive data for the role of H_2O_2 in mediating ABA-induced closure was provided by a study of Pei et al. (2000). H_2O_2 production in guard cells was stimulated by ABA and resulted in the activation of Ca^{2+}-permeable channels required for stomatal closure. No activation of calcium channels by H_2O_2 was observed in the *gca2-1* mutant, which is insensitive to ABA-triggered stomatal closing. A subsequent study revealed an altered Ca^{2+}-oscillation pattern in the *gca2-1* mutant that is responsible for the observed ABA-insensitivity (Allen et al. 2001). Thus, the altered H_2O_2 and ABA sensitivity of *gca2-1* may reflect a modified Ca^{2+}-signature, which seems to be regulated by several input signals including elicitor and pathogen interaction (Klüsener et al. 2002). Recent analyses of knockout *Arabidopsis* lines with impaired functionality of NADPH-oxidases support the role of H_2O_2 in mediating ABA responses (Kwak et al. 2003). In addition, H_2O_2 seems to mediate ABA-relayed inhibition of K^+ inward channels (Zhang et al. 2001b) and also interferes with key regulators of ABA signalling by enzymatic inactivation of the protein phosphatases ABI1 and ABI2 (Meinhard and Grill 2001, Meinhard et al. 2002). *In vitro* studies revealed rapid inactivation of the two PP2Cs probably by oxidation of critical cysteine residue(s). Transient inactivation of ABI1 and ABI2 acting as negative regulators of ABA signalling (Gosti et al. 1999, Merlot et al. 2001) would result in stimulation of ABA signalling while in case of the recently described positive regulatory function of ABI1 (Wu et al. 2003) inactivation by H_2O_2 would generate a downregulation of the signalling pathway. Apart from a role in regulating ion currents, H_2O_2 has also been assigned a role in cellular responses to ABA at the level of gene regulation (Guan et al. 2000). Nitric oxide (NO) is an important effector in animal cells. In plants, NO affects processes related to growth and development (Beligni and Lamattina 2001) and it seems to play a role in adjustments of physiological processes to external abiotic influences (Durner et al. 1998). Recently, it has been reported that in guard cells of *Vicia faba* (Garcia-Mata and Lamattina, 2002) or *Pisum sativum* (Neill et al. 2002a), NO is a component of ABA signalling. As depicted from pharmacological experiments NO induction of stomatal closure requires cADPR and cGMP and thus might act upstream of calcium. The analysis of *Arabidopsis* mutants defective in two nitrate reductases revealed the putative source for NO development (Desikan et al. 2002). In the ABA-treated guard cells from *nia1*, *nia2* double mutant plants both NO development and stomatal closure were greatly reduced. Interestingly, NO production was not impaired in *abi1* and *abi2* mutants, arguing for a role of NO and nitrate reductase upstream of or parallel to the action of the ABI1 and ABI2 protein phosphatases. A more detailed discussion covering the putative function of nitrate reductase and the chemistry of NO in plant cells can be found in a recent review by Garcia-Mata and Lamattina (2003).

2.4.2.2 Cyclic nucleotides

cADPR triggers ABA-mediated gene activation (Wu et al. 1997) as well as stomatal closure via release of Ca^{2+} from internal stores (Lecki et al. 1998). As stated above analyses argue for cGMP and cADPR to be components in a stomatal ABA signalling pathway involving NO production (Neill et al. 2002a). In animal systems, cADPR is generated by a specific Ca^{2+}-stimulated cyclase from NAD and the intracellular signal constitutes part of a calcium-induced-calcium release-system in that it increases cytosolic calcium levels by activation of ryanodine-sensitive Ca^{2+}-channels. Such ryanodine-sensitive Ca^{2+}-channels exist in the tonoplast of plants (Allen et al. 1995). In this context, it is intriguing that in lower animals, in sponges, ABA, cADPR and calcium-induced-calcium release form a signal pathway mediating thermo-induced responses (Zocchi et al. 2003).

2.4.2.3 Lipid-derived signals

In guard cells, phospholipases C (PLC) and D (PLD) are activated by ABA (PLD: Jacob et al. 1999, PLC: Lee et al. 1996, Staxen et al. 1999). Activation occurs at the enzymatic level and for AtPLC1 of *Arabidopsis* a transcriptional upregulation by ABA has been found (Hirayama et al. 1995). Phospholipase C (PLC) catalyzed cleavage of phosphatidylinositol 4,5-bisphosphate generates the second messengers diacylglycerol and $InsP_3$. $InsP_3$ mobilizes Ca^{2+}-release from internal stores into the cytosol via an $InsP_3$-receptor localized at the vacuolar membrane (Allen et al. 1995). The exact contribution of $InsP_3$ or diacylglycerol to ABA responses is still not clear, but reducing PLC expression limited ABA responses (Sanchez and Chua 2001, Hunt et al. 2003). A central role of $InsP_3$ is also supported by the *Arabidopsis* loss of function mutant *fry1*. FRY1 is an inositol polyphosphate 1-phosphatase that converts $InsP_3$ to $InsP_2$. *fry1* plants display an enhanced sensitivity to ABA obviously due to the impaired phosphoinositide catabolism (Xiong et al. 2001c). In agreement with that, overexpression of other $InsP_3$ degrading enzymes, two inositol 5-phosphatases (At5PTase 1 and 2), resulted in ABA insensitivity with respect to stomatal regulation and gene regulation in seedlings (At5PTase 1, Burnette et al. 2003) or seed germination (At5PTase 2, Sanchez and Chua 2001). In addition, other phosphoinositides like phosphatidylinositol 3- and 4-phosphate (Jung et al. 2002) and $InsP_6$ (Lemtiri-Chlieh et al. 2000) interfere with ABA signal transduction. $InsP_6$ is transiently increased by ABA and $InsP_6$

was found to be more potent than $InsP_3$ to mediate ABA-linked ion channel regulation (Lemtiri-Chlieh et al. 2000).

Taken all these facts into consideration, phosphoinositide metabolism seems to play a central role in ABA signal transduction, although the actual role of single compounds in relation to Ca^{2+} and other signalling cassettes are still largely unknown.

Phosphatidic acid (PA), a phospho-diacylglycerol, is generated by PLD activity from phospholipids. PA induced stomatal closure and inhibited stomatal opening in epidermal peels (Jacob et al. 1999). Several studies provide further evidence for a role of PLD in ABA signalling. ABA-mediated upregulation of gene expression in *Arabidopsis* was accompanied by a transient stimulation of PLD activity (Hallouin et al. 2002). Antisense expression of PLD reduced ABA-induced senescence (Fan et al. 1997). Treatment with 1-butanol, a presumed selective inhibitor of PLD inhibited both ABA-induced production of PA and partially ABA-induced stomatal closure (Jacob et al. 1999). Additionally, antisense lines for PLD showed ABA insensitivity with respect to stomatal closure, while overexpression of PLD resulted in increased drought resistance (Sang et al. 2001). These findings suggest that PA is involved in ABA responses as a secondary messenger. Since PA treatments did not increase guard cell $Ca^{2+}{}_{cyt}$, PLD must act either downstream of Ca^{2+} or in a parallel pathway (Jacob et al. 1999).

In the barley aleurone, too, ABA effects seem to be triggered by phosphatidic acid, which is released by an ABA triggered increase in PLD activity (Ritchie et al. 2002). The activation of PLD is claimed to be G-protein mediated and localized to the plasma membrane (Ritchie and Gilroy 2000). It is not yet clear, however, if stimulation of PLD activity by ABA in general involves G proteins or if ABA induced changes in Ca^{2+} oscillations are responsible for PLD activation.

Recently, another phospholipid, sphingosine-1-phosphate (S1P), was implicated as secondary messenger in drought and ABA signalling (Ng et al. 2002). The enzyme involved in S1P formation, sphingosine kinase, is activated by ABA and sphingosine kinase inhibition impairs stomatal regulation (Coursol et al. 2003).

2.4.2.4 Calcium

Ca^{2+} is a major component in many signalling pathways in plants and animals. It has been suggested that individual stimuli evoke increases of $Ca^{2+}{}_{cyt}$, which are unique in terms of their spatio-temporal characteristics (Evans et al. 2001) and such stimulus-specific Ca^{2+} signals have been referred to as "Ca^{2+} signatures". In general, two distinct but not exclusive types of Ca^{2+} increases can be observed. One type brings along a single increase in $Ca^{2+}{}_{cyt}$ that usually is of transient nature. This may in some cases be followed by a variable number of additional transients with a defined temporal distance to each other and usually declining amplitudes leading to the second type, Ca^{2+}oscillations. Specificity of a calcium signal depends on the compartment(s) from which Ca^{2+} is released and on the amplitude and frequency of the stimulus-induced $Ca^{2+}{}_{cyt}$ oscillations. Guard cells achieve an optimal aperture under a certain set of environmental conditions by integrating signals from differing stimuli. Since many of these stimuli use calcium as a second messenger (Evans et al. 2001), integration of the signals might take place on the levels of resulting "Ca^{2+} signatures". Ca^{2+}_{cyt} elevations can result in opposing reactions such as stomatal closure or opening, which are both preceded by a rise in Ca^{2+}_{cyt} (Irving et al. 1992).

In response to ABA, cytosolic-free Ca^{2+} increases in guard cells and this Ca^{2+} increase precedes stomatal closure (McAinsh et al. 1990). Furthermore, ABA apparently sensitizes Ca^{2+} influx to membrane potential (Grabov and Blatt 1998, Hamilton et al. 2000). In an elegant electrophysiological study, Allen et al. (2001) explored the specificity of Ca^{2+} signatures for regulation of the stomatal aperture. Using buffer changes they artificially superimposed Ca^{2+} oscillations in *Arabidopsis* guard cells and found optimal parameters with respect to frequency and duration of the Ca^{2+} transients, finally leading to reduced steady state stomatal apertures. These parameters where consistent with those observed after challenging guard cells with ABA. Moreover, *gca2*, an ABA insensitive mutant, displayed suboptimal oscillation parameters that turned out to be insufficient for prolonged stomatal closing. When wild type oscillations were experimentally imposed in *gca2* guard cells, stomatal closure was partially restored. Interestingly, while single Ca^{2+} transients already led to a rapid closure, they were not able to induce a long-term effect, since stomata reopened after the imposed program was stopped. This strongly argues for two distinct mechanisms driven by Ca^{2+}. A single, transient Ca^{2+} increase seems to be sufficient for short time closing of stomata, for example by inhibition of the plasma membrane H^+-ATPase (Kinoshita et al. 1995), while a long time response is dependent on oscillations with distinct frequency, duration and amplitude.

ABA-induced increases in Ca^{2+}_{cyt} by activation of plasma membrane calcium channels are reduced in the protein phosphatase mutants *abi1* and *abi2* (Allen et al. 1999, Murata et al. 2001). Different mechanism of Ca^{2+} release (Allen et al. 2000) are responsible for generation of Ca^{2+} oscillations in response to different stimuli as exemplified with the *det3* mutant devoid of the C-subunit of vacuolar ATPase (Allen et al. 2000). Ca^{2+} or H_2O_2 failed to generate Ca^{2+} oscillations and to induce stomata closure in *det3* whereas both ABA and cold induced Ca^{2+}_{cyt} oscillations and the proper stomatal response.

2.4.3 G-proteins

Heterotrimeric GTP-binding proteins (G-proteins) as well as small G-proteins are involved in modulating ABA responses. GPA1, the sole typical α subunit of trimeric G-proteins in *Arabidopsis*, is required for pH-independent ABA activation of slow anion channels (Wang et al. 2001). The activation was abolished in *gpa1* knockout lines not in wild type plants when ABA-induced pH changes of the cytosol were suppressed. In essence, the data suppose that different ABA signalling pathways are able to activate anion channels. Effects of S1P on stomatal regulation were diminished in *gpa1-1* and *gpa1-2* supporting an interaction of the Gα subunit with sphingosine-type signals (Coursol et al. 2003).

In the barley aleurone, G-protein mediates ABA induction of phospholipase D (PLD), an effect, which is localized to the plasma membrane (Ritchie and Gilroy 2000). Other studies support this notion (Pappan and Wang 1999) while in *Arabidopsis* cells a G protein-mediated activation of PLD was ruled out (Hallouin et al. 2002).

A monomeric G protein AtRac1 is involved in mediating ABA-triggered stomatal closure. AtRac1 inactivation is necessary for an ABA-induced reorganization of the actin skeleton in guard cells required for stomatal closure (Lemichez et al. 2001, Eun et al. 1997). Analyses of transgenic plant with deregulated expression of such monomeric GTPases employing constitutively active and dominant-negative forms have revealed altered ABA responses (Yang 2002). A null mutant of the small GTPase ROP10 (RHO-like small G protein of plants) of *Arabidopsis* revealed specifically enhanced ABA responses (Zheng et al. 2002). Baxter-Burrell et al. (2002) established a link between ROP signalling and H_2O_2 production. The proposed role of H_2O_2 as a regulator of ABA signalling supports a function of those GTPases as nodes of a regulatory network in which inputs from different signalling pathways including the ABA signal transduction chain are integrated.

2.4.4. Farnesyltransferase ERA1

The *ERA1* gene encodes a farnesyltranferase β-subunit with multifaceted roles, since the loss-of-function mutant *era1-2* displays a pleiotropic phenotype covering changes in development, auxin and ABA signalling (Pei et al. 1998, Yalovsky et al. 2000, Brady et al. 2003). The influence of ERA1 as a negative regulator of ABA signal transduction in guard cells is exerted on the level of $Ca^{2+}{}_{cyt}$ increases through plasma membrane Ca^{2+} channels (Allen et al. 2002).

2.4.5 Protein phosphatases

2.4.5.1 Protein phosphatases ABI1, ABI2

ABI1 and ABI2 are homologous type 2C protein phosphatases with overlapping yet distinct functions. The ABA-insensitive *Arabidopsis* mutants *abi1-1* and *abi2-1* characterized ABI1 and ABI2 as key regulators of ABA-invoked seed dormancy, stomatal closure, and vegetative growth inhibition (Koornneef et al. 1984). The mutant abi1-1 (abi1) and abi2-1 (abi2) proteins confer a genetically dominant ABA-insensitivity and both carry the same amino acid substitution at equivalent positions of the catalytic PP2C domain resulting in a reduced phosphatase activity (Leung et al. 1997, Rodriguez et al. 1998a). Despite major efforts to elucidate the mechanism of ABI1 and ABI2 action, the precise role is still a conundrum, partly due to the functional redundancy and to the 'gain-of-function' action of abi1 and abi2. Ectopic (over)-expression of abi1 and ABI1 in transient system generated ABA-insensitivity in agreement with a negative regulatory role of the PP2C on ABA signalling (Sheen 1998). In line with this conclusion, intragenic revertants of

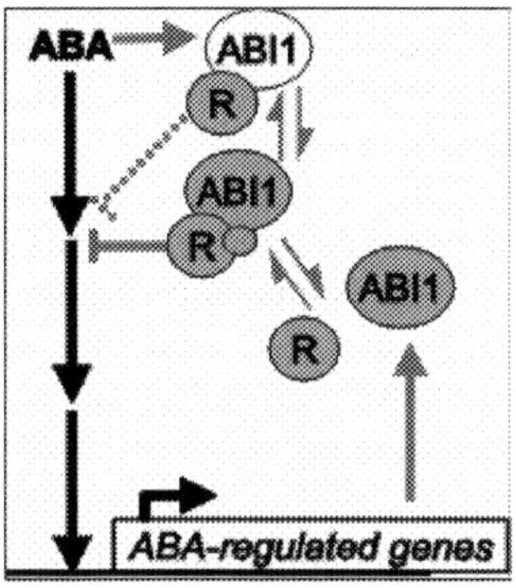

Fig. 3. Model of dual ABI1 action as a positive and negative regulator. Binding of ABA to the ABA receptor mediates inactivation of the repressor R of ABA signal transduction by activation of ABI1 (positive role). This step requires dephosphorylation, which is impaired in the phosphatase-deficient abi1-repressor complex resulting in a genetically dominant failure to activate ABA-responsive genes including induction of ABI1 expression. ABI1 is required to form the active repressor as a repressor R-ABI1 complex that is stabilized by protein phosphorylation (negative role). Induction of ABI1 expression by ABA results in increased formation of repressor protein that offsets the balance towards active repressor that results in a ABA desensitizing. Alternatively, ABI1 released from the complex exerts a second negative control of the signalling pathway. Thus, ABI1 exerts both a positive regulatory role in ABA signalling as well as a negative feedback requiring ABA-induced gene expression of ABI1. Transient analyses by microinjected ABI1 forms interfered with the activation of ABA signalling whereas ectopic expression of ABI1 primarily generated the ABA-desensitized phenotype.

abi1-1 exhibited an ABA hypersensitive recessive phenotype (Gosti et al. 1999, Merlot et al. 2001). The revertant genes were assumed to be loss-of-function alleles though the secondary mutations in all instances inactivated only the catalytic domain. There seems to be a clear bias against inactivation of the aminoterminal domain that is supposed to represent a regulatory or interaction domain (Leube et al. 1998). In addition, transient expression analyses documented the requirement for phosphatase activity of ABI1 for mediating ABA-insensitivity, however, a distinct mutated form of ABI1 that lacked PP2C activity still inhibited ABA-inducible transcription (Sheen 1998), suggesting interference of ABI1 with interaction partners of ABA signalling.

Recent protein microinjection data shed new light on ABI1 action (Wu et al. 2003). Tomato hypocotyls cells revealed an ABA-insensitive phenotype after injection of abi1 protein while coinjection of ABI1 at a two- to threefold excess over the mutant form rescued ABA-inducible transcription. Thus, ABI1 and abi1 compete for common binding sites and the wild-type protein is capable to restore proper ABA signal relay in agreement with a *positive* regulatory function of the specific PP2C. The different assay systems, ectopic gene expression and transient protein introduction, probably reveal different natures of ABI1 action. Consistent high expression levels of ABI1 generate an ABA-insensitive phenotype, possibly by forming more active repressor complex (Fig. 3) or *via* re-addressing gene expression in a feed-back loop. Short-transients of introduced abi1/ABI1 protein interferes with primary ABA-signalling events in which the PP2C activity is required for ABA-signal propagation.

To reconcile the findings, the emerging mechanism of ABI1 (and possibly ABI2) action includes a first step characterized by the requirement for ABI1 phosphatase activity to relay the ABA signal into the nucleus, and a second step in which induction of ABI1 results in the desensitizing of the plant against ABA. In this scenario, the abi1/abi2 mutant forms inhibit the first step possibly by preventing dephosphorylation of a negative regulator of ABA-signal transfer within a protein complex (Himmelbach et al. 1998). Consistent with this view, microinjection of a phosphatase-inactive ABI1 blocked signal transduction (Wu et al. 2003).The second action of ABI1 probably involves nuclear localization and interaction with transcriptional regulators (Himmelbach et al. 2002, see below)

The components of the proposed ABI1-regulatory protein complex are still at large though the association of ABI1 and ABI2 with the protein kinase PKS3 that interacts with the Ca^{2+}-binding proteins SCaBP5 (Ishitani et al. 2000, Guo et al. 2002) provides an intriguing paradigm of a physical interaction of a putative Ca^{2+}-sensor and its associated protein kinase with the PP2C. Analysis of knockout and RNAi lines of SCaBP5 and PKS3 revealed alteration of ABA responses. In addition, studies of double mutants support the notion that the proposed complex negatively interferes with ABA signal transduction.

2.4.5.2 Other protein phosphatases

Apart from ABI1 and ABI2, two other PP2Cs of *Arabidopsis*, AtPP2CA and AtP2C-HA, have been implicated to regulate ABA signal transduction. Both are transcriptionally upregulated by ABA (Tahtiharju and Palva 2001, Rodriguez et al. 1998b). AtPP2CA inhibited ABA action in maize protoplasts comparable to ABI1 (Sheen 1998) and interacts with an inward rectifying K^+ channel (Cherel et al. 2002). Silencing AtPP2CA expression in *Arabidopsis* by an antisense approach resulted in accelerated freezing tolerance (Tahtiharju and Palva 2001). Thus, AtPP2CA probably acts in a negative regulatory feedback circuit.

A positive regulatory role in ABA signalling is assigned to the protein phosphatase 2A, RCN1 (Kwak et al. 2002). *Arabidopsis rcn1* mutant plants display ABA insensitivity with respect to stomatal opening and germination in line with disruption of ABA induced activation of anion channels in guard cells and reduced Ca^{2+} increases. Interestingly, RCN1 was formerly described to be involved in auxin transport, gravitropic responses, and lateral root growth (Garbers et al. 1996, Rashotte et al. 2001). These findings point to interference of regulatory networks controlled by differing signals and reflect pleiotropic phenotypic alterations. Alternatively, these regulators could exert differential roles in distinct signalling pathways.

2.4.6 Protein kinases

The action of protein phosphatases such as ABI1 is counterbalanced by protein kinases. Several protein kinases have been implicated in ABA responses including Ca^{2+}-calmodulin regulated protein kinases (Sheen 1996) and SNF1-like protein kinases such as PKABA1 (Anderberg and Walker-Simmons 1992).

2.4.6.1 Protein kinases AAPK, PKABA1 and OST1/SRK2E

PKABA1 of wheat and barley (Anderberg and Walker-Simmons 1992, Yamauchi et al. 2002) as well as AAPK from *Vicia* (Li and Assmann 1996) and OST1/SRK2E from *Arabidopsis* (Mustilli et al. 2002, Yoshida et al. 2002) belong to the protein family of SNF1-like protein kinases. All three kinases contain an N-terminal domain similar to SNF1/AMP-regulated protein kinase of yeast (Hardie et al. 1998) and a C-terminal domain with putative regulatory functions.

PKABA1 of wheat is ABA-induced and antagonizes GA-induced gene expression in seeds and germinating seedlings (Anderberg and Walker-Simmons 1992, Yamauchi et al. 2002) while PKABA1 of barley aleurone cells is only involved in GA-regulated gene expression, which suggests functional separation of paralogues (Shen et al. 2001). PKABA1 of wheat physically interacts with the putative phosphorylation substrate TaABF, a basic leucine zipper transcription factor that has high structural homology to ABI5 of *Arabidopsis* (Johnson et al. 2002, see below). AAPK from *Vicia* is a positive regulator of ABA-induced stomatal closure presumably by mediating the activation of plasma membrane anion channels (Li et al. 2000). It can physically interact with an RNA binding protein, AKIP1 (Li et al. 2002). AAPK shares high homology in primary structure with OST1/SRK2E of *Arabidopsis* (Mustilli et al. 2002, Yoshida et al. 2002) and both are rapidly activated by ABA. The recessive *ost1* mutants were isolated by screening for thermal surface differences due to altered transpiration rates (Mustilli et al. 2002), a technique elegantly applied to identify an ABA insensitive barley mutant (Raskin and Ladyman 1988). *ost1* mutants are impaired in both ABA-dependent stomatal closing and ABA-mediated inhibition of stomatal opening. Seed dormancy and the response of germinating seedlings to ABA are not altered (Mustilli et al. 2002, Yoshida et al. 2002), arguing for a specific role of OST1/SRK2E in guard cell regulation.

2.4.6.2 Mitogen-activated protein kinases

The involvement of mitogen-activated protein (MAP) kinase in ABA signal transfer is still not clearly substantiated despite an earlier report (Knetsch et al. 1996). However, MAPK cascades are involved in stress signalling throughout eukaryots including H_2O_2 and pathogen signalling in plants (Kovtun et al. 2000, Jonak et al. 2002). An interference of such pathways with ABA signalling is evident considering the convergence of such pathways at the level of Ca^{2+}-oscillations (Klüsener et al. 2002). Recently, the MAPK AtMPK3 has been reported to participate in ABA-evoked postgermination arrest (Lu et al. 2002).

2.4.6.3 Calcium-regulated protein kinases

The calcium-dependent protein kinases (CDPKs) comprise a family of Ca^{2+} sensors consisting of a protein kinase domain and a carboxyterminal Ca^{2+}-binding calmodulin-like domain (Cheng et al. 2002b). Ca^{2+}-mediated stimulation of kinase activity is the proposed action of CDPKs (Huang et al. 1996, Romeis et al. 2000). Transient expression analysis supports the involvement of a CDPK in ABA-signalling as a positive regulator (Sheen 1996). In guard cells, CDPKs can phosphorylate the inward K^+ channel protein KAT1 (Li et al. 1998), but the link between ABA-mediated stomatal closure and Ca^{2+}-triggered CDPK action is not resolved which could involve regulation of the endoplasmatic Ca^{2+}-pump ACA2 (Hwang et al. 2000). In addition, two other protein kinases that associate with Ca^{2+}-sensors regulate or modulate ABA-responses: the already mentioned PKS3, which interacts with the Ca^{2+}-binding proteins SCaBP5 and binds to ABI1 and ABI2 (Ishitani et al. 2000, Guo et al. 2002), as well as CIPK3, a Ser/Thr kinase that associates with a calcineurin B-like calcium sensor (Kim et al. 2003).

2.4.7 Transcriptional regulators

Targets of ABA signalling are preformed elements such as ion channels, the cytoskeleton (Eun and Lee 1997, Hwang and Lee 2001), the vesicle trafficking machinery (Leyman et al. 1999, Geelen et al. 2002), or transcription factors. The transcription factors control ABA-regulated genes, possibly including secondary transcription factors that activate a set of ABA-responsive genes further downstream in the signalling cascade. The ABA-signal massively readdresses genomic expression as revealed by transcriptome analyses (Hoth et al. 2002, Seki et al. 2002). By random massive sequencing of transcripts more than 1300 ABA-regulated genes were identified in *Arabidopsis* seedlings, approximately half of them were upregulated and the other downregulated (Hoth et al. 2002). ABA regulation of the majority of the genes (more than 90%) was impaired in the *abi1* mutant emphasizing the central role of this locus in ABA signal transduction. Several *cis*-acting elements are known that confer regulation of gene expression by ABA and represent interaction sites of transcriptional regulators including VP1/ABI3, basic region/leucine zipper (bZIP), homeodomain-containing, as well as MYB- and MYC-type transcription factors. Unfortunately, nothing is known about transcriptional regulators conferring ABA-mediated downregulation. Among the ABA regulated genes, transcripts encoding ABA signalling components like ABI1, ABI2 and AtHB6 are upregulated, obviously reflecting adjustment of the signalling machinery by negative feedback loops (Hoth et al. 2002, Himmelbach et al. 2002).

2.4.7.1 VP1/ABI3

ABI3 from *Arabidopsis* and its putative orthologue VIVIPAROUS 1 (VP1) from maize contain four highly conserved domains, an acidic domain (A1) and three basic domains capable to mediate DNA (B2, B3) or protein binding (B1) (Nakamura et al. 2001, Suzuki et al. 1997). VP1/ABI3 interfere with ABRE-type *cis*-elements by binding to the bZIP transcription factor ABI5/TRAB1 (see below) and this interaction is required to maintain ABA-mediated seed dormancy (Hobo et al. 1999, Nakamura et al. 2001). Moreover, VP1 and the bZIP factor EmBP1 form a DNA binding complex together with a member of the highly conserved 14-3-3 protein family (Schultz et al. 1998). 14-3-3 proteins have been suggested to be fine-tuners of their targets by binding to specific phosphorylated serine residues. Such interaction of 14-3-3 proteins with transcription factors may reflect an additional mechanism to couple ABA regulated phosphorylation/dephosphorylation events to gene expression.

2.4.7.2 Basic region/leucine zipper (bZIP) transcription factors

The bZIP transcription factors constitutes a prominent group of ABA-response regulators that interact with ABA-response elements (ABRE), ACGT-containing 'G-boxes' of promoter elements (Hattori et al. 2002), and usually require a second related motif, also called coupling element, to achieve optimal ABA responsiveness (Busk and Pages 1998, Rock 2000). Interestingly, a dehydration-responsive element (DRE) could serve as a coupling element (Narusaka et al. 2003). The bZIP transcription factors with a functional role in ABA or stress signalling represent a clade within the 75 bZIP transcription factors of *Arabidopsis* (Jacoby et al. 2002). The subclass includes the bZIP transcription factors ABI5/TRAB1 (Finkelstein and Lynch 2000, Hobo et al. 1999), and the ABA-responsive element binding factors AREB1/ABF2, AREB2/ABF4, AREB3, ABF1 and ABF3 (Choi et al. 2000, Uno et al. 2000). AREB1/ABF2 and AREB2/ABF4 transactivate ABA-regulated gene expression in dependence on ABA probably after phosphorylation of the aminoterminal domain mediated by a specific ABA-activated protein kinase (Uno et al. 2000). Similarly, rice TRAB1 becomes phosphorylated in response to ABA at a specific serine residue that is required for activation of the preformed and idle bZIP transcriptional regulator (Kagaya et al. 2002). While AREBs/ABFs seem to play a regulatory function predominantly in vegetative tissues, ABI5 is involved as a positive regulator in ABA signal transduction during seed and early seedling development (Finkelstein and Lynch 2000). ABA exerts a dual effect on ABI5, it enhances ABI5 gene expression and stabilizes the protein by mediating ABI5 phosphorylation (Finkelstein and Lynch 2000, Lopez-Molina et al. 2001). The proteolytic degradation of ABI5 by the 26S proteasome (Lopez Molina et al. 2003, Smalle et al. 2003) is required to overcome postembryonic growth arrest. ABA blocks the degradation, which leads to accumulation of ABI5 and reactivation of embryonic genes (Lu et al. 2002). ABI5 is able to bind to the ABA transcriptional activator ABI3 thereby possibly recruiting ABI3 to target promoters such as the ABA-regulated late embryo abundant genes (Nakamura et al. 2001). Deficiency of ABI3 results in the repression of ABI5 expression (Lopez-Molina et al. 2001). These findings provide a basis to explain the ABA-insensitive phenotype of the loss-of-function mutants *vp1* of maize and *Arabidopsis abi3* and *abi5*.

2.4.7.3 Homeodomain-leucine zipper transcriptional regulators

ABA induces expression of several homeodomain- and leucine zipper-containing (HD-Zip) transcriptional regulators including ATHB5 (Johannesson et al. 2003), ATHB6 (Söderman et al. 1999), ATHB7 (Söderman et al. 1996) and ATHB12 (Lee et al. 2001). Functional characterization of ATHB5 (Johannesson et al. 2003) and ATHB6 (Himmelbach et al. 2002) support a role in ABA signal transduction as a positive and negative regulator, respectively. ATHB6 targets the AT-rich *cis*-element *CAATTATTA* and physically interacts with the protein phosphatase 2C ABI1 (Himmelbach et al. 2002). The nuclear-localised ATHB6 requires translocation of ABI1 into the nuclear compartment for interaction. Upregulation of ATHB-6 gene expression is ABI1-dependent and results in desensitizing of guard cells against ABA.

2.4.7.4 AP2-type transcription factors

EREBPs (ethylene-responsive element binding proteins) and AP2 (APETALA2) are the prototypic members of a family of transcription factors unique to plants, whose distinguishing characteristic is that they contain the so-called AP2 DNA-binding domain. Members of this family are the dehydration-responsive element binding proteins (DREBs, Liu et al. 1998), which recognise the drought-responsive element (DRE, Yamaguchi-Shinozaki and Shinozaki 1994) in target promoters. DREBs are intermediates in an ABA-independent pathway, which relays drought, cold, and pathogen stress signals (Kizis et al. 2001, Park et al. 2001). Recently, one maize DRE-binding protein has been shown to be induced by ABA (Kizis and Pagès 2002) implying that not all DRE-binding factors function independently from ABA. Furthermore, there exists interference between ABRE and DRE and their transcriptional regulators in that DRE/DRE motifs are able to serve as a coupling element of ABRE (Narusaka et al. 2003) and the maize DRE-binding factor DBF1 is an activator of ABA-induced transcription while DBF2 antagonizes the action (Kizis and Pagès 2002). ABI4 is another member of this family, which has been implicated in a seed-specific signalling pathway (Finkelstein 1994). ABI4 interferes with ABA signal transduction by interacting with ABI3 and ABI5 (Söderman et al. 2000). The maize orthologue ZmABI4 specifically binds to a coupling element of ABA-responsive genes (Niu et al. 2002).

2.4.7.5 Other transcription factors

133 basic helix-loop-helix (bHLH) transcriptional regulator genes have been determined in *Arabidopsis* (Heim et al. 2003). Among them, rd22BP1 (renamed AtMYC2) has been shown to activate ABA-inducible gene expression under drought stress (Abe et al. 2003), while expression of three other members of the bHLH family, BEE1, BEE2, and BEE3 which are positive regulators in brassinosteroid signalling is repressed by ABA.

Plant MYB transcription factors, a family with more than 100 members in *Arabidopsis* with probably distinct functions, show structural similarity to the vertebrate cellular proto-oncogene c-MYB (Martin and Paz-Ares 1997). ABA and drought induce the expression of three specific MYB family members (Abe et al. 1997). While the bZIP and HD-Zip proteins seem to work as preformed targets, *de novo* synthesis is required for MYB/MYC action necessitating a primary transcriptional regulator of ABA action (Shinozaki and Yamaguchi-Shinozaki 2000).

Single C2H2 zinc finger protein genes comprise a gene family with approximately 30 genes in *Arabidopsis* (Dinkins et al. 2002). SCOF-1 is a C2H2-type zinc finger protein from soybean, which is induced by low temperature and abscisic acid (ABA) but not by dehydration or high salinity (Kim et al. 2001). SCOF-1 does not bind to an ABA responsive element (ABRE) directly but greatly enhanced the DNA binding activity of SGBF-1, a soybean G-box binding bZIP transcription factor, to ABRE *in vitro.*

2.5 RNA and protein turnover during ABA response

Recent findings uncovered a novel facet of ABA-relayed control of gene expression at the posttranscriptional level. The control seems to be exerted by affecting maturation of mRNA as well as stability of transcripts and proteins. In addition, genome-wide expression analysis unravelled downregulation of transcript abundance for ribosomal proteins paralleled by a concomitant upregulation of genes involved in proteolysis (Hoth et al. 2002).

RNA-binding proteins frequently regulate turnover or access of transcripts to the translational machinery (Fedoroff 2002b). The maize ABA-regulated and glycine-rich RNA-binding protein MA16 preferentially interacts with uridine- and guanosine-rich stretches and is associated with RNAs and several other proteins in a sort of ribonucleoprotein (RNP) complex (Ludevid et al. 1992, Freire et al. 1995). Interestingly, in the presence of ABA the ABA-response regulator AAPK phosphorylates the RNA-binding protein AKIP1 which thereupon partitions into subnuclear foci and interacts with dehydrin mRNA (Li et al. 2002).

Both the mRNA cap-binding protein ABH1 (Hugovieux et al. 2001, 2002) as well as SAD1 (Xiong et al. 2001a), likely to act as a SM-like snRNP protein in mRNA maturation or turnover, revealed a negative regulatory role on ABA responses. The recessive *sad1* and *abh1* mutants are characterized by ABA hypersensitivity of germination and stomatal closure to exogenous ABA. Interestingly, the mutations affect expression of ABA-regulated genes in a targeted manner. The molecular basis for the phenomenon is unclear but points to an ABA-regulated posttranscriptional control of distinct RNA transcripts. Early posttranscriptional regulation might also involve HYL1, a dsRNA binding protein (Lu and Fedoroff 2000), which may act as an integrator of auxin, cytokinin and ABA signalling at the transcriptional or posttranscriptional level. At the posttranslational level, protein turnover establishes an additional regulatory mechanism that has gained interest from the susceptibility to ABA5 to proteolytic degradation *via* the ubiquitin pathway (Lopez-Molina et al. 2003, Smalle et al. 2003). The *Arabidopsis* SNF1-like kinases AKIN10 and AKIN11 are complexed in the proteasome (Farras et al.

2001) but also interact with an importin-binding nuclear protein PRL1 that regulates pleiotropic responses to sugars and hormones, including abscisic acid (Nemeth et al. 1998). In addition, stability of ABA signal components or targets might be regulated by the small ubiquitin-like modifier (SUMO; Lois et al. 2003).

These examples probably reflect only "the tip of an iceberg" (Fedoroff 2002b) and illustrate that posttranscriptional regulatory mechanisms may address ABA specific targets (dehydrin mRNA, ABI5) as well as knots or integrators of several signal transduction pathways (HYL1, AKIN10/11).

2.6 Cross-talk

The different facets of ABA action such as regulation of ion status and metabolism, as well as regulation of gene expression at the transcriptional and posttranscriptional level just mirror the complexity and cybernetic challenges of a sessile plant to adjust to stress situations. External signals like cold, drought, or salt stress trigger the generation of ABA as an internal signal in addition to the initiation of an ABA-independent cascade required for stress-optimised adaptation (Shinozaki and Yamaguchi-Shinozaki 1997, Fedoroff 2002a).

The necessity of interference between ABA signalling and other signalling pathways is obvious just considering the growth-inhibitory action of ABA and ethylene that antagonize the growth-promotive effect of auxin, cytokinin, and gibberellic acid. The cross-talk occurring between signalling of ABA and ethylene (Ghassemian et al. 2000), auxin (Suzuki et al., 2001; Brady et al. 2003), gibberellin (Gómez-Cadenas et al. 2001), pathogen interaction, and wounding (Pena-Cortes et al. 1995, Audenaert et al. 2002; Neill et al. 2002a), or sugar sensing (Finkelstein et al. 2002) emphasizes the tight interaction of regulatory circuits. In light of this situation, we should be aware that our major tools for analysis of signalling, mutants and phenocopies generated by interfering gene expression, are prone to pleiotropic alteration *via* cross-talk and feedback loops.

Acknowledgements

We thank the Deutsche Forschungsgemeinschaft and the "Fonds der Chemischen Industrie" for financial support.

References

Abe H, Yamaguchi-Shinozaki K, Urao T, Iwasaki T, Hosokawa D, Shinozaki K (1997) Role of *Arabidopsis* MYC and MYB homologs in drought- and abscisic acid-regulated gene expression. Plant Cell 9:1859-1868

Abe H, Urao T, Ito T, Seki M, Shinozaki K, Yamaguchi-Shinozaki K (2003) *Arabidopsis* AtMYC2 (bHLH) and AtMYB2 (MYB) function as transcriptional activators in abscisic acid signaling. Plant Cell 15:63-78

Allan AC, Fricker MD, Ward JL, Beale MH, Trewavas AJ (1994) Two transduction pathways mediate rapid effects of abscisic acid in *Commelina* guard cells. Plant Cell 6:1319-1328

Allen GJ, Muir SR, Sanders D. (1995) Release of Ca^{2+} from individual plant vacuoles by both $InsP_3$ and cyclic ADP-ribose. Science 268:735-737

Allen GJ, Kuchitsu K, Chu SP, Murata Y, Schroeder JI (1999) *Arabidopsis abi1-1* and *abi2-1* phosphatase mutations reduce abscisic acid-induced cytoplasmic calcium rises in guard cells. Plant Cell 11:1785-1798

Allen GJ, Chu SP, Schumacher K, Shimazaki CT, Vafeados D, Kemper A, Hawke SD, Tallman G, Tsien RY, Harper JF, Chory J, Schroeder JI (2000) Alteration of stimulus-specific guard cell calcium oscillations and stomatal closing in *Arabidopsis det3* mutant. Science 289:2338-2342

Allen, GJ, Chu SP, Harrington CL, Schumacher K, Hoffmann T, Tang YY, Grill E Schroeder JI (2001): A defined range of guard cell calcium oscillation parameters encodes stomatal movements. Nature 411:1053-1057

Allen GJ, Murata Y, Chu SP, Nafisi M, Schroeder JI. (2002) Hypersensitivity of abscisic acid-induced cytosolic calcium increases in the *Arabidopsis* farnesyltransferase mutant *era1-2*. Plant Cell 14:1649-1662

Anderberg RJ, Walker-Simmons MK (1992) Isolation of a wheat cDNA clone for an abscisic acid-inducible transcript with homology to protein kinases. Proc Natl Acad Sci USA 89:10183-10187

Anderson BE, Ward JM, Schroeder JI (1994) Evidence for an extracellular reception site for abscisic acid in *Commelina* guard cells. Plant Physiol 104: 1177-1183

Arigoni D, Sagner S, Latzel C, Eisenreich W, Bacher A, Zenk MH (1997) Terpenoid biosynthesis from 1-deoxy-D-xylulose in higher plants by intramolecular skeletal rearrangement. Proc Natl Acad Sci USA 94:10600-10605

Audenaert K, De Meyer GB, Hofte MM (2002) Abscisic acid determines basal susceptibility of tomato to Botrytis cinerea and suppresses salicylic acid-dependent signaling mechanisms. Plant Physiol 128:491-501

Audran C, Borel C, Frey A, Sotta B, Meyer C, Simonneau T, Marion-Poll A (1998) Expression studies of the zeaxanthin epoxidase gene in Nicotiana plumbaginifolia. Plant Physiol 118:1021-1028

Baxter-Burrell A, Yang Z, Springer PS, Bailey-Serres J (2002) RopGAP4-dependent Rop GTPase rheostat control of *Arabidopsis* oxygen deprivation tolerance. Science 296:2026-2028

Beligni MV, Lamattina L (2001) Nitric oxide: a non-traditional regulator of plant growth. Trends Plant Sci 6:508-509

Bittner F, Oreb M, Mendel RR (2001) ABA3 is a molybdenum cofactor sulfurase required for activation of aldehyde oxidase and xanthine dehydrogenase in *Arabidopsis thaliana*. J Biol Chem 276:40381-40384

Blatt MR, Armstrong F (1993) Potassium channels of stomatal guard cells: abscisic acid-evoked control of the outward rectifier mediated by cytoplasmic pH. Planta 191:330-341

Bouvier F, D'Harlingue A, Backhaus RA, Kumagai MH, Camara B. (2000) Identification of neoxanthin synthase as a carotenoid cyclase paralog. Eur J Biochem 267:6346-6352

Brady SM, Sarkar SF, Bonetta D, McCourt P (2003) The ABSCISIC ACID INSENSITIVE 3 (ABI3) gene is modulated by farnesylation and is involved in auxin signaling and lateral root development in *Arabidopsis*. Plant J 34:67-75

Burnette RN, Gunesekera BM, Gillaspy GE (2003) An *Arabidopsis* inositol 5-phosphatase gain-of-function alters abscisic acid signaling. Plant Physiol 132:1011-1019

Busk PK, Pages M (1998) Regulation of abscisic acid-induced transcription. Plant Mol Biol 37:425-435

Cheng WH, Endo A, Zhou L, Penney J, Chen HC, Arroyo A, Leon P, Nambara E, Asami T, Seo M, Koshiba T, Sheen J (2002a) A unique short-chain dehydrogenase/reductase in *Arabidopsis* glucose signaling and abscisic acid biosynthesis and functions. Plant Cell 14:2723-2743

Cheng SH, Willmann MR, Chen HC, Sheen J (2002b) Calcium signaling through protein kinases. The *Arabidopsis* calcium-dependent protein kinase gene family. Plant Physiol 129:469-485

Cheong YH, Chang HS, Gupta R, Wang X, Zhu T, Luan S (2002) Transcriptional profiling reveals novel interactions between wounding, pathogen, abiotic stress, and hormonal responses in *Arabidopsis*. Plant Physiol 129:661-677

Cherel I, Michard E, Platet N, Mouline K, Alcon C, Sentenac H, Thibaud JB (2002) Physical and functional interaction of the *Arabidopsis* K^+ channel AKT2 and phosphatase AtPP2CA. Plant Cell 14:1133-1146

Choi H, Hong J, Ha J, Kang J, Kim SY (2000) ABFs, a family of ABA-responsive element binding factors. J Biol Chem 275:1723-1730

Coursol S, Fan L-M, Le Stunff H, Spiegel S, Gilroy S, Assmann SM (2003) Sphingolipid signaling in *Arabidopsis* guard cells involves heterotrimeric G proteins. Nature 423:651-654

Cowan AK (2000) Is abscisic aldehyde really the immediate precursor to stress-induced ABA? Trends Plant Sci 5:191-192

Cunningham FX Jr, Gantt E (1998) Genes and enzymes of carotenoid biosynthesis in plants. Annu Rev Plant Physiol Plant Mol Biol 49:557-583

Desikan R, Griffiths R, Hancock J, Neill S (2002) A new role for an old enzyme: Nitrate reductase-mediated nitric oxide generation is required for abscisic acid-induced stomatal closure in *Arabidopsis thaliana*. Proc Natl Acad Sci USA 99:16314-16318

Dinkins R, Pflipsen C, Thompson A, Collins GB (2002) Ectopic expression of an *Arabidopsis* single zinc finger gene in tobacco results in dwarf plants. Plant Cell Physiol 43:743-750

Duckham SC, Linforth RST, Taylor IB (1991) Abscisic acid-deficient mutants at the *aba* gene locus of *Arabidopsis thaliana* are impaired in the epoxidation of zeaxanthin. Plant Cell Environment 14:601-606

Durner J, Wendehenne D, Klessig DF (1998) Defense gene induction in tobacco by nitric oxide, cyclic GMP, and cyclic ADP-ribose. Proc Natl Acad Sci USA 95:10328-10333

Eun SO, Lee Y (1997) Actin filaments of guard cells are reorganized in response to light and abscisic acid. Plant Physiol 115:1491-1498

Evans NH, McAinsh MR, Hetherington AM (2001) Calcium oscillations in higher plants. Curr Opin Plant Biol 4:415-420

Fan L, Zheng S, Wang X (1997) Antisense suppression of phospholipase D alpha retards abscisic acid- and ethylene-promoted senescence of postharvest *Arabidopsis* leaves. Plant Cell 9:2183-2196

Farras R, Ferrando A, Jasik J, Kleinow T, Okresz L, Tiburcio A, Salchert K, del Pozo C, Schell J, Koncz C (2001) SKP1-SnRK protein kinase interactions mediate proteasomal binding of a plant SCF ubiquitin ligase. EMBO J 20:2742-2756

Fedoroff NV (2002a) Cross-talk in abscisic acid signaling. Sci STKE 2002/140/re10

Fedoroff NV (2002b) RNA-binding proteins in plants: the tip of an iceberg? Curr Opin Plant Biol 5:452-459

Finkelstein RR (1994) Mutations at two new *Arabidopsis* ABA response loci are similar to the *abi3* mutations. Plant J 5:765-771

Finkelstein RR, Lynch TJ (2000) The *Arabidopsis* abscisic acid response gene ABI5 encodes a basic leucine zipper transcription factor. Plant Cell 12:599-609

Finkelstein RR, Wang ML, Lynch TJ, Rao S, Goodman HM (1998) The *Arabidopsis* abscisic acid response locus ABI4 encodes an APETALA 2 domain protein. Plant Cell 10:1043-1054

Finkelstein RR, Gampala SS, Rock CD (2002) Abscisic acid signaling in seeds and seedlings. Plant Cell 14 Suppl:S15-S45

Frey A, Audran C, Marin E, Sotta B, Marion-Poll A (1999) Engineering seed dormancy by the modification of zeaxanthin epoxidase gene expression. Plant Mol Biol 39:1267-1274

Freire MA, Pages M (1995) Functional characteristics of the maize RNA-binding protein MA16. Plant Mol Biol 29:797-807

Garbers C, DeLong A, Deruere J, Bernasconi P, Soll D (1996) A mutation in protein phosphatase 2A regulatory subunit A affects auxin transport in *Arabidopsis*. EMBO J 15:2115-2124

Garcia-Mata C, Lamattina L (2002) Nitric oxide and abscisic acid cross talk in guard cells. Plant Physiol. 128:790-792

Garcia-Mata C, Lamattina L (2003) Abscisic acid, nitric oxide and stomatal closure - is nitrate reductase one of the missing links? Trends Plant Sci 8:20-26

Geelen D, Leyman B, Batoko H, Di Sansebastiano GP, Moore I, Blatt MR, Di Sansabastiano GP (2002) The abscisic acid-related SNARE homolog NtSyr1 contributes to secretion and growth: evidence from competition with its cytosolic domain. Plant Cell 14:387-406

Ghassemian M, Nambara E, Cutler S, Kawaide H, Kamiya Y, McCourt P (2000) Regulation of abscisic acid signaling by the ethylene response pathway in *Arabidopsis*. Plant Cell 12:1117-1126

Gomez-Cadenas A, Zentella R, Walker-Simmons MK, Ho TH (2001) Gibberellin/abscisic acid antagonism in barley aleurone cells: site of action of the protein kinase PKABA1 in relation to gibberellin signaling molecules. Plant Cell 13:667-679

Gonzalez-Guzman M, Apostolova N, Belles JM, Barrero JM, Piqueras P, Ponce MR, Micol JL, Serrano R, Rodriguez PL (2002) The short-chain alcohol dehydrogenase ABA2 catalyzes the conversion of xanthoxin to abscisic aldehyde. Plant Cell 14:1833-1846

Gosti F, Beaudoin N, Serizet C, Webb AA, Vartanian N, Giraudat J (1999) ABI1 protein phosphatase 2C is a negative regulator of abscisic acid signaling. Plant Cell 11:1897-1910

Grabov A, Blatt MR (1998) Membrane voltage initiates Ca^{2+} waves and potentiates Ca^{2+} increases with abscisic acid in stomatal guard cells. Proc Natl Acad Sci USA 95:4778-4783

Grill E, Ziegler H (1998) A plant's dilemma. Science 282: 252-253

Guan LM, Zhao J, Scandalios JG (2000) Cis-elements and trans-factors that regulate expression of the maize Cat1 antioxidant gene in response to ABA and osmotic stress: H_2O_2 is the likely intermediary signaling molecule for the response. Plant J 22:87-95

Guo Y, Xiong L, Song CP, Gong D, Halfter U, Zhu JK (2002) A calcium sensor and its interacting protein kinase are global regulators of abscisic acid signaling in *Arabidopsis*. Dev Cell 3:233-244.

Hallouin M, Ghelis T, Brault M, Bardat F, Cornel D, Miginiac E, Rona JP, Sotta B, Jeannette E (2002) Plasmalemma abscisic acid perception leads to RAB18 expression via phospholipase D activation in *Arabidopsis* suspension cells. Plant Physiol 130:265-272

Hamilton DW, Hills A, Kohler B, Blatt MR (2000) Ca^{2+} channels at the plasma membrane of stomatal guard cells are activated by hyperpolarization and abscisic acid. Proc Natl Acad Sci USA 97:4967-4972

Hardie DG, Carling D, Carlson M (1998) The AMP-activated/SNF1 protein kinase subfamily: metabolic sensors of the eukaryotic cell? Annu Rev Biochem 67:821-855

Hattori T, Totsuka M, Hobo T, Kagaya Y, Yamamoto-Toyoda A (2002) Experimentally determined sequence requirement of ACGT-containing abscisic acid response element. Plant Cell Physiol 43:136-140

Heim MA, Jakoby M, Werber M, Martin C, Weisshaar B, Bailey PC (2003) The basic helix-loop-helix transcription factor family in plants: a genome-wide study of protein structure and functional diversity. Mol Biol Evol 20:735-747

Hetherington AM (2001) Guard cell signaling. Cell 107:711-714

Himmelbach A, Iten M, Grill E (1998) Signaling of abscisic acid to regulate plant growth. Philos Trans R Soc Lond B Biol Sci 353:1439-1444

Himmelbach A, Hoffmann T, Leube M, Höhener B, Grill E (2002) Homeodomain protein ATHB6 is a target of the protein phosphatase ABI1 and regulates hormone responses in *Arabidopsis*. EMBO J 21:3029-3038

Hirai N, Yoshida R, Todoroki Y, Ohigashi H (2000) Biosynthesis of abscisic acid by the non-mevalonate pathway in plants, and by the mevalonate pathway in fungi. Biosci Biotechnol Biochem 64:1448-1458

Hirayama T, Ohto C, Mizoguchi T, Shinozaki K (1995) A gene encoding a phosphatidylinositol-specific phospholipase C is induced by dehydration and salt stress in *Arabidopsis thaliana*. Proc Natl Acad Sci USA 92:3903-3907

Hobo T, Kowyama Y, Hattori T (1999) A bZIP factor, TRAB1, interacts with VP1 and mediates abscisic acid-induced transcription. Proc Natl Acad Sci USA 96:15348-15353

Hornberg C, Weiler E (1984) High-affinity binding sites for abscisic acid on plasmalemma of *Vicia faba* guard cells. Nature 310:321-324

Hoth S, Morgante M, Sanchez JP, Hanafey MK, Tingey SV, Chua NH (2002) Genome-wide gene expression profiling in *Arabidopsis thaliana* reveals new targets of abscisic acid and largely impaired gene regulation in the abi1-1 mutant. J Cell Sci 115:4891-4900

Hsu TC, Liu HC, Wang JS, Chen RW, Wang YC, Lin BL (2001) Early genes responsive to abscisic acid during heterophyllous induction in *Marsilea quadrifolia*. Plant Mol Biol 47:703-715

Huang JF, Teyton L, Harper JF (1996) Activation of a Ca(2+)-dependent protein kinase involves intramolecular binding of a calmodulin-like regulatory domain. Biochem 35:13222-13230

Hugouvieux V, Kwak JM, Schroeder JI (2001). An mRNA cap binding protein, ABH1, modulates early abscisic acid signal transduction in *Arabidopsis*. Cell 106: 477-487

Hugouvieux V, Murata Y, Young JJ, Kwak JM, Mackesy DZ, Schroeder JI (2002) Localization, ion channel regulation, and genetic interactions during abscisic acid signaling of the nuclear mRNA cap-binding protein, ABH1. Plant Physiol 130:1276-1287

Hunt L, Mills LN, Pical C, Leckie CP, Aitken FL, Kopka J, Mueller-Roeber B, McAinsh MR, Hetherington AM, Gray JE (2003) Phospholipase C is required for the control of stomatal aperture by ABA. Plant J 34:47-55

Hwang I, Sze H, Harper JF (2000) A calcium-dependent protein kinase can inhibit a calmodulin-stimulated Ca^{2+} pump (ACA2) located in the endoplasmic reticulum of *Arabidopsis*. Proc Natl Acad Sci USA 97:6224-6229

Hwang JU, Lee Y (2001) Abscisic acid-induced actin reorganization in guard cells of dayflower is mediated by cytosolic calcium levels and by protein kinase and protein phosphatase activities. Plant Physiol 125:2120-2128

Irving HR, Gehring CA, Parish RW (1992) Changes in cytosolic pH and calcium of guard cells precede stomatal movements. Proc Natl Acad Sci USA 89:1790-1794

Ishitani M, Liu J, Halfter U, Kim CS, Shi W, Zhu JK (2000) SOS3 function in plant salt tolerance requires N-myristoylation and calcium binding. Plant Cell 12:1667-1678

Iuchi S, Kobayashi M, Yamaguchi-Shinozaki K, Shinozaki K (2000) A stress-inducible gene for 9-cis-epoxycarotenoid dioxygenase involved in abscisic acid biosynthesis under water stress in drought-tolerant cowpea. Plant Physiol 123:553-562

Iuchi S, Kobayashi M, Taji T, Naramoto M, Seki M, Kato T, Tabata S, Kakubari Y, Yamaguchi-Shinozaki K, Shinozaki K (2001) Regulation of drought tolerance by gene manipulation of 9-cis-epoxycarotenoid dioxygenase, a key enzyme in abscisic acid biosynthesis in *Arabidopsis*. Plant J 27:325-333

Jacob T, Ritchie S, Assmann SM, Gilroy S (1999) Abscisic acid signal transduction in guard cells is mediated by phospholipase D activity. Proc Natl Acad Sci USA 96:12192-12197

Jakoby M, Weisshaar B, Droge-Laser W, Vicente-Carbajosa J, Tiedemann J, Kroj T, Parcy F; bZIP Research Group. (2002) bZIP transcription factors in *Arabidopsis*. Trends Plant Sci 7:106-111

Jeannette E, Rona JP, Bardat F, Cornel D, Sotta B, Miginiac E (1999) Induction of RAB18 gene expression and activation of K^+ outward rectifying channels depend on an extracellular perception of ABA in *Arabidopsis thaliana* suspension cells. Plant J 18:13-22

Johannesson H, Wang Y, Engstrom P (2001) DNA-binding and dimerization preferences of *Arabidopsis* homeodomain-leucine zipper transcription factors in vitro. Plant Mol Biol 45:63-67

Johnson RR, Wagner RL, Verhey SD, Walker-Simmons MK (2002) The abscisic acid-responsive kinase PKABA1 interacts with a seed-specific abscisic acid response element-binding factor, TaABF, and phosphorylates TaABF peptide sequences. Plant Physiol 130:837-846

Jonak C, Okresz L, Bogre L, Hirt H (2002) Complexity, cross talk and integration of plant MAP kinase signalling. Curr Opin Plant Biol 5:415-424

Jung JY, Kim YW, Kwak JM, Hwang JU, Young J, Schroeder JI, Hwang I, Lee Y (2002) Phosphatidylinositol 3- and 4-phosphate are required for normal stomatal movements. Plant Cell 14:2399-2412

Kagaya Y, Hobo T, Murata M, Ban A, Hattori T (2002) Abscisic acid-induced transcription is mediated by phosphorylation of an abscisic acid response element binding factor, TRAB1. Plant Cell 14:3177-3189

Kim JC, Lee SH, Cheong YH, Yoo CM, Lee SI, Chun HJ, Yun DJ, Hong JC, Lee SY, Lim CO, Cho MJ (2001) A novel cold-inducible zinc finger protein from soybean, SCOF-1, enhances cold tolerance in transgenic plants. Plant J 25:247-259

Kim KN, Cheong YH, Grant JJ, Pandey GK, Luan S (2003) CIPK3, a calcium sensor-associated protein kinase that regulates abscisic acid and cold signal transduction in *Arabidopsis*. Plant Cell 15:411-423

Kinoshita T, Nishimura M, Shimazaki K-I (1995) Cytosolic concentration of Ca^{2+} regulates the plasma membrane H^{+}-ATPase in guard cells of fava bean. Plant Cell 7:1333-1342

Kizis D, Pagès M (2002) Maize DRE-binding proteins DBF1 and DBF2 are involved in rab17 regulation through the drought-responsive element in an ABA-dependent pathway. Plant J 30:679-689

Kizis D, Lumbreras V, Pagès M (2001) Role of AP2/EREBP transcription factors in gene regulation during abiotic stress. FEBS Lett. 498:187-189

Klüsener B, Young JJ, Murata Y, Allen GJ, Mori IC, Hugouvieux V, Schroeder JI (2002) Convergence of calcium signaling pathways of pathogenic elicitors and abscisic acid in *Arabidopsis* guard cells. Plant Physiol 130:2152-2163

Knetsch M, Wang M, Snaar-Jagalska BE, Heimovaara-Dijkstra S (1996) Abscisic acid induces mitogen-activated protein kinase activation in barley aleurone protoplasts. Plant Cell 8:1061-1067

Koornneef M, Reuling G, Karssen CM (1984) The isolation and characterization of abscisic acid-insensitive mutants of *Arabidopsis thaliana*. Physiol Plant 61:377-383

Kovtun Y, Chiu WL, Tena G, Sheen J (2000) Functional analysis of oxidative stress-activated mitogen-activated protein kinase cascade in plants. Proc Natl Acad Sci USA 97:2940-2945

Kruse T, Tallman G, Zeiger E (1989) Isolation of guard cell protoplasts from mechanically prepared epidermis of Vicia faba leaves. Plant Physiology 90:1382-1386

Kushiro T, Nambara E, McCourt P (2003) Hormone evolution: The key to signalling. Nature 422:122

Kwak JM, Moon JH, Murata Y, Kuchitsu K, Leonhardt N, DeLong A, Schroeder JI (2002) Disruption of a guard cell-expressed protein phosphatase 2A regulatory subunit, RCN1, confers abscisic acid insensitivity in *Arabidopsis*. Plant Cell 14:2849-2861

Kwak JM, Mori IC, Pei ZM, Leonhardt N, Torres MA, Dangl JL, Bloom RE, Bodde S, Jones JD, Schroeder JI (2003) NADPH oxidase *AtrbohD* and *AtrbohF* genes function in ROS-dependent ABA signaling in *Arabidopsis*. EMBO 22:2623-2633.

Lång V, Mäntylä E, Welin B, Sundberg B, Palva ET (1994) Alterations in water status, endogenous abscisic acid content, and expression of rab18 gene during the development of freezing tolerance in *Arabidopsis thaliana*. Plant Physiol 104:1341-1349

Leckie CP, McAinsh MR, Allen GJ, Sanders D, Hetherington AM (1998) Abscisic acid induced stomatal closure mediated by cyclic ADP-ribose. Proc Natl Acad Sci USA 95:15837-15842

Lee Y, Choi YB, Suh S, Lee J, Assmann SM, Joe CO, Kelleher JF, Crain RC (1996) Abscisic acid-induced phosphoinositide turnover in guard cell protoplasts of *Vicia faba*. Plant Physiol 110:987-996

Lee Y, Oh H, Cheon C, Hwang I, Kim Y Chun J (2001) Structure and expression of the *Arabidopsis thaliana* homeobox gene Athb-12. Biochem Biophys Res Commun 284:133–141

Lemichez E, Wu Y, Sanchez JP, Mettouchi A, Mathur J, Chua NH (2001) Inactivation of AtRac1 by abscisic acid is essential for stomatal closure. Genes Dev 15:1808-1816

Lemtiri-Chlieh F, MacRobbie EA, Brearley CA (2000) Inositol hexakisphosphate is a physiological signal regulating the K^+-inward rectifying conductance in guard cells. Proc Natl Acad Sci USA 97:8687-8692

Leube MP, Grill E, Amrhein N (1998): ABI1 of *Arabidopsis* is a protein serine/threonine phosphatase highly regulated by the proton and magnesium ion concentration. FEBS Lett 424:100-104

Leung J, Merlot S, Giraudat J (1997) The *Arabidopsis* abscisic acid-insensitive2 (ABI2) and ABI1 genes encode homologues protein phosphatases 2C involved in abscisic acid signal transduction. Plant Cell 9:759-771

Leyman B, Geelen D, Quintero FJ, Blatt MR (1999) A tobacco syntaxin with a role in hormonal control of guard cell ion channels. Science 283:537-540

Li J, Assmann SM (1996) An abscisic acid-activated and calcium-independent protein kinase from guard cells of fava bean. Plant Cell 8:2359-2368

Li J, Lee YR, Assmann SM (1998) Guard cells possess a calcium-dependent protein kinase that phosphorylates the KAT1 potassium channel. Plant Physiol 116:785-795

Li J, Wang XQ, Watson MB, Assmann SM (2000) Regulation of abscisic acid-induced stomatal closure and anion channels by guard cell AAPK kinase. Science 287:300-303

Li J, Kinoshita T, Pandey S, Ng CK, Gygi SP, Shimazaki K, Assmann SM (2002) Modulation of an RNA-binding protein by abscisic-acid-activated protein kinase. Nature 418:793-797

Lichtenthaler HK (1999) The 1-deoxy-D-xylulose-5-phosphate pathway of isoprenoid biosynthesis in plants. Annu Rev Plant Physiol Plant Mol Biol 50:47-65

Lichtenthaler HK, Schwender J, Disch A, Rohmer M (1997) Biosynthesis of isoprenoids in higher plant chloroplasts proceeds via a mevalonate-independent pathway. FEBS Lett 400:271-274

Liu Q, Kasuga M, Sakuma Y, Abe H, Miura S, Yamaguchi-Shinozaki K, Shinozaki K.(1998) Two transcription factors, DREB1 and DREB2, with an EREBP/AP2 DNA binding domain separate two cellular signal transduction pathways in drought- and low-temperature-responsive gene expression, respectively, in *Arabidopsis*. Plant Cell. 10:1391-1406

Lois LM, Lima CD, Chua NH (2003) Small ubiquitin-like modifier modulates abscisic acid signaling in *Arabidopsis*. Plant Cell 15:1347-1359

Lopez-Molina L, Mongrand S, Chua NH (2001) A postgermination developmental arrest checkpoint is mediated by abscisic acid and requires the ABI5 transcription factor in *Arabidopsis*. Proc Natl Acad Sci USA 98:4782-4787

Lopez-Molina L, Mongrand S, Kinoshita N, Chua NH (2003) AFP is a novel negative regulator of ABA signaling that promotes ABI5 protein degradation. Genes Dev 17:410-418

Lu C, Fedoroff N (2000) A mutation in the *Arabidopsis* HYL1 gene encoding a dsRNA binding protein affects responses to abscisic acid, auxin, and cytokinin. Plant Cell 12:2351-2366

Lu C, Han MH, Guevara-Garcia A, Fedoroff NV (2002) Mitogen-activated protein kinase signaling in postgermination arrest of development by abscisic acid. Proc Natl Acad Sci USA 99:15812-15817

Luan, S (2002): Signaling drought in guard cells. Plant Cell Environ 25:229-237

Ludevid MD, Freire MA, Gomez J, Burd CG, Albericio F, Giralt E, Dreyfuss G, Pages M (1992) RNA binding characteristics of a 16 kDa glycine-rich protein from maize. Plant J 2:999-1003

Marin E, Nussaume L, Quesada A, Gonneau M, Sotta B, Hugueney P, Frey A, Marion-Poll A (1996) Molecular identification of zeaxanthin epoxidase of *Nicotiana plumbaginifolia*, a gene involved in abscisic acid biosynthesis and corresponding to the *ABA* locus of *Arabidopsis thaliana*. EMBO J 15:2331-2342

Martin C, Paz-Ares J (1997) MYB transcription factors in plants. Trends Genet 13:67-73

McAinsh MR, Brownlee C, Hetherington AM (1990) Abscisic acid-induced elevation of guard cell cytosolic Ca^{2+} precedes stomatal closure. Nature 343:186-188

Meinhard M, Grill E (2001): Hydrogen peroxide is a regulator of ABI1, a protein phosphatase 2C from *Arabidopsis*. FEBS Lett 508:443-446

Meinhard M, Rodriguez PL, Grill E (2002) The sensitivity of ABI2 to hydrogen peroxide links the abscisic acid-response regulator to redox signaling. Planta 214:775-782

Merlot S, Gosti F, Guerrier D, Vavasseur A, Giraudat J (2001) The ABI1 and ABI2 protein phosphatases 2C act in a negative feedback regulatory loop of the abscisic acid signaling pathway. Plant J. 25:295-303

Milborrow BV (2001) The pathway of biosynthesis of abscisic acid in vascular plants: a review of the present state of knowledge of ABA biosynthesis. J Exp Bot 52:1145-1164

Morriss-Kay GM, Ward SJ (1999) Retinoids and mammalian development. Int Rev Cytol. 188:73-131

Murata Y, Pei ZM, Mori IC, Schroeder J (2001): Abscisic acid activation of plasma membrane $Ca(^{2+})$ channels in guard cells requires cytosolic NAD(P)H and is differentially disrupted upstream and downstream of reactive oxygen species production in *abi1-1* and *abi2-1* protein phosphatase 2C mutants. Plant Cell 13: 2513-2523

Mustilli AC, Merlot S, Vavasseur A, Fenzi F, Giraudat J (2002): *Arabidopsis* OST1 protein kinase mediates the regulation of stomatal aperture by abscisic acid and acts upstream of reactive oxygen species production. Plant Cell 14:3089-3099

Nakamura S, Lynch TJ, Finkelstein RR (2001) Physical interactions between ABA response loci of *Arabidopsis*. Plant J 26:627-635

Nambara E, Suzuki M, Abrams S, McCarty DR, Kamiya Y, McCourt P (2002) A screen for genes that function in abscisic acid signaling in *Arabidopsis thaliana*. Genetics 161: 1247-1255

Narusaka Y, Nakashima K, Shinwari ZK, Sakuma Y, Furihata T, Abe H, Narusaka M, Shinozaki K, Yamaguchi-Shinozaki K (2003) Interaction between two cis-acting elements, ABRE and DRE, in ABA-dependent expression of *Arabidopsis* rd29A gene in response to dehydration and high-salinity stresses. Plant J 34:137-148

Neill SJ, Desikan R, Clarke A, Hancock JT (2002a) Nitric oxide is a novel component of abscisic acid signaling in stomatal guard cells. Plant Physiol 128:13-16

Neill SJ, Desikan R, Clarke A, Hurst RD, Hancock JT (2002b) Hydrogen peroxide and nitric oxide as signaling molecules in plants. J Exp Bot 53:1237-1247

Nemeth K, Salchert K, Putnoky P, Bhalerao R, Koncz-Kalman Z, Stankovic-Stangeland B, Bako L, Mathur J, Okresz L, Stabel S, Geigenberger P, Stitt M, Redei GP, Schell J,

Koncz C (1998) Pleiotropic control of glucose and hormone responses by PRL1, a nuclear WD protein, in *Arabidopsis*. Genes Dev 12:3059-3073

Ng CK, Carr K, McAinsh MR, Powell B, Hetherington AM (2001) Drought-induced guard cell signal transduction involves sphingosine-1-phosphate. Nature 410:596-599

Niu X, Helentjaris T, Bate NJ (2002) Maize ABI4 binds coupling element 1 in abscisic acid and sugar response genes. Plant Cell 11:2565-2575

Pappan K, Wang X (1999) Molecular and biochemical properties and physiological roles of plant phospholipase D. Biochim Biophys Acta 1439:151-166

Park JM, Park CJ, Lee SB, Ham BK, Shin R, Paek KH (2001) Overexpression of the tobacco Tsi1 gene encoding an EREBP/AP2-type transcription factor enhances resistance against pathogen attack and osmotic stress in tobacco. Plant Cell. 13:1035-1046

Parry AD, Babiano MJ, Horgan R (1990) The role of cis-carotenoids in abscisc acid biosynthesis. Planta 182:118-128

Pedron J, Brault M, Nake C, Miginiac E (1998) Detection of abscisic acid-binding proteins in the microsomal protein fraction of *Arabidopsis thaliana* with abscisic acid-protein conjugates used as affinity probes. Eur J Biochem 252:385-391

Pei ZM, Ghassemian M, Kwak CM, McCourt P, Schroeder JI (1998) Role of farnesyltransferase in ABA regulation of guard cell anion channels and plant water loss. Science 282: 287-290

Pei ZM, Murata Y, Benning G, Thomine S, Klüsener B, Allen GJ, Grill E, Schroeder JI (2000) Calcium channels activated by hydrogen peroxide mediate abscisic acid signaling in guard cells. Nature 406:731-734

Pena-Cortes H, Fisahn J, Willmitzer L (1995) Signals involved in wound-induced proteinase inhibitor II gene expression in tomato and potato plants. Proc Natl Acad Sci USA. 92:4106-4113

Qin X, Zeevaart JA (1999) The 9-cis-epoxycarotenoid cleavage reaction is the key regulatory step of abscisic acid biosynthesis in water-stressed bean. Proc Natl Acad Sci USA 96:15354-15361

Qin X, Zeevaart JA (2002) Overexpression of a 9-cis-epoxycarotenoid dioxygenase gene in *Nicotiana plumbaginifolia* increases abscisic acid and phaseic acid levels and enhances drought tolerance. Plant Physiol 128:544-551

Rashotte AM, DeLong A, Muday GK (2001) Genetic and chemical reductions in protein phosphatase activity alter auxin transport, gravity response, and lateral root growth. Plant Cell. 13:1683-1697

Raskin I and Ladyman JAR (1988) Isolation and characterization of a barley mutant with abscisic acid-insensitive stomata. Planta 173:73-78

Ritchie S, Gilroy S (2000) Abscisic acid stimulation of phospholipase D in the barley aleurone is G-protein-mediated and localized to the plasma membrane. Plant Physiol 124:693-702

Ritchie SM, Swanson SJ, Gilroy S (2002) From common signalling components to cell specific responses: insights from the cereal aleurone. Physiol Plant 115: 342-351

Roberts SK, Snowman BN (2000) The effects of ABA on channel-mediated K^+ transport across higher plant roots. J Exp Bot 51:1585-1594

Rock CD (2000) Pathways to ABA-regulated gene expression. New Phytol 148:357-396

Rodriguez PL, Benning G, Grill E. (1998a) ABI2, a second protein phosphatase 2C involved in abscisic acid signal transduction in *Arabidopsis*. FEBS Lett. 421:185-190

Rodriguez PL, Leube MP, Grill E (1998b) Molecular cloning in *Arabidopsis thaliana* of a new protein phosphatase 2C (PP2C) with homology to ABI1 and ABI2. Plant Mol Biol 38:879-883.

Romeis T, Piedras P, Jones JD (2000) Resistance gene-dependent activation of a calcium-dependent protein kinase in the plant defense response. Plant Cell 12:803-816

Sanchez JP, Chua NH (2001) *Arabidopsis* PLC1 is required for secondary responses to abscisic acid signals. Plant Cell 13:1143-1154

Sang Y, Zheng S, Li W, Huang B, Wang X (2001) Regulation of plant water loss by manipulating the expression of phospholipase Dα. Plant J 28:135-144

Schroeder JI, Allen GJ, Hugouvieux V, Kwak JM, Waner D (2001) Guard cell signal transduction. Annu Rev Plant Physiol Plant Mol Biol 52:627-658

Schultz TF, Medina J, Hill A, Quatrano RS (1998) 14-3-3 proteins are part of an abscisic acid-VIVIPAROUS1 (VP1) response complex in the Em promoter and interact with VP1 and EmBP1. Plant Cell 10:837-847

Schwartz A, Wu WH, Tucker EB, Assmann SM (1994) Inhibition of inward K^+ channels and stomatal response by abscisic acid: an intracellular locus of phytohormone action. Proc Natl Acad Sci USA. 91:4019-4023

Schwartz SH, Tan BC, Gage DA, Zeevaart JA, McCarty DR (1997) Specific oxidative cleavage of carotenoids by VP14 of maize. Science 276:1872-1874

Seki M, Ishida J, Narusaka M, Fujita M, Nanjo T, Umezawa T, Kamiya A, Nakajima M, Enju A, Sakurai T, Satou M, Akiyama K, Yamaguchi-Shinozaki K, Carninci P, Kawai J, Hayashizaki Y, Shinozaki K (2002) Monitoring the expression pattern of around 7,000 *Arabidopsis* genes under ABA treatments using a full-length cDNA microarray. Funct Integr Genomics 2:282-291

Seo M, Koiwai H, Akaba S, Komano T, Oritani T, Kamiya Y, Koshiba T (2000a) Abscisic aldehyde oxidase in leaves of *Arabidopsis thaliana.* Plant J 23:481-488

Seo M, Peeters AJ, Koiwai H, Oritani T, Marion-Poll A, Zeevaart JA, Koornneef M, Kamiya Y, Koshiba T (2000b) The *Arabidopsis* aldehyde oxidase 3 (AAO3) gene product catalyzes the final step in abscisic acid biosynthesis in leaves. Proc Natl Acad Sci USA 97:12908-12913

Sharp RE (2002) Interaction with ethylene: changing views on the role of abscisic acid in root and shoot growth responses to water stress. Plant Cell Environ 25:211-222

Sheen J (1996) Ca^{2+}-dependent protein kinases and stress signal transduction in higher plants. Science 274:1900-1902

Sheen J (1998) Mutational analysis of protein phosphatase 2C involved in abscisic acid signal transduction in higher plants. Proc Natl Acad Sci USA 95: 975-980

Sheen J (2001) Signal transduction in maize and *Arabidopsis* mesophyll protoplasts. Plant Physiol 127:1466-1475

Shen Q, Gomez-Cadenas A, Zhang P, Walker-Simmons MK, Sheen J, Ho TH (2001) Dissection of abscisic acid signal transduction pathways in barley aleurone layers. Plant Mol Biol. 47:437-448

Shinozaki K, Yamaguchi-Shinozaki K (1997) Gene expression and signal transduction in water-stress response.Plant Physiol 115:327-334

Shinozaki K, Yamaguchi-Shinozaki K (2000) Molecular responses to dehydration and low temperature: differences and cross-talk between two stress signaling pathways. Curr Opin Plant Biol 3:217-223

Siefermann-Harms D (1977) The xanthophyll cycle in higher plants. In Tevini M, Lichtenthaler HK, eds, Lipid and lipid polymers in higher plants. Springer, Berlin, pp 218-230

Slovik S, Daeter W, Hartung W (1995): Compartmental redistribution and long-distance transport of abscisic acid (ABA) in plants as influenced by environmental changes in the rhizosphere - a biomathematical model. J. Exp. Bot. 46:881-894

Smalle J, Kurepa J, Yang P, Emborg TJ, Babiychuk E, Kushnir S, Vierstra RD (2003) The pleiotropic role of the 26S proteasome subunit RPN10 in *Arabidopsis* growth and development supports a substrate-specific function in abscisic acid signaling. Plant Cell 15:965-980

Söderman E, Mattsson J, Engstrom P (1996) The *Arabidopsis* homeobox gene ATHB-7 is induced by water deficit and by abscisic acid. Plant J 10:375-381

Söderman E, Hjellström M, Fahleson J, Engström P (1999) The HD-Zip gene *ATHB6* in *Arabidopsis* is expressed in developing leaves, roots and carpels and up-regulated by water deficit conditions. Plant Mol Biol 40:1073–1083

Söderman EM, Brocard IM, Lynch TJ, Finkelstein RR (2000) Regulation and function of the *Arabidopsis* ABA-insensitive4 gene in seed and abscisic acid response signaling networks. Plant Physiol 124:1752-1765

Staxen II, Pical C, Montgomery LT, Gray JE, Hetherington AM, McAinsh MR (1999) Abscisic acid induces oscillations in guard-cell cytosolic free calcium that involve phosphoinositide-specific phospholipase C. Proc Natl Acad Sci USA 96:1779-1784

Sutton F, Paul SS, Wang XQ, Assmann SM (2000) Distinct abscisic acid signaling pathways for modulation of guard cell versus mesophyll cell potassium channels revealed by expression studies in *Xenopus laevis* oocytes. Plant Physiol 124:223-230

Suzuki M, Kao CY, McCarty DR (1997) The conserved B3 domain of VIVIPAROUS1 has a cooperative DNA binding activity. Plant Cell 9:799-807

Suzuki M, Kao CY, Cocciolone S, McCarty DR (2001) Maize VP1 complements *Arabidopsis abi3* and confers a novel ABA/auxin interaction in roots. Plant J 28:409-418

Tahtiharju S, Palva T (2001) Antisense inhibition of protein phosphatase 2C accelerates cold acclimation in *Arabidopsis thaliana*. Plant J 26:461-470

Takahashi S, Katagiri T, Hirayama T, Yamaguchi-Shinozaki K, Shinozaki K (2001) Hyperosmotic stress induces a rapid and transient increase in inositol 1,4,5-trisphosphate independent of abscisic acid in *Arabidopsis* cell culture. Plant Cell Physiol 42:214-222

Tan BC, Cline K, McCarty DR (2001) Localization and targeting of the VP14 epoxycarotenoid dioxygenase to chloroplast membranes. Plant J 27:373-382

Tan BC, Joseph LM, Deng WT, Liu L, Li QB, Cline K, McCarty DR (2003) Molecular characterization of the *Arabidopsis* 9-cis epoxycarotenoid dioxygenase gene family. Plant J 35:44-56

Taylor IB, Burbidge A, Thompson AJ (2000) Control of abscisic acid synthesis. J Exp Bot 51:1563-1574

Thompson AJ, Jackson AC, Parker RA, Morpeth DR, Burbidge A, Taylor IB (2000a) Abscisic acid biosynthesis in tomato: regulation of zeaxanthin epoxidase and 9-cis-epoxycarotenoid dioxygenase mRNAs by light/dark cycles, water stress and abscisic acid. Plant Mol Biol 42:833-845

Thompson AJ, Jackson AC, Symonds RC, Mulholland BJ, Dadswell AR, Blake PS, Burbidge A, Taylor IB (2000b) Ectopic expression of a tomato 9-cis-epoxycarotenoid dioxygenase gene causes over-production of abscisic acid. Plant J 23:363-374

Uno Y, Furihata T, Abe H, Yoshida R, Shinozaki K, Yamaguchi-Shinozaki K (2000) *Arabidopsis* basic leucine zipper transcription factors involved in an abscisic acid-dependent signal transduction pathway under drought and high-salinity conditions. Proc Natl Acad Sci USA 97:11632-11637

Walker-Simmons MK, Holappa LD, Abrams GD and Abrams SR (1997) ABA metabolites induce group 3 LEA mRNA and inhibit germination in wheat. Physiol Plant 7:125-134

Wang XQ, WuW-H, Assmann SM (1998) Differential responses of abaxial and adaxial guard cells of broad bean to abscisic acid and calcium. Plant Physiol 118:1421-1429

Wang XQ, Ullah H, Jones AM, Assmann SM (2001) G protein regulation of ion channels and abscisic acid signaling in *Arabidopsis* guard cells. 292:2070-2072

Webb AA, Larman MG, Montgomery LT, Taylor JE, Hetherington AM (2001) The role of calcium in ABA-induced gene expression and stomatal movements. Plant J 26:351-362

Wilkinson S, Davies WJ (1997) Xylem sap pH increase: a drought signal received at the apoplastic face of the guard cell that involves the suppression of saturable abscisic acid uptake by the epidermal symplast. Plant Physiol 113:559-573

Wu Y, Kuzma J, Marechal E, Graeff R, Lee HC, Foster R, Chua NH (1997) Abscisic acid signaling through cyclic ADP-ribose in plants. Science 278:2126-2130

Wu Y, Sanchez JP, Lopez-Molina L, Himmelbach A, Grill E, Chua NH (2003) The *abi1-1* mutation blocks ABA signaling downstream of cADPR action. Plant J. 34:307-315

Xiong L, Gong Z, Rock CD, Subramanian S, Guo Y, Xu W, Galbraith D, Zhu JK (2001a) Modulation of abscisic acid signal transduction and biosynthesis by an Sm-like protein in *Arabidopsis*. Dev Cell 1:771-781

Xiong L, Ishitani M, Lee H, Zhu JK (2001b) The *Arabidopsis* LOS5/ABA3 locus encodes a molybdenum cofactor sulfurase and modulates cold stress- and osmotic stress-responsive gene expression. Plant Cell 13:2063-2083

Xiong L, Lee BH, Ishitani M, Lee H, Zhang C, Zhu JK (2001c) FIERY1 encoding an inositol polyphosphate 1-phosphatase is a negative regulator of abscisic acid and stress signaling in *Arabidopsis*. Genes Dev 15:1971-1984

Xiong L, Lee H, Ishitani M, Zhu JK (2002) Regulation of osmotic stress-responsive gene expression by the LOS6/ABA1 locus in *Arabidopsis*. J Biol Chem 277:8588-8596

Yalovsky S, Kulukian A, Rodriguez-Concepcion M, Young CA, Gruissem W (2000) Functional requirement of plant farnesyltransferase during development in *Arabidopsis*. Plant Cell. 12:1267-1278

Yamaguchi-Shinozaki K, Shinozaki K (1994) A novel cis-acting element in an *Arabidopsis* gene is involved in responsiveness to drought, low-temperature, or high-salt stress. Plant Cell 6:251-264

Yamauchi D, Zentella R, Ho D (2002) Molecular analysis of the barley (*Hordeum vulgare* L.) gene encoding the protein kinase PKABA1 capable of suppressing gibberellin action in aleurone layers. Planta 215:319-326

Yang Z (2002) Small GTPases: Versatile signaling switches in plants. Plant Cell 14 Suppl:S375-388

Yoshida R, Hobo T, Ichimura K, Mizoguchi T, Takahashi F, Aronso J, Ecker JR, Shinozaki K (2002) ABA-activated SnRK2 protein kinase is required for dehydration stress signaling in *Arabidopsis*. Plant Cell Physiol 43:1473-1483

Zeevaart JAD, Creelman RA (1988) Metabolism and physiology of abscisic acid. Annu Rev Plant Physiol Plant Mol Biol 39:439-473

Zhang DP, Chen SW, Peng YB, Shen YY (2001a) Abscisic acid-specific binding sites in the flesh of developing apple fruit. J Experimental Botany 52:2097-2103

Zhang DP, Wu ZY, Li XY, Zhao ZX (2002): Purification and identification of a 42-kilodalton abscisic acid-specific-binding protein from epidermis of broad bean leaves. Plant Physiol 128:714-725

Zhang X, Miao YC, An GY, Zhou Y, Shangguan ZP, Gao JF, Song CP (2001b) K^+ channels inhibited by hydrogen peroxide mediate abscisic acid signaling in Vicia guard cells. Cell Res. 11:195-202.

Zheng ZL, Nafisi M, Tam A, Li H, Crowell DN, Chary SN, Schroeder JI, Shen J, Yang Z (2002) Plasma membrane-associated ROP10 small GTPase is a specific negative regulator of abscisic acid responses in *Arabidopsis*. Plant Cell. 14:2787-2797

Zocchi E, Basile G, Cerrano C, Bavestrello G, Giovine M, Bruzzone S, Guida L, Carpaneto A, Magrassi R, Usai C (2003) ABA- and cADPR-mediated effects on respiration and filtration downstream of the temperature-signaling cascade in sponges. J Cell Sci 116:629-636

3 Plant responses to heat stress

Priti Krishna

Abstract

The heat stress response is characterized by inhibition of normal transcription and translation, higher expression of heat shock proteins (hsps) and induction of thermotolerance. If stress is too severe, signaling pathways leading to apoptotic cell death are also activated. As molecular chaperones, hsps provide protection to cells against the damaging effects of heat stress and enhance survival. The enhanced expression of hsps is regulated by heat shock transcription factors (HSFs). Recent advances in molecular genetic approaches have provided new insights into the plant heat stress response. A striking characteristic of plants is that they contain highly complex multigene families encoding HSFs and hsps. This review outlines our current knowledge of the functions of plant hsps and the regulation of HSFs, and offers a comparative view of heat stress responses in plants and other organisms. Recent observations indicating that heat stress response overlaps with other stress responses are also discussed.

3.1 Introduction

Heat stress response is invoked in organisms as diverse as bacteria, fungi, plants, and animals by sudden increases in temperature, and is characterized by elevated synthesis of a set of proteins called heat shock proteins (hsps). Hsps comprise several evolutionarily conserved protein families, such as hsp100, hsp90, hsp70, hsp60, and small hsps (shsps). A common feature of the heat stress response is that an initial exposure to mild heat stress provides resistance against a subsequent usual lethal dose of heat stress. This phenomenon is referred to as 'acquired thermotolerance'. Since thermotolerant cells express high levels of hsps, these proteins have been associated with the development of thermotolerance (reviewed in Parsell and Lindquist 1993). High temperature stress causes extensive denaturation and aggregation of cellular proteins, which, if unchecked, lead to cell death. Through their chaperoning activity, hsps help cells to cope with heat-induced damage to cellular proteins. During stress, hsps function primarily to prevent aggregation and promote proper refolding of denatured proteins, but because protein conformation is important right from the time a protein is synthesized, hsps play important roles under normal conditions as well. The principal role of hsps under non-stress conditions is to assist in the synthesis, transport, and proper folding of the target proteins. Although all hsps function as molecular chaperones, each hsp

Topics in Current Genetics, Vol. 4
H. Hirt, K. Shinozaki (Eds.) Plant Responses To Abiotic Stress

family has a unique mechanism of action. The relative importance of individual hsp families in stress tolerance varies from one organism to another.

In nature, temperature changes are likely to occur more rapidly than other stress-causing factors. Due to their inability to translocate, plants are subject to wide variations in temperature both diurnally and seasonally, and must therefore adapt to temperature stress quickly and efficiently. Indeed, it appears from published experimental data and *in silico* analyses of the fully sequenced *Arabidopsis thaliana* model that plants have evolved extensive and complex mechanisms to combat detrimental effects of heat stress. While modest knowledge about hsp expression and functions has been gained, our understanding of the regulatory mechanisms controlling the heat shock response in plants is limited. In this review, I focus on the recent progress made in understanding the molecular mechanism of the heat shock response in plants.

3.2 Major families of heat shock proteins

3.2.1 Hsp100

The hsp100 family of proteins is present in both prokaryotes and eukaryotes, with sizes ranging from 75 to 100 kDa. The bacterial hsp100 proteins, referred to as Clp proteins, have been studied extensively as components of a 2-subunit protease system (Squires and Squires 1992). The large subunit ClpA functions as a chaperone, while the small subunit ClpP is the protease. Hsp100 proteins are divided into 2 major classes; class 1 proteins contain 2 ATP-binding sites, and class 2 proteins contain only 1 ATP-binding site (Miernyk 1999; Schirmer et al. 1996). An interesting feature of the hsp100 proteins is their ability to promote dissociation of aggregated proteins in an ATP-dependent manner as opposed to mainly preventing unfolding and aggregation of proteins, as is attributed to other chaperones (Parsell et al. 1994).

Hsp100 proteins have been identified in a number of plant species, and an analysis of their expression has revealed that they are both developmentally regulated and stress-induced (reviewed in Agarwal et al. 2001). The *A. thaliana* hsp100 family of proteins consists of 8 members, of which 5 proteins have predicted plastidial localization signals (Agarwal et al. 2001). Several studies have established that Athsp101 of the *A. thaliana* hsp100 family is essential for thermotolerance (Hong and Vierling 2000, 2001; Queitsch et al. 2000). The yeast homolog hsp104 is also required for induced thermotolerance in yeast (Sanchez and Lindquist 1990). Since the thermotolerance defect in yeast caused by the deletion of *hsp104* gene can be complemented by plant hsp100 proteins, it is concluded that the function of yeast and plant hsp100 proteins in thermotolerance is conserved (Lee et al. 1994; Schirmer et al. 1994). In addition to its role in heat stress, plant hsp101 has been demonstrated to function as a RNA-binding protein for mediating the translational enhancement of tobacco mosaic virus RNA and ferredoxin mRNA (Ling et al. 2000; Wells et al. 1998).

3.2.2 Hsp90

Hsp90 is an essential molecular chaperone in eukaryotic cells, with key functions in signal transduction networks, cell-cycle control, protein degradation, and protein trafficking. The number of identified protein targets whose functions are facilitated by hsp90 continues to grow, especially in animal systems. A critical dependence on hsp90 has been established for steroid hormone receptors, several serine/threonine and tyrosine kinases, and other distinct proteins. Two key features regarding the mechanism of animal and yeast hsp90 have emerged over the last decade: 1) hsp90 is an ATP-dependent chaperone, and 2) hsp90 functions in cooperation with other chaperones and co-chaperones that together constitute the hsp90 chaperone complex (reviewed in Buchner 1999; Young et al. 2001). Due to its critical role in signal transduction, hsp90 has emerged as a promising drug target (reviewed in Neckers 2002). The hsp90 inhibitor geldanamycin (GA) binds with high specificity within the ATP-binding pocket of hsp90, inhibiting its function and resulting in the degradation of client proteins via the ubiquitin proteasome pathway. A derivative of GA with cancer selectivity is now in Phase I clinical trial as an anticancer drug. New hsp90 inhibitors are being engineered to target specific functions of hsp90 that are applicable to medical conditions other than cancer. Further significance of hsp90 was shown in recent work in insect (Rutherford and Lindquist 1998) and plant (Queitsch et al. 2002), suggesting that hsp90 links the response to environmental stresses with development in a way that could influence evolutionary change. The hypothesis is that hsp90 buffers genetic variation in nature by maintaining mutant proteins in their wild type conformations. When this buffering is compromised, for example, by temperature stress, hsp90 is diverted from its normal clients to other partially denatured proteins and variations become exposed, thereby allowing selection to remodel developmental processes.

Hsp90 genes have been isolated from several plant species (reviewed in Krishna and Gloor 2001). The *A. thaliana* hsp90 family of proteins consists of 7 members, of which 4 constitute the cytoplasmic subfamily and the remaining 3 are predicted to be within the plastidial, mitochondrial and endoplasmic reticulum (ER) compartments. Although the occurrence of multiple hsp90 proteins in the cytoplasm and of other family members in various subcellular compartments suggests a range of specific functions for these proteins, our understanding of hsp90 in plants remains relatively limited. The study of the hsp90-based chaperone complex in animals has revealed that hsp90 and hsp70 and their co-chaperones Hop (hsp70 and hsp90 organizing protein), p23, and hsp40 participate in the conformational regulation of client proteins. High molecular weight immunophilins are also recovered in hsp90 complexes, and the co-chaperone Cdc37/$p50^{cdc37}$ is predominantly found in hsp90-kinase complexes (reviewed in Pratt and Toft 1997; Richter and Buchner 2001). An hsp90-based chaperone system is present also in plants, and, to date, hsp90, hsp70, high molecular weight immunophilins and a Hop-like protein have been identified in plant hsp90 complexes (Owens-Grillo et al. 1996; Stancato et al. 1996; Pratt et al. 2001; Reddy et al. 1998; Zhang et al. 2003). Some observations suggest that the plant hsp90 complex has evolved some unique characteristics. For example, *Brassica napus* p23 appears to be expressed only under heat stress (Z. Zhang and P. Krishna, unpublished data), unlike animal and yeast p23 that are expressed under normal growth conditions. Furthermore, a Cdc37/p50^{cdc37}-related protein has not been identified in plants based on sequence comparisons. Currently there is no published report on any plant hsp90 client protein, but since reducing hsp90 function in *A. thaliana* through treatment with GA produces an array of morphological phenotypes (Queitsch et al. 2002), it appears that hsp90 chaperones signaling proteins in plants that control plant growth and development.

Though hsp90 is an abundant protein under non-stress conditions, its increased expression in response to elevated temperatures suggests a protective role for hsp90 under heat stress conditions (Krishna and Gloor 2001; Parsell and Lindquist 1993). Indeed, mammalian and yeast hsp90 can promote refolding of thermally denatured proteins *in vitro* (Schumacher et al. 1994; Wiech et al. 1992), and stabilize early unfolding intermediates in thermal unfolding pathway of proteins, thereby preventing their irreversible aggregation (Jakob et al. 1995). Also, transiently expressed *B. napus* hsp90 in *A. thaliana* cell suspension culture containing stably integrated firefly luciferase as a reporter, can accelerate luciferase renaturation during recovery (Forreiter et al. 1997). Thus, hsp90 has chaperoning activity that is linked to denaturing stress. Several lines of evidence suggest that hsp90 functions as a regulator of the heat shock response; this aspect is discussed in section 3.3.2 of this review.

3.2.3 Hsp70

Members of the hsp70 family exist in the cytosol of all eubacteria and eukaryotes, and some archae, as well as within mitochondria, ER and plastids of eukaryotic cells (Lin et al. 2001). In higher eukaryotes, including plants, some hsp70 family members are expressed constitutively (Hsc70) while others are stress-inducible (reviewed in Boston et al. 1996; Hartl and Hayer-Hartl 2002). The domain structure of hsp70 comprises of a ~45 kDa NH_2-terminal ATPase domain and a ~25 kDa COOH-terminal peptide-binding domain. In higher eukaryotes, hsp70 functions both co- and post-translationally. The various functions of hsp70 have been reviewed previously (Hendrick and Hartl 1993; Parsell and Lindquist 1993). Substrates in their non-native states are bound by hsp70 and successive cycles of binding and releasing coupled with ATP hydrolysis promotes protein folding. The mechanistic details of hsp70 function are best understood for the prokaryotic homolog DnaK, which requires 2 accessory proteins: DnaJ (eukaryotic counterparts are referred to as hsp40) and the nucleotide exchange factor GrpE (Bukau and Horwich 1998). DnaJ regulates the ATPase activity of DnaK, and GrpE facilitates the release of ADP. In comparison to the bacterial DnaK, relatively little is known about the hsp70 protein folding machinery in plants. A total of 89 J-domain proteins were identified in the genome of *A. thaliana* (Miernyk 2001). Seven of these are closely related to hsp40 and are likely to perform functions analogous to hsp40/DnaJ proteins. A GrpE-like protein appears to be absent in eukaryotic cytosol, although a structurally unrelated protein Bag-1 acts as a nucleotide exchange factor and a regulator of hsp70 (Hohfeld and Jentsch 1997). Bag-domain proteins are also present in plants but a description of these proteins has yet to appear in literature. Other co-chaperones of eukaryotic hsp70, such as Hop and Hip (hsp70-interacting protein) that affect the nucleotide-bound state of hsp70 (Frydman and Hohfeld 1997), may also regulate the functional cycle of hsp70 proteins. Residues critical for interaction with hsp70 are conserved in the tetratricopeptide repeat domains of plant orthologs of Hip and Hop (Webb et al. 2001; Zhang et al. 2003), but the details of these interactions remain to be elucidated.

The hsp70 protein family is relatively large in plants; 18 members have been assigned to the hsp70 family in *A. thaliana* with distribution in the cytoplasm as well as plastid, ER and mitochondria (Lin et al. 2001). The expression of hsp70 is both developmentally regulated and stress-induced (Boston et al. 1996; Sung et al. 2001). A comprehensive analysis of the expression profile of the *A. thaliana* hsp70 gene family indicates that individual members differ in their response to different conditions and stimuli (Sung et al. 2001). Based on the expression patterns, functions for members of the hsp70 family can be ascribed to heat and cold stress, and to seed maturation and germination. Together, these data suggest that plant hsp70 proteins interact with diverse substrates and take part in a plethora of cellular processes. Clearly, the challenge for the future is to obtain detailed knowledge of the specific functions of individual members. Like hsp90, hsp70 has also been implicated in the regulation of the heat shock response. This aspect is discussed in section 3.3.2 of this review.

3.2.4 Small hsps

Shsps are a group of proteins ranging in size from 15 to 42 kDa that are ubiquitously produced in prokaryotic and eukaryotic cells in response to heat stress. In plants, shsps are the most dominant proteins produced in response to heat stress. A total of 13 shsps belonging to 6 classes, defined on the basis of sequence relatedness and intracellular localization, and an additional 6 open reading frames encoding distantly-related proteins have been identified in the *A. thaliana* genome (Scharf et al. 2001). The unusual abundance and complexity of these proteins in plants suggest their unique significance, but a comprehensive understanding of their roles and mechanisms awaits further revelation. Detailed descriptions of the classification, structure, and functions of plant shsps have been provided in several recent reviews (Boston et al. 1996; Scharf et al. 2001; Sun et al. 2002; Waters et al. 1996). The NH_2-terminal end of shsps belonging to different classes are quite divergent, but all shsps share a conserved COOH-terminal domain referred to as the α-crystallin domain. A common feature of shsps is their ability to form large oligomers and changes in the oligomeric state are associated with the chaperone activity of shsps. The molecular chaperone activity of plant shsps has been demonstrated both *in vitro* (Lee et al. 1995a, 1997) and *in vivo* (Forreiter et al. 1997; Low et al. 2000). In contrast to hsp60 and hsp70 proteins, the chaperone activity of shsps is ATP-independent (Lee et al. 1995a). The current model for shsp chaperone function is that shsps bind to unfolding intermediates to protect them from irreversible aggregation and to maintain them in a refolding competent state. The captured proteins are then refolded by other molecular chaperones in an ATP-dependent manner (Lee and Vierling 2000). The expression patterns and chaperone function of shsps suggest that shsp production is correlated with thermotolerance. Experimental data in support of an important role of shsps in thermotolerance as well as other stresses has been derived from genetic modification of the expression of shsps in transgenic plants (Harndahl et al. 1999; Malik et al. 1999; Sun et al. 2001). The functions of shsps extend beyond those associated with protection against heat stress as shsps are also synthesized in the absence of any environmental stress at specific stages of development, such as germination, embryogenesis, pollen development and fruit maturation (reviewed in Sun et al. 2002). Getting a clear picture of the distinct functions of shsps during stress and developmental processes is a task for the future.

3.2.5 The Chaperonins

Chaperonins comprise a diverse family of molecular chaperones that are present in the cytoplasm, plastids, and mitochondria of eukaryotes and eubacteria (reviewed in Boston et al. 1996; Bukau and Horwich 1998; Hill and Hemmingsen 2001). They occur as two distinct subgroups, type I and type II, that are distantly related in sequence. The eukaryotic type I chaperonins are localized in the chloroplast and mitochondria and are referred to as chaperonin 60 (Cpn60). The bacterial type I chaperonin, referred to as GroEL, has been studied extensively, and serves as a paradigm for the type I system. Typically, these proteins consist of 2 stacked rings, each comprised of 7 subunits (Saibil 2002), and they act in concert with co-chaperonins GroES in bacteria, and Cpn21 and Cpn10 in eukaryotes (Saibil and Ranson 2002). Through ATP hydrolysis, the type I chaperonins assist a large variety of proteins to reach their native states. A recent examination of the *A. thaliana* genome for chaperonin sequences indicates that the chaperonin family in plants is more diverse than previously described (Hill and Hemmingsen 2001). Nine plastidic and mitochondrial Cpn60 proteins, and a previously undescribed 10 kDa potential plastidic co-chaperonin were identified in this search. The presence of the 10 kDa co-chaperonin together with the previously recognized Cpn21 is intriguing and raises the possibility of coexistence of 2 chaperonin systems in the chloroplast (Schlicher and Soll 1996).

The plastidic Cpn60 is composed of 2 distinct subunit types, α and β. The expression of *Cpn60* gene members is developmentally regulated, and some members are heat shock-inducible while others are not (Zabaleta et al. 1994a). The chloroplast Cpn60 plays a prominent role in the folding of plastid proteins, including Rubisco. In accordance with this, reduced expression of Cpn60β in transgenic tobacco resulted in abnormal phenotypes (Zabaleta et al. 1994b), and a functionless *Cpn60α* gene led to defects in *A. thaliana* embryo development (Apuya et al. 2001). The results of Apuya et al. (2001) clearly demonstrate that Cpn60α is required for chloroplast development and that proper embryo development is dependent on functional chloroplasts, but do not preclude the possibility that Cpn60α may also be involved in the folding of some cytoplasmic proteins.

The cytoplasmic counterparts of type I chaperonins are referred to as type II chaperonins and they too function in double-ring oligomeric complex. The eukaryotic complex is referred to as TCP-1 (T complex protein-1), CCT (chaperonin containing TCP-1), or TRiC (TCP-1 ring complex). In *A. thaliana*, 9 coding regions are predicted to encode CCT-related proteins (Hill and Hemmingsen 2001). Study of the type II chaperonins in plants is still in its infancy.

3.3 Transcriptional regulation of hsps

The stress-inducible expression of hsps is regulated primarily at the transcriptional level by heat shock transcription factors (Hsfs) that bind to conserved regulatory elements located in the promoters of hsp genes, referred to as heat shock elements (HSEs). The eukaryotic HSEs with the palidromic consensus sequence (nGAAn) (nTTCn) are located within a few hundred base pairs of the 5' flanking regions of heat shock genes (reviewed in Schoffl et al. 1998). Upon activation, the Hsf binds to HSEs and interacts with proteins of the basal transcription machinery. Features of heat shock gene transcription in other organisms involving Hsfs and components of the chromatin remodeling and basal transcription machinery have been reviewed (Morimoto 1998; Wu 1995), and are not discussed here. In plants, the interaction of *A. thaliana* Hsf1 with the TATA box binding transcription factor is the only observation reported in this direction (Reindl and Schoffl 1998).

The first plant Hsf was cloned in 1990 from *Lycopersicon peruvianum* (Scharf et al. 1990). Further analysis of the tomato Hsf system revealed that in contrast to a single Hsf in *Saccharomyces cerevisiae* and *Drosophila melanogaster*, plants contain multiple Hsfs. An idea of the unusual complexity of the Hsf network in plants can be obtained by examination of the *A. thaliana* genome sequence that appears to encode 21 such proteins (Nover et al. 2001). This number far exceeds the 4 Hsfs found in vertebrates (Pirkkala et al. 2001). The plant Hsf gene family is also unique in that some of its members are heat stress-inducible, and others have structural peculiarities. Typically, the Hsfs are composed of a conserved DNA binding domain, an oligomerization domain, and an activation domain. The Hsfs of higher eukaryotes are converted from a monomeric to a trimeric form in response to stress. The activated trimeric form acquires DNA binding and transcriptional activation properties. In yeast, the Hsf is bound to DNA in a trimeric form even in the absence of stress, but becomes competent for transactivation function during heat stress following temperature-dependent phosphorylation and conformational change (Jakobsen and Pelham 1988; Sorger 1991). In contrast, human Hsf1 is phosphorylated as an inactive monomer and becomes activated after binding to DNA in its trimeric state by heat-induced hyperphosphorylation (Cotto et al. 1996).

3.3.1 Structure of plant Hsfs

A detailed description of the Hsf family of proteins in plants and their structural characteristics has recently been provided by Nover et al. (2001). At the NH_2-terminal end of plant Hsfs is a conserved helix-turn-helix DNA binding domain (DBD) that specifically recognizes the palindromic HSE. Connected to the DBD by a linker region of variable length and sequence is the oligomerization domain (HR-A/B). The heptad pattern of hydrophobic residues in the HR-A/B region suggests a coiled-coil structure that is likely involved in trimerization of Hsfs. Based on differences in their flexible linker and HR-A/B regions, the plant Hsfs have been divided into 3 classes: A, B, and C (Fig. 1). Similar to non-plant Hsfs, the HR-A and B parts of class B Hsfs are separated by 7 amino acid residues, whereas class A and class C Hsfs have insertions of an additional 21 and 7 amino acid residues, respectively (Nover et al. 2001). It is interesting that class C Hsfs were identified only after examination of the *A. thaliana* genome sequence, although EST analysis suggests that members of this new class are well expressed in different plant tissues. Hsfs also contain sequence motifs essential for both nuclear import and export. The nuclear localization signals (NLS) of class A and C Hsfs lie adjacent to the HR-A/B region, but for most class B Hsfs, NLS is located towards the COOH-terminal end of the protein (Fig. 1). The nucleocytoplasmic distribution of Hsfs is also influenced by the nuclear export signal (NES). The predominant cytoplasmic distribution of the tomato HsfA2 is dependent on its NES (Heerklotz et al. 2001). The activation domains of Hsfs, located at the COOH-terminus, are least conserved in size and sequence. The transcription activation function of Hsfs is correlated with short peptide motifs referred to as AHA motifs, usually found in the centre of the activation domain of most *A. thaliana* class A Hsfs (Doring et al. 2000). It is believed that these motifs constitute the sites for interaction with components of the basal transcription machinery (Nover et al. 2001). Interestingly, class B Hsfs lack the AHA motifs (Fig. 1). This has raised the question of whether these proteins function in cooperation with other Hsfs. Nover and co-workers have evidence suggesting that HsfB1 acts as a synergistic partner of HsfA1 in activating gene transcription (Nover et al. 2001).

3.3.2 Regulation of plant Hsfs

3.3.2.1 Positive and negative regulation by Hsfs

The multiplicity of Hsfs in plants raises the question of whether different members are functionally redundant or diverse. Experimental data describing the structural and functional characteristics of plant Hsfs has been obtained primarily for tomato Hsfs. Although 17 Hsfs have been identified in tomato thus far, only the constitutively expressed HsfA1 has been ascribed the unique role as a master regulator of thermotolerance in tomato (Mishra et al. 2002). In transgenic tomato plants with strongly reduced expression of HsfA1, no other constitutively expressed Hsf could substitute for HsfA1 as master regulator of the heat shock response. These plants

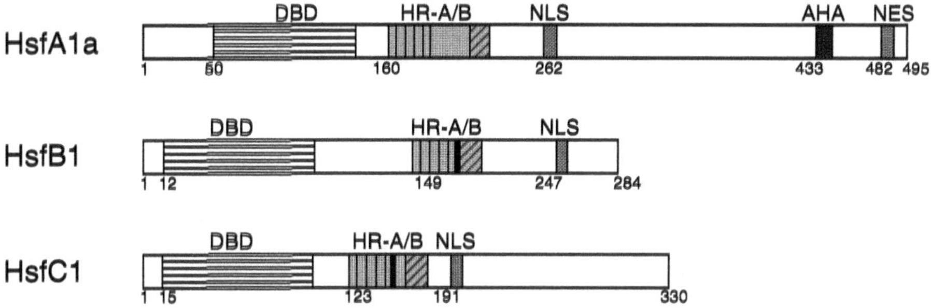

Fig. 1. Structures of *A. thaliana* Hsfs belonging to classes A, B, and C. Only a single member of each class is shown. The functional domains include: DBD, DNA binding domain; HR-A/B, oligomerization domain (the plain grey area in HsfA1a and HsfC1 represents insertion of additional amino acid residues between parts A and B); NLS, nuclear localization signal; AHA, short motifs in the activator region that are rich in aromatic, hydrophobic, and acidic amino acid residues; NES, nuclear export signal. Reproduced with permission from Nover et al. (2001) Cell Stress Chaperones 6:177-189.

were highly sensitive to elevated temperature and showed reduced or no synthesis of hsps and heat stress-inducible Hsfs. In contrast, no comparable defects in the heat shock response were noticed in plants with strongly reduced or no expression of the heat stress-inducible HsfA2 and HsfB1 (Mishra et al. 2002). The authors speculate a sequential model of heat stress-induced gene expression in which the heat-inducible HsfA2 and HsfB1 act as HsfA1-dependent enhancer or modifier of hsp synthesis. This model is supported by the observation that the nuclear retention and activator function of HsfA2 is markedly influenced by heteroligomerization with HsfA1 (Heerklotz et al. 2001), and that HsfB1, which has no transcription activator function itself, can act as a strong synergistic co-activator of HsfA1 (Nover et al. 2001).

As an alternative to an activator function, it is proposed that some plant Hsfs, in particular those belonging to class B, could function as negative regulators of the heat stress response (Czarnecka-Verner et al. 2000). One possibility is that class B Hsfs occupy HSEs under non-stress conditions and maintain heat shock genes in a repressed state. The constitutive nuclear localization of tomato HsfB1 (Scharf et al. 1998) is compatible with this model. The negative regulatory effects of these proteins may also be exerted by forming non-functional complexes with components of the basal transcriptional machinery or even with class A Hsfs. The complexity of class B Hsfs suggests important biological functions for these proteins and the possibility that they have evolved as co-activators or repressors of class A Hsfs is supported by some experimental evidence. Elaborate analysis in the future of the interactions between different Hsfs and between Hsfs and proteins of the transcription complex will reveal how protein:protein interactions regulate the plant heat shock response.

3.3.2.2 Negative regulation by hsps

With the exception of yeast, eukaryotic Hsf is maintained as an inert monomer under non-stress conditions, and assembles into trimers only upon activation by stress. How Hsf of higher eukaryotes is maintained in the monomeric form has been a question of much interest. Studies from various organisms have led to the belief that Hsf is negatively regulated, at least in part, by hsp70 (reviewed in Wu 1995). For example, mutations in yeast hsp70 result in overexpression of hsps (Craig and Gross 1991), and results of biochemical and overexpression studies indicate that hsp70 interacts with Hsf (Abravaya et al. 1992; Mosser et al. 1993; Rabindran et al. 1994). These observations, together with the fact that the heat shock response is transient in nature, led to the proposal that there is feedback regulation of Hsf by hsp70. According to this model, during heat shock, when protein unfolding increases and non-native proteins accumulate, hsp70 and other molecular chaperones are recruited to prevent unfolding and stabilizing partially unfolded intermediates. The sequestration of hsp70 by denatured proteins results in the activation of Hsf and transcription of hsps, including hsp70. As the levels of hsp70 increase during heat stress, hsp70 and Hsf reassociate. This leads to repression of Hsf transcriptional activity and attenuation of the heat shock response. In metazoans, activation of Hsf1 is a multistep process, including trimerization, DNA binding and inducible phosphorylation. Any of these events could be affected by hsp70. Shi et al. (1998) observed that hsp70 and its co-chaperone Hdj-1/hsp40 interact directly with the activation domain of Hsf1, leading to repression of heat shock gene transcription. Since neither the DNA binding activity nor inducible phosphorylation of Hsf1 is affected in hsp70 overexpressing cells, it is concluded that the primary function of hsp70 as a regulator of Hsf1 is to repress its transcriptional activity during attenuation of the heat shock response (Shi et al. 1998). Presumably, other events and factors are involved in dissociating trimers and converting them to the inert monomeric form. A protein that associates with Hsf1, termed as heat shock factor binding protein 1 (HSBP1), has properties of a negative regulator of Hsf1 (Satyal et al. 1998). HSBP1 interacts with only the trimeric form of Hsf1 and with hsp70. It is proposed that during attenuation of the heat shock response, interaction of Hsf1 with hsp70, Hdj-1/hsp40 and HSBP1 eventually leads to dissociation of Hsf1 trimers to inert monomers (Morimoto 1998). The details of this model remain to be worked out.

Several lines of evidence indicate that the hsp90 chaperone machinery also regulates Hsf activity. Yeast and rat Hsfs can physically associate with hsp90 (Nadeau et al. 1993). Genetic tests in yeast revealed that hsp90 and Cpr7, a CyP-40-type cyclophilin required for full hsp90 function *in vivo*, act synergistically to repress gene expression from Hsf-dependent promoters (Duina et al. 1998). Furthermore, trimeric human Hsf1 dynamically associates with an hsp90-immunophilin (FKBP52)-p23 complex through the regulatory domain, and formation of this heterocomplex results in the repression of Hsf1 transcriptional activity (Guo et al. 2001). It is clear from these observations that molecular chaperones have pivotal roles in regulating the heat shock response, but the specific steps of the Hsf inactivation/dissociation pathway at which individual chaperones act re-

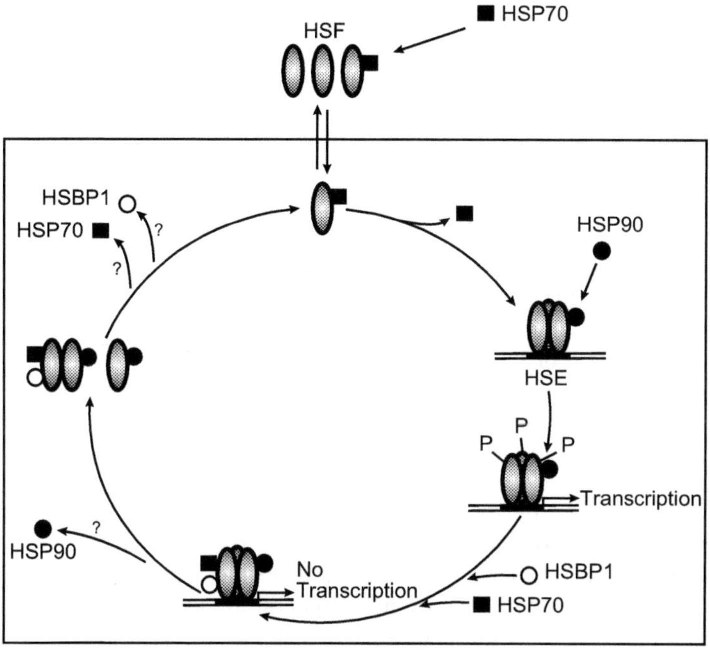

Fig. 2. Model of Hsf regulation by hsps and HSBP1. Hsf exists in the inactive monomeric state through interaction with hsp70 or another protein(s). Following activation, Hsf localizes to the nucleus where it trimerizes, acquires DNA binding ability, and upon phosphorylation becomes transcriptionally competent. Hsp90 and its co-chaperones (not shown) bind to Hsf trimers and maintain them in an activable state. During attenuation of the heat stress response, hsp70 and its co-chaperones (not shown), and HSBP1 bind to Hsf, repressing its transcriptional activity and leading to the dissociation of Hsf trimers to inert monomers. Hsp90 and its co-chaperones may also be involved in the disassembly of Hsf trimers.

main to be worked out. Several possibilities exist for how they may exert their effects. Because hsp70 and hsp90 can be found together in the same heterocomplexes (Pratt and Toft 1997), it is possible that they jointly affect aspects of Hsf regulation. Alternatively, it may be that repression of activated Hsf is mediated by the binding of hsp70 to the activation domain of Hsf, and that the hsp90 complex keeps the inactive Hsf in an activable state, as in the case of steroid receptors (Pratt and Toft 1997). The repression mechanism must involve controls to ensure that Hsf activation occurs in proportion to the severity of stress (Guo et al. 2001). Thus, the involvement of molecular chaperones in 'controlling' Hsf hyperphosphorylation or disassembly also remain as attractive possibilities. The recent demonstration that hsp90 and p23 help disassemble transcriptional regulatory complexes (Freeman and Yamamoto 2002) warrants further studies of the effects of these chaperones on Hsf trimer disassembly. The model shown in Figure 2 illustrates both possible scenarios by which hsp90 may regulate Hsf: 1) keep the trimers in an activable state and 2) participate in the disassembly of the trimers. It is possible that hsp90 participates in only one of these steps, or alternatively, in the presence of different co-chaperones, regulates both Hsf activation and deactivation.

In plants, circumstantial evidence suggests the presence of a negative regulator(s) of Hsfs. For example, overexpression of *A. thaliana* Hsf1 fused to β-glucuronidase (GUS) in transgenic *A. thaliana* plants derepressed the heat shock response, leading to constitutive synthesis of hsps and an increase in basal thermotolerance (Lee et al. 1995b). Since overexpression of unfused Hsf1 did not produce this effect, a model was put forth suggesting that derepression may be due to impaired recognition of Hsf1-GUS by a repressing factor. In contrast to Hsf1, overexpression of either Hsf3 or Hsf3-GUS in transgenic plants derepressed the heat shock response (Prandl et al. 1998), suggesting that either titration of a negative regulator of Hsf3 or just the intrinsic constitutive activity of Hsf3 is responsible for derepression. To date, very few studies have investigated the involvement of molecular chaperones in regulating the activity of plant Hsfs. In a transgenic approach, the time required to turn off Hsf activity during recovery from heat stress was significantly prolonged in *A. thaliana* plants expressing antisense *hsp70* gene as compared to wild type plants (Lee and Schoffl 1996). These results are consistent with the involvement of hsp70 at an early stage of plant Hsf inactivation. However, since reduced levels of hsp70 in these transgenic plants did not correlate with derepressed Hsf at normal temperature; heat stress was required for activation, a role for hsp70 in the disassembly of Hsf trimers is favored by the authors. A role for hsp70 as a regulator of Hsf activity in plant cells is further corroborated by the demonstration that *A. thaliana* Hsf1 and hsp70 interact *in vitro* as well as in yeast 2-hybrid assays (Kim and Schoffl 2002).

The recent cloning of the maize *empty pericarp2* (*emp2*) has led to the identification of a putative plant ortholog of HSBP1 (Fu et al. 2002). Null mutations in *emp2* result in increased expression of hsps, retarded embryo development and early-stage abortion of embryogenesis. *emp2* is a loss of function mutation of a negative regulator of the plant heat shock response, and, therefore, shows unattenuated heat shock response. The conservation of the hydrophobic heptad repeat domains in EMP2 suggests that it functions as a negative regulator by binding to Hsf. The developmental retardation of *emp2* mutant kernels before the onset of the heat shock response indicates that EMP2 has an additional role during embryo development, distinct from its role in heat stress response. Further investigation of EMP2 will expand our understanding of the roles of EMP2 during heat stress response and embryogenesis.

3.3.2.3 Heat stress granules

During prolonged heat stress in plants, shsps aggregate to produce highly ordered cytoplasmic complexes of 40 mM diameter. The so-called heat stress granules (HSGs) are unique to plants and are produced in all plant species and tissues examined so far (Nover et al. 1989). HSGs comprise mainly of cytosolic shsps belonging to both classes I and II. An examination of the structural prerequisites of hsps for their assembly into HSGs revealed that they first assemble into dodecamers of 210-280 kDa, which, upon heat stress, become incorporated into HSG complexes (Kirschner et al. 2000). Class II dodecamers alone can form HSG complexes but class I dodecamers require the presence of class II proteins. An intact COOH-terminus of class II shsp is critical for oligomerizing, aggregating into HSGs, and recruiting class I proteins (Kirschner et al. 2000). Experimental data suggests that HSGs represent storage and protection sites for housekeeping mRNPs, which are released following removal of stress (Nover et al. 1989). It is also proposed that during long-term heat stress, unfolded proteins bound to shsps exceed the capacity of the hsp70/hsp40 refolding machinery, and that these denatured protein-shsp complexes are stored transiently in HSGs (Low et al. 2000).

The heat-inducible tomato HsfA2 accumulates to high levels in the course of prolonged heat stress and recovery periods. A study of the intracellular localization of tomato Hsfs showed that following heat shock, HsfA2 is present as a high salt-resistant nuclear form and as a stored form in cytoplasmic HSGs (Scharf et al. 1998). The binding of HsfA2 to HSG is specific as HsfA1 and other cytosolic proteins examined did not associate with the HSG fraction. Both the HSG-bound and the nuclear high salt-resistant forms were reversible to the soluble cytoplasmic form after removal of stress. Incorporation of HsfA2 into HSG and its release during recovery could be an aspect of Hsf regulation during plant heat stress response. It remains to be seen if HsfA2 is incorporated into pre-HSG particles or becomes associated with HSG during heat stress-induced aggregation of the dodecamers.

3.3.2.4 Regulation of Hsfs by phosphorylation

In mammalian model systems, the DNA binding and transactivating capabilities of Hsf1 are uncoupled and serine phosphorylation is an important determinant of the transactivating potency of Hsf1 (reviewed in Pirkkala et al. 2001). The latent form of mammalian Hsf1 under normal conditions is also constitutively phosphorylated. Multisite phosphorylation-mediated regulation of Hsfs is not surprising given that the heat shock response is both fast and flexible, and that Hsf activation has to vary in proportion to the severity of the stress and in response to stimuli other than heat stress in a constantly changing environment. Details of Hsf1 phosphorylation have been slow in forthcoming, presumably due to the many potential sites of phosphorylation that can be associated with either increases or decreases in activity. A subset of phosphorylated sites involved in Hsf1 repression have been identified, but to date only a single site, Ser230, has been linked with inducible transcriptional activity (Holmberg et al. 2001). Several signaling pathways involving glycogen synthase kinase 3β (GSK-3β), protein kinase C (PKC), extracellular signal-regulated kinase (ERK), and c-Jun N-terminal kinase (JNK) have been implicated in the repression of mammalian Hsf1 (Chu et al. 1996, 1998; Dai et al. 2000; He et al. 1998; Kim et al. 1997; Knauf et al. 1996). However, an in vivo demonstration of these kinases in Hsf1 regulation, as well as an understanding of how these different kinases interact to affect Hsf1 repression are currently lacking. Since Ser230 lies within a consensus site for calcium/calmodulin-dependent protein kinase II (CaMKII), and overexpression of CaMKII enhances both the level of Ser230 phosphorylation and transactivation of Hsf1, Holmberg et al. (2001)

propose that CaMKII signaling is involved in the positive regulation of the trans-activating capability of Hsf1. In this regard, the yeast Hsf is also inducibly serine phosphorylated (Cotto et al. 1996), and the *Drosophila* Hsf undergoes phosphorylation at some sites and dephosphorylation at others in response to heat stress, with no net increase in the steady state level of Hsf phosphorylation (Fritsch and Wu 1999). Since in all cases the DNA binding activity of Hsfs remains unaffected by inducible phosphorylation, it is likely that the influence is on the transcriptional activity of the Hsfs. It is clear from these results that Hsf regulation by phosphorylation is a fairly complex process.

Relatively little is known about the role of phosphorylation in plant heat stress response. Due to the pivotal roles of mitogen-activated protein kinases (MAPKs) in signal transduction of extracellular stimuli, such as hormone regulators and environmental stresses, in various organisms, MAPKs have also received attention in plants. In a recent analysis of the *A. thaliana* genome, 20 MAPKs, 10 MAPK kinases (MAPKK), and 60 MAPKK kinases (MAPKKK) were identified (Ichimura et al. 2002). MAPK cascades in plants are activated in response to different biotic and abiotic stresses (reviewed in Jonak et al. 2002), including cold and drought (Jonak et al. 1996), high salt (Munnik et al. 1999), wounding (Bogre et al. 1997), and pathogen infection (Nuhse et al. 2000). Until recently, no heat shock-activated MAPK was reported in plants. Sangwan et al. (2002) provided the first demonstration that a MAPK immunologically related to the ERK superfamily of protein kinases is activated by heat stress in alfalfa (*Medicago sativa*) cells. A study of the mechanism leading to activation of this heat shock-activated MAPK (HAMK) indicated that heat is sensed by changes in membrane fluidity that occur directly, rapidly, and reversibly in response to temperature and that translate the signal via cytoskeleton, Ca^{2+} fluxes and Ca^{2+}-dependent protein kinases (CDPKs) into activation of HAMK. While these results suggest that a temporal sequence of events is involved in the activation of HAMK, further characterization of this enzyme revealed that HAMK can be heat-activated in cell-free extracts (Sangwan and Dhindsa 2002). The integrity of cellular membranes and the partitioning of Ca^{2+} are likely disrupted in cell-free extracts, which raises the intriguing question of how HAMK is activated in cell-free extracts. Direct effects of temperature on the conformation and activity of either HAMK or its upstream activator(s), leading to temperature perception directly by a protein, are possibilities that need to be explored in the future. The inability of tobacco cells in which HAMK activity is blocked to launch a heat stress response suggests that HAMK is an important regulator of the heat stress response in plants (RS Dhindsa, personal communication).

Further support for a role of MAPKs in plant heat stress response comes from studies in tomato. Link et al. (2002) observed a 50 kDa MAPK is activated by heat stress in tomato cells in a Ca^{2+}-dependent manner. A partially purified preparation of the heat-activated MAPK could phosphorylate tomato HsfA3 but not HsfA1 even though both proteins contain several copies of consensus MAPK phosphorylation sites. Whether or not this substrate specificity is biologically relevant remains to be seen. Heat stress induces several events in cells, including cell cycle arrest. Thus it is not surprising that a cyclin-dependent CDC2a kinase forms a stable complex with *A. thaliana* Hsfl and phosphorylates it on multiple serine residues (Reindl et al. 1997). Phosphorylation by CDC2a results in reduced DNA binding of AtHsfl to HSEs *in vitro*. Although a link between heat stress response and cell-cycle control is suggested by these results, no further evidence confirming this interaction or its significance has emerged following this study.

In addition to MAPKs, the GSK-3-like kinases in plants are emerging as important regulators of development, stress, and hormone signaling (reviewed by Jonak and Hirt 2002). In view of the fact that mammalian Hsfl is phosphorylated by GSK-3β (He et al. 1998), the involvement of plant GSK-3-like kinases in heat stress response should also be explored. The role of phosphorylation in heat stress response is not limited to Hsf activation or deactivation. Heat shock causes significant reduction in normal transcription and translation processes, and affects the cell at the level of proteins, nucleic acids, membrane, and the cytoskeleton. These changes are likely to correlate with altered phosphorylation of several cellular proteins. A comprehensive study of protein phosphorylation in plant heat stress response using high throughput proteomic analysis and computer assisted methodology should be undertaken to formulate a systematic working hypothesis for the future. Such an approach in mammalian cell lines allowed proteins to be grouped based on kinetic analysis of phosphorylation by heat shock (Kim et al. 2002). Identifying what protein kinases are activated and what proteins are phosphorylated in response to heat stress will serve only as a starting point in understanding heat stress signaling. Defining the upstream and downstream components of different protein kinases as well as the mechanisms of cross-talk between different cascades are the real challenges for the future.

3.4 Ca^{2+} and heat shock response

There is considerable evidence that Ca^{2+}-mediated second messenger systems are involved in the perception and reaction of plants to different environmental signals (Bush 1995). Changes in cytosolic Ca^{2+} levels occur in response to cold stress (Knight et al. 1996), mechanical stimulation (Haley et al. 1995), oxidative stress (Price et al. 1994), salinity stress (Lynch et al. 1989), and hypo-osmotic shock (Takahashi et al. 1997) suggesting that signaling by Ca^{2+} is one of the primary mechanisms leading to molecular changes that help plants to adapt to new conditions. The involvement of Ca^{2+} in thermotolerance in plants has also been documented. Braam (1992) observed that expression of calmodulin-related *TCH* genes is regulated in response to heat shock and that external Ca^{2+} is required for maximal heat shock induction of *TCH* genes but not of *hsp70* gene. Pretreatment of tobacco seedlings with Ca^{2+} increased thermotolerance of seedlings and led to a transient increase, lasting 10 to 20 minutes, in cytoplasmic but not chloroplastic Ca^{2+} (Gong et al. 1998b). A single heat stress treatment initiated a refractory period during which additional heat treatments failed to increase cytosolic Ca^{2+} levels, but responsiveness, measured as cytosolic Ca^{2+} increases, to other stimuli such as cold shock and touch was maintained. These observations suggest that plants can distinguish between different stimulus-induced increases in cytosolic Ca^{2+}. That Ca^{2+} improves thermotolerance of seedlings is further supported by studies in *A. thaliana*. Treatment of *A. thaliana* plants with calcium chloride improved survival after severe heat treatment, whereas treatment with Ca^{2+} channel blockers and calmodulin inhibitors reduced survival (Larkindale and Knight 2002). Contrary to the observations made in tobacco seedlings (Gong et al. 1998b), no increase in cytosolic Ca^{2+} was detected in *A. thaliana* during heat treatment, although transient elevation in levels were noted during recovery from heat stress. Plants that had acquired thermotolerance through a mild heat treatment produced larger Ca^{2+} peaks when exposed to a lethal heat treatment as compared to thermosensitive plants (without mild heat treatment), in particular during initiation of recovery. Together, these results suggest a role for Ca^{2+} in some aspect of plant heat stress response. Whether Ca^{2+} elevation during heat stress affects Hsf activity and hsp synthesis is not clear. Some investigations on animal Hsf activity in response to Ca^{2+} treatments have been made (Mosser et al. 1990; Soncin et al. 2000), but a clear picture of the effects has not emerged. Future questions are: how Ca^{2+} signal produced during heat stress is different from Ca^{2+} signals triggered by other stimuli, what components sense and transduce Ca^{2+} signal during heat stress, and what are the downstream responses?

3.5 Hormones and heat stress response

Hormones control virtually all aspects of plant physiology, including stress response. The roles of abscisic acid (ABA) in cold, salt, and drought stresses (Chandler and Robertson 1994; Zhu 2002) and those of ethylene and salicylic acid (SA) in plant defense responses (Johnson and Ecker 1998; Wang et al. 2002) are well documented. In comparison, the effects of plant growth regulators on heat stress response are less studied, and investigations in this direction appear to be limited to examining hsp gene expression in response to different hormones. Hsp90 transcripts or protein are induced in response to indoleacetic acid (Yabe et al. 1994), ABA (Pareek et al. 1995) and brassinosteroid (Wilen et al. 1995). Elevated expression of hsp100 by ABA has been similarly observed (Campbell et al. 2001; Pareek et al. 1995). The expression of a subset of shsps during embryogenesis has generated much interest in the regulation of shsp expression by hormones, especially ABA (Almoguera and Jordano 1992; Coca et al. 1996; Kaukinen et al. 1996; Wehmeyer et al. 1996).

Convincing evidence exists for ABA and brassinosteroid action in increasing thermotolerance in plants. A bromegrass (*Bromus inermis*) cell suspension culture, without a prior mild heat treatment, had significant increase in survival rate when it was pretreated with 75 μM ABA. The heat tolerance provided by ABA treatment was first observed after 4 days of culture in the presence of ABA and reached a maximum after 11 days of culture (Robertson et al. 1994). Typically, thermotolerance has been associated with the accumulation of hsps during heat stress. In case of bromegrass cell culture, ABA-responsive heat stable proteins, a subset of which cross-reacted with anti-dehydrin and anti-Wcs120 (a cold-responsive winter wheat protein) antibodies, were demonstrated to confer thermo-stability in *in vitro* protection assays. The thermotolerance enhancing effect of ABA has also been noted in maize (Gong et al. 1998a) and in *A. thaliana* (Larkindale and Knight 2002). Pretreatment of *A. thaliana* seedlings with SA, 1-aminocyclopropane-1-carboxylic acid (a precursor to ethylene) and ABA prior to a usual lethal heat treatment enhanced survival by approximately 5-, 3-, and 2-fold, respectively, and reduced heat stress-induced oxidative damage (Larkindale and Knight 2002). In contrast to the 4 day treatment of bromegrass cell suspension culture with ABA (Robertson et al. 1994), a 1 hour treatment of *A. thaliana* plants with the growth regulators before exposure to heat stress was sufficient to increase survival (Larkindale and Knight 2002). While it is a natural assumption that both early and late events related to thermotolerance could be influenced by hormones, the difference in the treatment time leading to thermotolerance in the two plant systems is remarkable. *A. thaliana* mutants insensitive to ethylene and ABA, as well as transgenic lines inhibited in SA production, showed increased susceptibility to heat, confirming the involvement of these growth regulators in plant heat stress response. However, molecular changes mediated by these hormones that lead to thermotolerance are presently unknown. Some progress has been made towards understanding the molecular mechanism by which 24-epibrassinolide (EBR), a brassinosteroid, promotes thermotolerance in plants. Dhaubhadel et al. (1999) demonstrated that *B. napus* and tomato seedlings grown in the presence of EBR are significantly resistant to a heat treatment that is lethal to untreated seedlings. Since a mild heat treatment of seedlings prior to their exposure to the usual lethal heat stress was not required to observe this effect, it is concluded that EBR treatment increases the basic thermotolerance of seedlings. An examination of hsp expression before, during, and after heat stress treatment revealed that hsps accumulate to higher levels in EBR-treated than in untreated seedlings following heat stress, but not at the control temperature. Surprisingly though, the higher hsp levels in treated seedlings did not correlate with higher hsp mRNA levels during the recovery period. Further investigation into the mechanism leading to higher accumulation of hsps in EBR-treated seedlings revealed that this is a result of higher hsp synthesis in these seedlings, even when the mRNA levels are lower than in untreated seedlings. Consistent with this finding, several translation initiation and elongation factors were detected at significantly higher levels in treated seedlings as compared to untreated seedlings (Dhaubhadel et al. 2002). These results suggest that EBR-treatment limits the loss of some of the components of the translational apparatus during a prolonged heat stress and increases the level of expression of some of the components of the translational machinery during recovery, which correlates with higher hsp synthesis during heat stress, a more rapid resumption of cellular protein synthesis following heat stress, and a higher survival rate. Although higher level of hsps must contribute to increased thermotolerance in EBR-treated seedlings, factors other than hsps that may directly or indirectly contribute to EBR-mediated increase in stress tolerance were searched for using differential display. Four cDNAs characterized thus far that were upregulated in treated seedlings encode 3-ketoacyl CoA thiolase, myrosinase, glycine rich protein

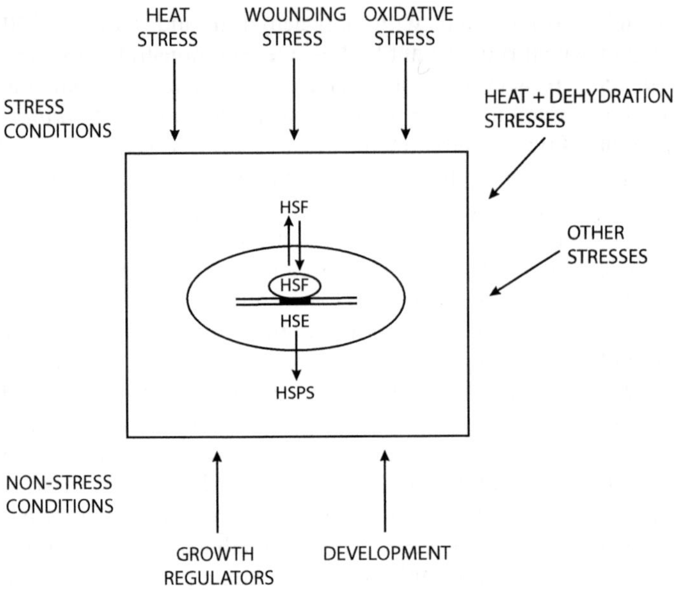

Fig. 3. Multiple stress and non-stress conditions that induce synthesis of hsps. Although heat shock gene expression is shown here to occur through activation of Hsf, this has not been confirmed for all conditions; other pathways may also induce hsp synthesis.

22 (GRP22), and a hypothetical protein (S. Dhaubhadel and P. Krishna, submitted manuscript). The thiolase transcript levels were higher in treated seedlings as compared to untreated seedlings during heat stress, but transcripts of the other three cDNAs were present at higher levels in treated seedlings prior to any stress. Higher expression of 3-ketoacyl thiolase, myrosinase, and GRP22 can be linked, at least hypothetically, to an increase in the general stress resistance of plants (S. Dhaubhadel and P. Krishna, submitted manuscript). Thus, modified translational machinery, coupled with increased expression of genes involved in a variety of physiological responses, and other as yet unidentified factors in treated seedlings, may contribute to increased overall stress tolerance in these seedlings.

3.6 Relationship between heat and other stresses

Surveys of global gene expression in response to different stresses are beginning to point to the interaction and overlap between signaling pathways initiated by one stress and other stimuli (Fig. 3). In a study focused on transcriptional profiling of genes regulated by wounding it was noted that many pathogen-responsive and osmotic and heat stress-regulated genes are highly responsive to wounding (Cheong et al. 2002). The wound activation of hsps followed a time frame consistent with the normal heat shock response. Two Hsf genes were activated within 30 minutes after wounding and this activation was followed by transcription of several hsp genes. To determine if other stress conditions also regulate the expression of hsps, Cheong et al. (2002) searched the *A. thaliana* gene expression database of Torrey Mesa Research Institute. Their search revealed that genes encoding Hsf4, hsp70 and hsp17.6A are activated by several stresses, including osmotic stress, electric shock, pathogen attack, light, and plant hormones, whereas the gene encoding Hsf21 is specifically activated by wounding and pathogen elicitor. These results suggest that some hsps and Hsfs may be important for a broad spectrum of stress conditions, whereas others may be involved in specific stress responses. The overlap of heat stress response with other stress responses is also evident from analysis of differentially expressed genes during heat stress in cowpea (*Vigna unguiculata*) nodules. In addition to hsps, wound-induced and disease resistance proteins were upregulated (Simões-Araujo et al. 2002). The complexity of signaling events associated with sensing and acclimating to stresses is further seen when a combination of environmental conditions are simultaneously applied to plants. A combination of drought and heat stress on tobacco plants resulted in the suppression of photosynthesis, enhancement of respiration, induction of several defense genes, and changes in genes involved in sugar metabolism (Rizhsky et al. 2002). The expression of some of the genes induced solely under drought or heat stress was suppressed when the two stresses were combined, while the expression of others was specifically induced under combined stress conditions. These results demonstrate that the response of plants to a combination of stresses, similar to what may be encountered in the field, is different from the response to each of the stresses applied individually. Some assumptions regarding functions of stress proteins can be made on the basis of unique expression patterns during a combination of stresses. For instance, the finding that the expression of dehydrin, which is highly expressed during drought stress, is suppressed during a combination of heat and drought stress may suggest that hsps can replace the stabilizing function of dehydrin during this combination of stresses (Rizhsky et al. 2002).

3.7 Developmental regulation of shsps by Hsfs

A subset of shsps in plants are expressed during zygotic embryogenesis in the absence of any stress (reviewed in Schoffl et al. 1998). Deletion analyses of shsp promoters indicates that HSEs are also required for developmental regulation in embryos (Coca et al. 1996; Prandl and Schoffl 1996). This poses the question: if Hsfs are involved in the developmental regulation of shsps, what mechanism(s) is used to distinguish promoter activation during development and during heat stress response? Recently some interesting observations have been made in this direction. In a transient promoter activation assay performed in sunflower (*Helianthus annuus*) embryos, tomato HsfA2, but not HsfA1, could promote transcriptional activation of the developmentally regulated *Ha hsp 17.6 G1* promoter that is characterized by an unusual low-consensus HSE (Rojas et al. 2002). Mutational analyses of the *Ha hsp 17.6 G1* promoter combined with *in vitro* DNA binding assays suggest that the low-consensus HSE sequence is crucial for Hsf promoter selectivity, but that discrimination occurs after DNA binding and may involve preferential transcriptional activation. Specific interactions with different transcription factors could confer functional specificity to plant Hsfs. A preliminary but crucial observation in this direction is that ABI3, a seed-specific transcription factor from *A. thaliana*, and tomato HsfA1 can synergistically activate the *Ha hsp 17.7 G4* promoter when it contains intact proximal and distal HSEs (Rojas et al. 1999). In the absence of either functional Hsf or HSEs, substantial activation of the promoter by ABI3 does not occur. The activation domain of HsfA1 is necessary for promoter activation, and a truncated ABI3 is incapable of promoter activation by Hsf. Since ABI3 and HsfA1 can activate a minimal CaMV 35S promoter fused to *Ha hsp 17.7 G4* HSEs, it is proposed that ABI3 functions as a co-activator (indirectly binding DNA) rather than as a transcriptional activator (directly binding DNA) through Hsfs. Though ABI3 specificity towards plant Hsfs remains to be addressed, it is possible that the co-activator function of ABI3 is limited to only some Hsf(s). Another possibility is that one or more seed-specific Hsf is involved in the developmental regulation of shsps during embryogenesis. The recent cloning of a class A Hsf in sunflower, HaHsf9, which is specifically expressed during embryogenesis in the absence of environmental stress (Almoguera et al. 2002), gives much credibility to this possibility. The HaHsf9 can transactivate promoters with poor consensus HSEs, including that of the seed-specific *Ha hsp 17.6 G1* gene. Mutations that improve the HSE consensus of *Ha hsp 17.6 G1* promoter impair activation by HaHsf9 but do not affect heat shock-induced gene expression of this promoter. Thus, specific HSE sequences, specific expression patterns of Hsfs, and the interactions of Hsfs with embryo-specific factors are all possible mechanisms of developmental regulation of shsps, and are not mutually exclusive.

3.8 Future directions

From this overview, it is clear that many aspects of heat stress response and signaling in plants remain unexplored. Current observations of Hsfs and protein kinases activated during the heat stress response appear only as small windows into gaining an understanding of the complex mechanisms by which heat may be sensed and the signal transduced to the nucleus for regulation of gene expression. Limited data suggests that members of the extensive Hsf family in plants likely differ in their expression patterns, promoter recognition, oligomerization behavior, and potential as transcription activators, making regulation of the heat shock response through Hsfs a highly complex phenomenon. At this point, very little is known about what phosphorylation events lead to activation and attenuation of the heat shock response in plants, and virtually nothing is known about the phosphatases that may be involved in dephosphorylating protein kinases as well as Hsfs and other transcription factors involved in this response. A multitude of signaling cascades must coordinate the response during combination of heat and the other stresses that are often concurrent in the natural environment. Alongside the identification of the pathways operating in plant heat stress response, lies the enormous task of comprehending the cross-talk between different signaling pathways. Genomic approaches to genome-wide expression analysis have revealed that heat stress is not limited to the synthesis of hsps. One challenge for the future is to understand the overlap between different stresses and to elucidate the roles of all proteins where expression dramatically increases upon heat shock. Finally, the functions of some hsps, such as hsp90, are just beginning to be addressed. The identification of plant hsp90 interactors and of the signaling pathways controlled by them will undoubtedly lead to novel aspects of hsp90, advancing the link between stress, development and evolution. This knowledge may present new opportunities for agricultural biotechnology, as well as providing a better understanding of the mechanisms regulating plant growth and development, and plant responses to environmental stresses.

Acknowledgements

I thank Professor M. Perry for many helpful suggestions on the manuscript; Dr. R.S. Dhindsa for sharing unpublished observations; Dr. Z. Zhang for assistance in preparation of the manuscript. Support from the Natural Sciences and Engineering Research Council of Canada is gratefully acknowledged. Due to the extensiveness of the research area and space limitation, I regret not directly citing all contributions in the field.

References

Abravaya K, Myers MP, Murphy SP, Morimoto RI (1992) The human heat shock protein hsp70 interacts with HSF, the transcription factor that regulates heat shock gene expression. Genes Dev 6:1153-1164

Agarwal M, Katiyar-Agarwal S, Sahi C, Gallie DR, Grover A (2001) *Arabidopsis thaliana* Hsp100 proteins: kith and kin. Cell Stress Chaperones 6:219-224

Almoguera C, Jordano J (1992) Developmental and environmental concurrent expression of sunflower dry-seed-stored low-molecular-weight heat-shock protein and Lea mRNAs. Plant Mol Biol 19:781-792

Almoguera C, Rojas A, Díaz-Martin J, Prieto-Dapena P, Carranco R, Jordano J (2002) A seed-specific heat-shock transcription factor involved in developmental regulation during embryogenesis in sunflower. J Biol Chem 277:43866-43872

Apuya NR, Yadegari R, Fischer RL, Harada JJ, Zimmerma JL, Goldberg RB (2001) The Arabidopsis embryo mutant *schlepperless* has a defect in the *chaperonin-60α* gene. Plant Physiol 126:717-730

Bögre L, Ligterink W, Meskiene I, Barker PJ, Heberle-Bors E, Huskisson NS, Hirt H (1997) Wounding induces the rapid and transient activation of a specific MAP kinase pathway. Plant Cell 9:75-83

Boston RS, Viitanen PV, Vierling E (1996) Molecular chaperones and protein folding in plants. Plant Mol Biol 32:191-222

Braam J (1992) Regulated expression of the calmodulin-related *TCH* genes in cultured Arabidopsis cells: induction by calcium and heat shock. Proc Natl Acad Sci USA 89:3213-3216

Buchner J (1999) Hsp90 & Co.- a holding for folding. Trends Biochem Sci 24:136-141

Bukau B, Horwich AL (1998) The Hsp70 and Hsp60 chaperone machines. Cell 92:351-366

Bush DS (1995) Calcium regulation in plant cells and its role in signaling. Annu Rev Plant Physiol Plant Mol Biol 46:95-122

Campbell JL, Klueva NY, Zheng H, Nieto-Sotelo J, Ho T-HD, Nguyen HT (2001) Cloning of new members of heat shock protein Hsp101 gene family in wheat (*Triticum aestivum* (L.) *Moench*) inducible by heat, dehydration and ABA. Biochim Biophys Acta 1517:270-277

Chandler PM, Robertson M (1994) Gene expression regulated by abscisic acid and its relation to stress tolerance. Annu Rev Plant Physiol Plant Mol Biol 45:113-141

Cheong YH, Chang H-S, Gupta R, Wang X, Zhu T, Luan S (2002) Transcriptional profiling reveals novel interactions between wounding, pathogen, abiotic stress, and hormonal responses in Arabidopsis. Plant Physiol 129:661-677

Chu B, Soncin F, Price BD, Stevenson MA, Calderwood SK (1996) Sequential phosphorylation by mitogen-activated protein kinase and glycogen synthase kinase 3 represses transcriptional activation by heat shock factor 1. J Biol Chem 271:30847-30857

Chu B, Zhong R, Soncin F, Stevenson MA, Calderwood SK (1998) Transcriptional activity of heat shock factor 1 at 37°C is repressed through phosphorylation on two distinct serine residues by glycogen synthase kinase 3α and protein kinases Cα and Cζ. J Biol Chem 273:18640-18646

Coca MA, Almoguera C, Thomas TL, Jordano J (1996) Differential regulation of small heat-shock genes in plants: analysis of a water-stress-inducible and developmentally activated sunflower promoter. Plant Mol Biol 31:863-876

Cotto JJ, Kline M, Morimoto RI (1996) Activation of heat shock factor 1 DNA binding precedes stress-induced serine phosphorylation. J Biol Chem 271:3355-3358

Craig EA, Gross CA (1991) Is hsp70 the cellular thermometer? Trends Biochem Sci 16:135-140

Czarnecka-Verner E, Yuan C-X, Scharf K-D, Englich G, Gurley WB (2000) Plants contain a novel multi-member class of heat shock factors without transcriptional activator potential. Plant Mol Biol 43:459-471

Dai R, Frejtag W, He B, Zhang Y, Mivechi NF (2000) JNK targeting and phosphorylation of heat shock factor-1 suppress its transcriptional activity. J Biol Chem 275:18210-18218

Dhaubhadel S, Chaudhary S, Dobinson KF, Krishna P (1999) Treatment with 24-epibrassinolide, a brassinosteroid, increases the basic thermotolerance of *Brassica napus* and tomato seedlings. Plant Mol Biol 40:333-342

Dhaubhadel S, Browning KS, Gallie DR, Krishna P (2002) Brassinosteroid functions to protect the translational machinery and heat shock protein synthesis following thermal stress. Plant J 29:681-691

Doring P, Treuter E, Kistner C, Lyck R, Chen A, Nover L (2000) The role of AHA motifs in the activator function of tomato heat stress transcription factors HsfA1 and HsfA2. Plant Cell 12:265-278

Duina AA, Kalton HM, Gaber RF (1998) Requirement for Hsp90 and a CyP-40-type cyclophilin in negative regulation of the heat shock response. J Biol Chem 273:18974-18978

Forreiter C, Kirschner M, Nover L (1997) Stable transformation of an Arabidopsis cell suspension culture with firefly luciferase providing a cellular system for analysis of chaperone activity in vivo. Plant Cell 9:2171-2181

Freeman BC, Yamamoto KR (2002) Disassembly of transcriptional regulatory complexes by molecular chaperones. Science 296:2232-2235

Fritsch M, Wu C (1999) Phosphorylation of Drosophila heat shock transcription factor. Cell Stress Chaperones 4:102-117

Frydman J, Hohfeld J (1997) Chaperones get in touch: the Hip-Hop connection. Trends Biochem Sci 22:87-92

Fu S, Meeley R, Scanlon MJ (2002) *empty pericarp2* encodes a negative regulator of the heat shock response and is required for maize embryogenesis. Plant Cell 14:3119-3132

Gong M, Li Y-J, Chen SZ (1998a) Abscisic acid induced thermotolerance in maize seedlings is mediated by Ca^{2+} and associated with antioxidant systems. J Plant Physiol 153:488-496

Gong M, van der Luit AH, Knight MR, Trewavas AJ (1998b) Heat-shock-induced changes in intracellular Ca^{2+} level in tobacco seedlings in relation to thermotolerance. Plant Physiol 116:429-437

Guo Y, Guettouche T, Fenna M, Boellmann F, Pratt WB, Toft DO, Smith DF, Voellmy R (2001) Evidence for a mechanism of repression of heat shock factor 1 transcriptional activity by a multichaperone complex. J Biol Chem 276:45791-45799

Haley A, Russell AJ, Wood N, Allan AC, Knight MR, Campbell AK, Trewavas AJ (1995) Effects of mechanical signaling on plant cell cytosolic calcium. Proc Natl Acad Sci USA 92:4124-4128

Harndahl U, Hall RB, Osteryoung KW, Vierling E, Bornman JF, Sundby C (1999) The chloroplast small heat shock protein undergoes oxidation-dependent conformational changes and may protect plants from oxidative stress. Cell Stress Chaperones 4:129-138

Hartl FU, Hayer-Hartl M (2002) Molecular chaperones in the cytosol: from nascent chain to folded protein. Science 295:1852-1858

He B, Meng YH, Mivechi NF (1998) Glycogen synthase kinase 3β and extracellular signal-regulated kinase inactivate heat shock transcription factor 1 by facilitating the disappearance of transcriptionally active granules after heat shock. Mol Cell Biol 18:6624-6633

Heerklotz D, Doring P, Bonzelius F, Winkelhaus S, Nover L (2001) The balance of nuclear import and export determines the intracellular distribution and function of tomato heat stress transcription factor HsfA2. Mol Cell Biol 21:1759-1768

Hendrick JP, Hartl FU (1993) Molecular chaperone functions of heat-shock proteins. Annu Rev Biochem 62:349-384

Hill JE, Hemmingsen SM (2001) *Arabidopsis thaliana* type I and II chaperonins. Cell Stress Chaperones 6:190-200

Hohfeld J, Jentsch SL (1997) GrpE-like regulation of the Hsc70 chaperone by the anti-apoptotic protein BAG-1. EMBO J 16:6209-6216

Holmberg CI, Hietakangas V, Mikhailov A, Rantanen JO, Kallio M et al. (2001) Phosphorylation of serine 230 promotes inducible transcriptional activity of heat shock factor 1. EMBO J 20:3800-3810

Hong SW, Vierling E (2000) Mutants of *Arabidopsis thaliana* defective in the acquisition of tolerance to high temperature stress. Proc Natl Acad Sci USA 97:4392-4397

Hong SW, Vierling E (2001) Hsp101 is necessary for heat tolerance but dispensable for development and germination in the absence of stress. Plant J 27:25-35

Ichimura K, Tena G, Henry Y, Zhang Z, Hirt H, Wislon C, Morris P, Mundy J, Innes R, Ecker J et al. (2002) Mitogen-activated protein kinase cascades in plants: a new nomenclature. Trends Plant Sci 7:301-308

Jakob U, Lilie H, Meyer I, Buchner J (1995) Transient interaction of Hsp90 with early unfolding intermediates of citrate synthase. J Biol Chem 270:7288-7294

Jakobsen BK, Pelham HR (1988) Constitutive binding of yeast heat shock factor to DNA in vivo. Mol Cell Biol 8:5040-5042

Johnson PR, Ecker JR (1998) The ethylene gas signal transduction pathway: a molecular perspective. Annu Rev Genet 32:227-254

Jonak C, Hirt H (2002) Glycogen synthase kinase 3/SHAGGY-like kinases in plants: an emerging family with novel functions. Trends Plant Sci 7:457-461

Jonak C, Kiegerl S. Ligterink W, Barker PJ, Huskisson NS, Hirt H (1996) Stress signaling in plants: a mitogen-activated protein kinase pathway is activated by cold and drought. Proc Natl Acad Sci USA 93:11274-11279

Jonak C, Ökrész L, Bögre L, Hirt H (2002) Complexity, cross talk and integration of plant MAP kinase signaling. Curr Opin Plant Biol 5:415-424

Kaukinen KH, Tranbarger TJ, Misra S (1996) Post-germination-induced and hormonally dependent expression of low-molecular-weight heat shock protein genes in Douglas-fir. Plant Mol Biol 30:1115-1128

Kim B-H, Schoffl F (2002) Interaction between *Arabidopsis* heat shock transcription factor 1 and 70 kDa heat shock proteins. J Exp Bot 53:371-375

Kim H-J, Song EJ, Lee K-J (2002) Proteomic analysis of protein phosphorylations in heat shock response and thermotolerance. J Biol Chem 277:23193-23207

Kim J, Nueda A, Meng Y-H, Dynan WS, Mivechi NF (1997) Analysis of the phosphorylation of human heat shock transcription factor-1 by MAP kinase family members. J Cell Biochem 67:43-54

Kirschner M, Winkelhaus S, Thierfelder JM, Nover L (2000) Transient expression and heat-stress-induced co-aggregation of endogenous and heterologous small heat-stress proteins in tobacco protoplasts. Plant J 24:397-411

Knauf U, Newton EM, Kyriakis J, Kingston RE (1996) Repression of human heat shock factor 1 activity at control temperature by phosphorylation. Genes Dev 10:2782-2793

Knight H, Trewavas AJ, Knight MR (1996) Cold calcium signaling in Arabidopsis involves two cellular pools and a change in calcium signature after acclimation. Plant Cell 8:489-503

Krishna P, Gloor G (2001) The Hsp90 family of proteins in *Arabidopsis thaliana*. Cell Stress Chaperones 6:238-246

Larkindale J, Knight MR (2002) Protection against heat stress-induced oxidative damage in Arabidopsis involves calcium, abscisic acid, ethylene, and salicylic acid. Plant Physiol 128:682-695

Lee GJ, Vierling E (2000) A small heat shock protein cooperates with heat shock protein 70 systems to reactivate a heat-denatured protein. Plant Physiol 122:189-198

Lee GJ, Pokala N, Vierling E (1995a) Structure and in vitro molecular chaperone activity of cytosolic small heat shock proteins from pea. J Biol Chem 270:10432-10438

Lee GJ, Roseman AM, Saibil HR, Vierling E (1997) A small heat shock protein stably binds heat-denatured model substrates and can maintain a substrate in a folding-competent state. EMBO J 16:659-671

Lee JH, Schöffl F (1996) An *Hsp70* antisense gene affects the expression of HSP70/HSC70, the regulation of HSF, and the acquisition of thermotolerance in transgenic *Arabidopsis thaliana*. Mol Gen Genet 252:11-19

Lee JH, Hübel A, Schöffl F (1995b) Derepression of the activity of genetically engineered heat shock factor causes constitutive synthesis of heat shock proteins and increased thermotolerance in transgenic *Arabidopsis*. Plant J 8:603-612

Lee YJ, Nagao RT, Key JL (1994) A soybean 101-kDa heat stress protein complements yeast HSP104 deletion mutant in acquiring thermotolerance. Plant Cell 6:1889-1897

Lin B-L, Wang J-S, Liu H-C, Chen R-W, Meyer Y, Barakat A, Delseny M (2001) Genomic analysis of the Hsp70 superfamily in *Arabidopsis thaliana*. Cell Stress Chaperones 6:201-208

Ling J, Wells DR, Tanguay RL, Dickey LF, Thompson WF, Gallie DR (2000) Heat shock protein Hsp101 binds to the *Fed-1* internal light regulatory element and mediates its high translational activity. Plant Cell 12:1213-1227

Link V, Sinha AK, Vashista P, Hofmann MG, Proels RK, Ehness R, Roitsch T (2002) A heat-activated MAP kinase in tomato: a possible regulator of the heat stress response. FEBS Lett 531:179-183

Löw D, Brändle K, Nover L, Forreiter C (2000) Cytosolic heat-stress proteins Hsp17.7 class I and Hsp17.3 class II of tomato act as molecular chaperones in vivo. Planta 211:575-582

Lynch J, Polito VS, Lauchli A (1989) Salinity stress increases cytoplasmic Ca^{2+} activity in maize root protoplasts. Plant Physiol 90:1271-1274

Malik MK, Slovin JP, Hwang CH, Zimmerman JL (1999) Modified expression of a carrot small heat shock protein gene, Hsp17.7, results in increased or decreased thermotolerance. Plant J 20:89-99

Miernyk JA (1999) Protein folding in the plant cell. Plant Physiol 121:695-703

Miernyk JA (2001) The J-domain proteins of *Arabidopsis thaliana*: an unexpectedly large and diverse family of chaperones. Cell Stress Chaperones 6:209-218

Mishra SK, Tripp J, Winkelhaus S, Tschiersch B, Theres K, Nover L, Scharf K-D (2002) In the complex family of heat stress transcription factors, HsfA1 has a unique role as master regulator of thermotolerance in tomato. Genes Dev 16:1555-1567

Morimoto RI (1998) Regulation of the heat shock transcriptional response: cross talk between a family of heat shock factors, molecular chaperones, and negative regulators. Genes Dev 12:3788-3796

Mosser DD, Kotzbauer PT, Sarge KD, Morimoto RI (1990) In vitro activation of heat shock transcription factor DNA-binding by calcium and biochemical conditions that affect protein conformation. Proc Natl Acad Sci USA 87:3748-3752

Mosser DD, Duchaine J, Massie B (1993) The DNA-binding activity of the human heat shock transcription factor is regulated in vivo by hsp70. Mol Cell Biol 13:5427-5438

Munnik T, Ligterink W, Meskiene I, Calderini O, Beyerly J, Musgrave A, Hirt H (1999) Distinct osmosensing protein kinase pathways are involved in signaling moderate and severe hyperosmotic stress. Plant J 20:381-388

Nadeau K, Das A, Walsh CT (1993) Hsp90 chaperonins possess ATPase activity and bind heat shock transcription factors and peptidyl prolyl isomerases. J Biol Chem 268:1479-1487

Neckers L (2002) Hsp90 inhibitors as novel cancer chemotherapeutic agents. Trends Mol Med 8:S55-61

Nover L, Scharf K-D, Neumann D (1989) Cytoplasmic heat shock granules are formed from precursor particles and are associated with a specific set of mRNAs. Mol Cell Biol 9:1298-1308

Nover L, Bharti K, Döring P, Mishra SK, Ganguli A, Scharf K-D (2001) *Arabidopsis* and the heat stress transcription factor world: how many heat stress transcription factors do we need? Cell Stress Chaperones 6:177-189

Nühse T, Peck SC, Hirt H, Boller T (2000) Microbial elicitors induce activation and dual phosphorylation of the *Arabidopsis* MAP kinase AtMPK6. J Biol Chem 275:7521-7526

Owens-Grillo JK, Stancato LF, Hoffmann K, Pratt WB, Krishna P (1996) Binding of immunophilins to the 90 kDa heat shock protein (hsp90) via a tetratricopeptide repeat domain is a conserved protein interaction in plants. Biochemistry 35:15249-15255

Pareek A, Singla SL, Grover A (1995) Immunological evidence for accumulation of two high-molecular-weight (104 and 90 kDa) HSPs in response to different stresses in rice and in response to high temperature stress in diverse plant genera. Plant Mol Biol 29:293-301

Parsell DA, Lindquist S. (1993) The functions of heat-shock proteins in stress tolerance: degradation and reactivation of damaged proteins. Annu Rev Genet 27:437-496

Parsell DA, Kowal AS, Singer MA, Lindquist S (1994) Protein disaggregation mediated by heat stress protein 104. Nature 372:475-478

Pirkkala L, Nykanen P, Sistonen L (2001) Roles of the heat shock transcription factors in regulation of the heat shock response and beyond. FASEB J 15:1118-1131

Prändl R, Schöffl F (1996) Heat shock elements are involved in heat shock promoter activation during tobacco seed maturation. Plant Mol Biol 31:157-162

Prändl R, Hinderhofer K, Eggers-Schumacher G, Schöffl F (1998) HSF3, a new heat shock factor from *Arabidopsis thaliana,* derepresses the heat shock response and confers thermotolerance when overexpressed in transgenic plants. Mol Gen Genet 258:269-278

Pratt WB, Toft DO (1997) Steroid receptor interactions with heat shock protein and immunophilin chaperones. Endocr Rev 18:306-360

Pratt WB, Krishna P, Olsen LJ (2001) Hsp90-binding immunophilins in plants: the protein movers. Trends Plant Sci 6:54-58

Price AH, Taylor A, Ripley SJ, Griffiths A, Trewavas AJ, Knight MR (1994) Oxidative signals in tobacco increase cytosolic calcium. Plant Cell 6:1301-1310

Queitsch C, Hong S-W, Vierling E, Lindquist S (2000) Heat shock protein 101 plays a crucial role in thermotolerance in Arabidopsis. Plant Cell 12:479-492

Queitsch C, Sangster TA, Lindquist S (2002) Hsp90 as a capacitor of phenotypic variation. Nature 417:618-624

Rabindran SK, Wisniewski J, Li L, Li GC, Wu C (1994) Interaction between heat shock factor and hsp70 is insufficient to suppress induction of DNA-binding activity in vivo. Mol Cell Biol 14:6552-6560

Reddy RK, Kurek I, Silverstein AM, Chinkers M, Breiman A, Krishna P (1998) High molecular weight FK506-binding proteins are components of heat shock protein 90 heterocomplexes in wheat germ lysate. Plant Physiol 118:1395-1401

Reindl A, Schöffl F (1998) Interaction between the *Arabidopsis thaliana* heat shock transcription factor HSF1 and the TATA binding protein TBP. FEBS Lett 436:318-322

Reindl A, Schöffl F, Schell J, Koncz C, Bako L (1997) Phosphorylation by a cyclin-dependent kinase modulates DNA binding of the Arabidopsis heat-shock transcription factor HSF1 in vitro. Plant Physiol 115:93-100

Richter K, Buchner J (2001) Hsp90: chaperoning signal transduction. J Cell Physiol 188:281-290

Rizhsky L, Liang H, Mittler R (2002) The combined effect of drought stress and heat shock on gene expression in tobacco. Plant Physiol 130:1143-1151

Robertson AJ, Ishikawa M, Gusta LV, MacKenzie SL (1994) Abscisic acid-induced heat tolerance in *Bromus inermis* Leyss cell-suspension cultures. Heat-stable, abscisic acid-responsive polypeptides in combination with sucrose confer enhanced thermostability. Plant Physiol 105:181-190

Rojas A, Almoguera C, Jordano J (1999) Transcriptional activation of a heat shock gene promoter in sunflower embryos: synergism between ABI3 and heat shock factors. Plant J 20:601-610

Rojas A, Almoguera C, Carranco R, Scharf K-D, Jordano J (2002) Selective activation of the developmentally regulated *Ha hsp17.6 G1* promoter by heat stress transcription factors. Plant Physiol 129:1207-1215

Rutherford SL, Lindquist S (1998) Hsp90 as a capacitor for morphological evolution. Nature 396:336-342

Saibil H (2000) Molecular chaperones: containers and surfaces for folding, stabilizing or unfolding proteins. Curr Opin Struct Biol 10:251-258

Saibil HR, Ranson NA (2002) The chaperonin folding machine. Trends Biochem Sci 27:627-632

Sanchez Y, Lindquist S (1990) HSP104 is required for induced thermotolerance. Science 248:1112-1115

Sangwan V, Dhindsa RS (2002) In vivo and in vitro activation of temperature-responsive plant map kinases. FEBS Lett 531:561-564

Sangwan V, Orvar BL, Beyerly J, Hirt H, Dhindsa RS (2002) Opposite changes in membrane fluidity mimic cold and heat stress activation of distinct plant MAP kinase pathways. Plant J 31:629-638

Satyal SH, Chen D, Fox SG, Kramer JM, Morimoto RI (1998) Negative regulation of the heat shock transcriptional response by HSBP1. Genes Dev 12:1962-1974

Scharf K-D, Rose S, Zott W, Schoffl F, Nover L (1990) Three tomato genes code for heat stress transcription factors with a region of remarkable homology to the DNA-binding domain of the yeast HSF. EMBO J 9:4495-4501

Scharf K-D, Heider H, Höhfeld I, Lyck R, Schmidt E, Nover L (1998) The tomato Hsf system: HsfA2 needs interaction with HsfA1 for efficient nuclear import and may be localized in cytoplasmic heat stress granules. Mol Cell Biol 18:2240-2251

Scharf K-D, Siddique M, Vierling E (2001) The expanding family of *Arabidopsis thaliana* small heat stress proteins and a new family of proteins containing α-crystallin domains (Acd proteins). Cell Stress Chaperones 6:225-237

Schirmer EC, Lindquist S, Vierling E (1994) An *Arabidopsis* heat stress protein complements a thermotolerance defect in yeast. Plant Cell 6:1899-1909

Schirmer EC, Glover JR, Singer MA, Lindquist S (1996) HSP100/Clp proteins: a common mechanism explains diverse functions. Trends Biochem Sci 21:289-295

Schlicher T, Soll J (1996) Molecular chaperones are present in the thylakoid lumen of pea chloroplasts. FEBS Lett 379:302-304

Schöffl F, Prändl R, Reindl A (1998) Regulation of the heat-shock response. Plant Physiol 117:1135-1141

Schumacher RJ, Hurst R, Sullivan WP, McMahon NJ, Toft DO, Matts RI (1994) ATP-dependent chaperoning activity of reticulocyte lysate. J Biol Chem 269:9493-9499

Shi Y, Mosser DD, Morimoto RI (1998) Molecular chaperones as HSF1-specific transcriptional repressors. Genes Dev 12:654-666

Simoes-Araujo JL, Rodrigues RL, de A Gerhardt LB, Mondego JMC, Alves-Ferreira M, Rumjanek NG, Margis-Pinheiro M (2002) Identification of differentially expressed genes by cDNA-AFLP technique during heat stress in cowpea nodules. FEBS Lett 515:44-50

Soncin F, Asea A, Zhang X, Stevenson MA, Calderwood SK (2000) Role of calcium activated kinases and phophatases in heat shock factor-1 activation. Int J Mol Med 6:705-710

Sorger PK (1991) Heat shock factor and the heat shock response. Cell 65:363-366

Squires C, Squires CL (1992) The Clp proteins: proteolysis regulators or molecular chaperones? J Bacteriol 174:1081-1085

Stancato LF, Hutchison KA, Krishna P, Pratt WB (1996) Animal and plant cell lysates share a conserved chaperone system that assembles the glucocorticoid receptor into a functional heterocomplex with hsp90. Biochemistry 35:554-561

Sun W, Bernard C, van de Cotte B, van Montagu M, Verbruggen N (2001) *At-HSP17.6A*, encoding a small heat-shock protein in Arabidopsis, can enhance osmotolerance upon overexpression. Plant J 27:407-415

Sun W, van Montagu M, Verbruggen N (2002) Small heat shock proteins and stress tolerance in plants. Biochim Biophys Acta 1577:1-9

Sung DY, Vierling E, Guy CL (2001) Comprehensive expression profile analysis of the Arabidopsis Hsp70 gene family. Plant Physiol 126:789-800

Takahashi K, Isobe M, Knight MR, Trewavas AJ, Muto S (1997) Hypo-osomotic shock induces increases in cytosolic Ca^{2+} in tobacco suspension-culture cells. Plant Physiol 113:587-594

Wang KL-C, Li H, Ecker JR (2002) Ethylene biosynthesis and signaling networks. Plant Cell Supp S131-S151

Waters ER, Lee GJ, Vierling E. (1996) Evolution, structure and function of the small heat shock proteins in plants. J Exp Bot 47:325-338

Webb MA, Cavaletto JM, Klanrit P, Thompson GA (2001) Orthologs in *Arabidopsis thaliana* of the Hsp70 interacting protein Hip. Cell Stress Chaperones 6:247-255

Wehmeyer N, Hernandez LD, Finkelstein RR, Vierling E (1996) Synthesis of small heat shock proteins is part of the developmental program of late seed maturation. Plant Physiol 112:747-757

Wells DR, Tanguay RL, Le H, Gallie DR (1998) Hsp101 functions as a specific translational regulatory protein whose activity is regulated by nutrient status. Genes Dev 12:3236-3251

Wiech H, Buchner J, Zimmermann R, Jakob U (1992) Hsp90 chaperones protein folding in vitro. Nature 358:169-170

Wilen RW, Sacco M, Gusta LV, Krishna P. (1995) Effects of 24-epibrassinolide on freezing and thermotolerance of bromegrass (*Bromus inermis*) cell cultures. Physiol Plant 95:195-202

Wu C (1995) Heat shock transcription factors: structure and regulation. Annu Rev Cell Dev Biol 11:441-469

Yabe N, Takahashi T, Komeda Y (1994) Analysis of tissue-specific expression of *Arabidopsis thaliana* Hsp90-family gene *HSP81*. Plant Cell Physiol 35:1207-1219

Young JC, Moarefi I, Hartl FU (2001) Hsp90: a specialized but essential protein-folding tool. J Cell Biol 154:267-273

Zabaleta E, Assad N, Oropeza A, Salerno G, Herrera-Estrella L (1994a) Expression of one of the members of the *Arabidopsis* chaperonin 60β gene family is developmentally regulated and wound-repressible. Plant Mol Biol 24:195-202

Zabaleta E, Oropeza A, Assad N, Mandel A, Salerno G, Herrera-Estrella L (1994b) Antisense expression of chaperonin 60β in transgenic tobacco plants leads to abnormal phenotypes and altered distribution of photoassimilates. Plant J 6:425-432

Zhang Z, Quick MK, Kanelakis KC, Gijzen M, Krishna P (2003) Characterization of a plant homolog of Hop, a cochaperone of Hsp90. Plant Physiol 131:525-535

Zhu J-K (2002) Salt and drought stress signal transduction in plants. Annu Rev Plant Biol 53:247-273

4 Sensors of abiotic stress in *Synechocystis*

Koji Mikami, Iwane Suzuki and Norio Murata

Abstract

Systematic mutagenesis of histidine kinases in combination with DNA microarray technology has allowed us to identify sensors for cold, hyperosmotic stress, sodium chloride, phosphate, and metal ions in *Synechocystis* sp. PCC 6803. His kinase 33 (Hik33) senses both cold and hyperosmotic stress and regulates the expression of different sets of genes when cells are exposed to each respective stress. Salt stress is perceived by Hik33, Hik34, and Hik16, each of which regulates the expression of a different set of genes. Hik7 has been identified as a sensor of phosphate deficiency. Moreover, manganese deficiency and an excess of nickel ions are sensed by Hik27 (ManS) and Hik30 (NrsS), respectively. The genomes of six strains of cyanobacteria all include genes for Hik33, Hik34, and Hik7, suggesting that these three histidine kinases might be important for the perception of stress in many or even all cyanobacteria.

4.1 Introduction

Abiotic stresses, such as an abnormally high or low temperature, high osmolarity, high salinity and limited or excessive nutrient availability have negative effects on the growth and development of cells. Cells perceive a particular stress and react to it by expressing a specific set of stress-inducible genes, which appear to play important roles in the acclimation of the cells to the various types of stress (Shinozaki and Yamaguchi-Shinozaki 2000; Knight and Knight 2001; Sakamoto and Murata 2002; Xiong et al. 2002).

In photosynthetic organisms, such as cyanobacteria and plants, various two-component systems contribute to the perception and transduction of environmental signals (Mizuno et al. 1996; Hwang et al. 2002). A His kinase (Hik) perceives the stress and a response regulator (Rre) transduces the stress signal. The genome of *Synechocystis* sp. PCC 6803 (hereafter, *Synechocystis*) appears to include 44 genes for Hiks and 42 genes for Rres (Kaneko et al. 1996; Mount and Chang 2002; also see http://www.kazusa.or.jp/cyanobase/Synechocystis/index.html), whereas that of *Arabidopsis thaliana* contains 11 genes for Hiks and 22 genes for Rres (the Arabidopsis Genome Initiative 2000). Moreover, Hiks that perceive ethylene are found both in *A. thaliana* (Chang et al. 1993; Hua et al. 1995) and in *Synechocystis* (Mount and Chang 2002), suggesting the possibility that the Hiks and Rres in *A. thaliana* might have originated from cyanobacteria (Mount and Chang 2002).

Topics in Current Genetics, Vol. 4
H. Hirt, K. Shinozaki (Eds.) Plant Responses To Abiotic Stress

However, little is known about the two-component systems in plants and cyanobacteria that are involved in the perception and transduction of abiotic stresses, such as high and low temperatures, high osmolarity, and high salinity.

Gene-targeted mutagenesis and gene transfer have become routine techniques in studies of *Synechocystis* because of the capacity of this microorganism for homologous recombination. Moreover, the genome of *Synechocystis* is relatively small (3.7 Mbp) and the proportion of non-coding regions to coding regions in the genome is also small (12%). Thus, it is rather easy to generate mutant libraries in which each gene for a Hik or an Rre has been separately inactivated by the insertion of an antibiotic-resistance gene cassette into the coding region of the respective gene. Such mutant libraries can then be screened for stress sensors or signal transducers by monitoring the effects of each mutation on gene expression (Suzuki et al. 2000). In addition, the development of DNA microarray techniques, using microarrays based on the complete sequence of the genome of *Synechocystis*, has provided a new tool for the analysis of genome-wide patterns of transcription. Application of this technique to the screening of mutant libraries is rapidly increasing our understanding of sensors of abiotic stress and signal transducers. This review describes the characterization of Hiks as sensors of cold, hyperosmotic stress, salt stress, and levels of phosphate and metal ions in *Synechocystis*.

4.2 Hik33 as a cold sensor

The cold-inducible genes of *Synechocystis* include the *des* genes for fatty acid desaturases and these genes have been studied in great detail (Murata and Wada 1995; Los and Murata 1998). In this cyanobacterium, there are four desaturases, namely, the Δ12, Δ15, Δ9, and Δ6 desaturases, which are encoded, respectively, by the *desA, desB, desC*, and *desD* genes (Murata and Wada 1995; Los and Murata 1998). The expression of the *desD, desA*, and *desB* genes is induced by cold stress (Los et al. 1997) and a cold sensor, Hik33 (Sll0698) was originally identified as a positive regulator of the expression of these genes by monitoring the cold-induced activity of luciferase (LuxAB) that resulted from the cold-induced expression of a *desB* promoter-*luxAB* gene fusion in a library of mutants with inactivated Hiks (Suzuki et al. 2000). DNA microarray analysis using a mutant of the *hik33* gene indicated that Hik33 regulates, either fully or to a limited extent, the expression of 28 of 45 cold-inducible genes (Suzuki et al. 2001; Mikami et al. 2002). As shown in Figure 1, Hik33 regulates the expression of genes that are involved in the regulation of gene expression, in the regulation of photosynthesis and in the maintenance of the structure and function of the cell wall and cell membranes. Since 17 of the 45 cold-inducible genes are not regulated by Hik33 (Fig. 1), we can assume that *Synechocystis* must also have at least one other cold sensor.

The probable contribution of membrane fluidity to the perception of cold stress was suggested several years ago (Murata and Los 1997; Vigh et al. 1998; Los and Murata 2000) and, recently, we obtained evidence for such a contribution from

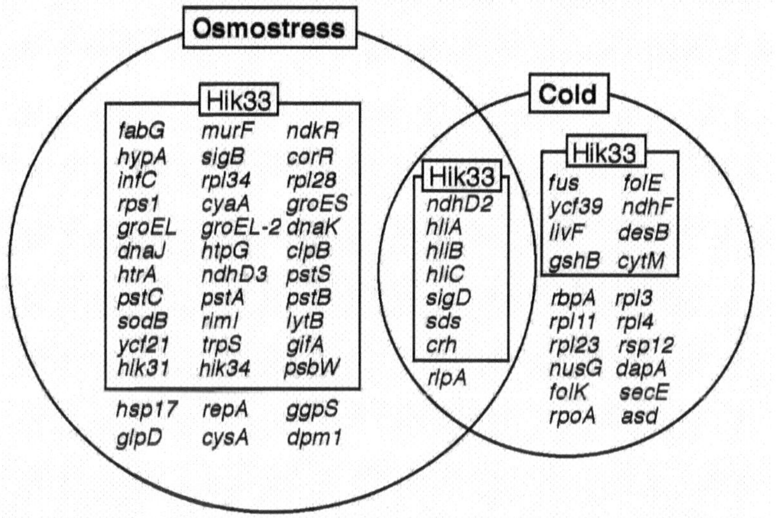

Fig. 1. Involvement of Hik33 in the regulation of expression of cold-inducible and hyperosmotic stress-inducible genes. Circles include genes whose expression is induced by cold stress or hyperosmotic stress, as indicated, and the overlapping region of the two circles encloses genes whose expression is induced by both cold stress and hyperosmotic stress. Rectangles with the circles include genes whose stress-inducible expression is regulated by Hik33. It is likely that genes shown outside rectangles are regulated by as yet unidentified sensors. Products of genes can be found in the Cyanobase (http://www.kazusa.or.jp/cyano/).

experiments in which membrane fluidity was manipulated by genetic engineering of the fatty acid desaturases in *Synechocystis*. We inactivated both the *desA* and the *desD* gene by targeted mutagenesis, generating *desA*⁻/*desD*⁻ cells that contained only monounsaturated lipid molecules (Tasaka et al. 1996). Fourier transform infrared (FTIR) spectrometory demonstrated that the replacement of polyunsaturated membrane lipids by monounsaturated membrane lipids, as a result of the *desA*⁻/*desD*⁻ mutation, rigidified the membrane lipids (Szalontai et al. 2000). Subsequent DNA microarray analysis demonstrated that, in *desA*⁻/*desD*⁻ mutant cells, a larger number of genes were induced by cold than in wild-type cells (Inaba et al. 2003). These observations suggested that the rigidification of membrane lipids might enhance the cold inducibility of gene expression. Indeed, the cold inducibility of some genes, such as the *hliA, hliB* and *sigD* genes, was inhibited in a triple mutant, namely, the *desA*⁻/*desD*⁻/*hik33*⁻ (Inaba et al. 2003). Thus, it seems likely that Hik33 senses the rigidification of membrane lipids to regulate the expression of cold-inducible genes, as postulated previously (Los and Murata 2000; Sakamoto and Murata 2002). Taken together, our observations indicate that a change in membrane fluidity is a primary signal when cells are exposed to cold stress and that this change is perceived by cold sensors.

4.3 Hik33 as a sensor of hyperosmotic stress

Screening of a *Synechocystis* library with mutations in Hiks allowed us to identify Hik33 as an osmosensor (Mikami et al. 2002). Hik33 acts positively to regulate, either fully or partially, the expression of 210 of 257 osmostress-inducible genes, whose products are involved in the structural maintenance of the cell wall and cell membranes, in photosynthesis, in gene expression, in phosphate-transport system and in the folding and turnover of proteins (Fig. 1; Mikami et al. 2002). Despite the major contribution of Hik33 to the regulation of osmostress-inducible genes, Hik33 does not regulate all the osmostress-inducible genes in *Synechocystis* (Fig. 1) so it is very likely that this cyanobacterium has additional osmosensors (Mikami et al. 2002).

An increase in the anisotropy of fluorescence polarization with 1,6-diphenyl-1,3,5-hexatriene was observed when *Escherichia coli* and *Saccharomyces cerevisiae* were exposed to hyperosmotic stress (Yamazaki et al. 1989; Laroche et al. 2001), suggesting a decrease in membrane fluidity as a result of hyperosmotic stress. The activity of two sensors of hyperosmotic stress in *E. coli*, KdpD and EnvZ (Mizuno et al. 1982; Voelkner et al 1993), is affected by the changes in the physical state of membrane lipids that occur upon exposure of cells to procaine, a membrane fluidizer (Sugiura et al. 1994; Lu et al. 1996; Rampersaud and Inouye 1991). Moreover, membrane-bound Hiks, such as Sln1, DocA, and MokA, have been identified as osmosensors in *S. cerevisiae*, *Dictyostelium discoidium,* and *Myxococcus xanthus*, respectively (Maeda et al. 1994, 1995; Schuster et al. 1996; Kimura et al. 2001). A membrane-bound Hik, AtHK1, was identified as a candidate for an osmosensor in *A. thaliana* (Urao et al. 1999). These findings suggest that the sensor of hyperosmotic stress, specifically Hik33, might recognize changes in membrane fluidity as the primary signal that occurs when cells are exposed to hyperosmotic stress. However, experimental evidence is required to prove this possibility.

4.4 Perception of multiple stresses by Hik33

Results from DNA microarray analysis indicate that cold stress and hyperosmotic stress might induce different sets of genes but a small group of genes is induced by both kinds of stress (Fig. 1). There are three distinct sets of genes whose cold-inducible or hyperosmotic stress-inducible expression is regulated by Hik33 (Fig. 1). These observations suggest that Hik33 might sense cold stress and hyperosmotic stress in different ways and might regulate the expression of distinct genes in a stress-specific manner (Mikami et al. 2002).

A putative homolog of Hik33 was recently identified in *Synechococcus elegatus* PCC 7942 as a sensor of strong light and nutrient stress (NblS; van Waasbergen et al. 2002). Moreover, Hik33 was originally identified as DspA, a chemical sensor of certain drugs, such as inhibitors of photosynthesis (Bartsevich and Shestakov 1995). These results suggest that Hik33 might be a “multi-stress” sen-

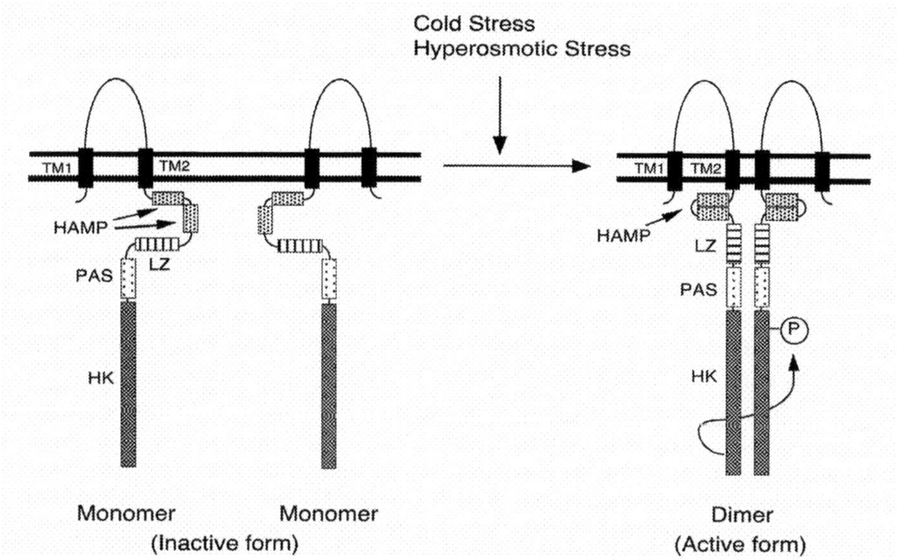

Fig. 2. A hypothetical model for the activation of Hik33 under stress conditions. Under normal conditions, Hik33 is inactive and exists as monomers. Cold stress and hyperosmotic stress induce the dimerization of Hik33 via intramolecular structural changes in the HAMP region, with the resultant activation of Hik33. TM, Transmembrane region; HAMP, the HAMP region; LZ, leucine zipper domain; PAS, the PAS domain; HK, histidine kinase domain.

sor, recognizing strong light, nutrient stress, and certain chemicals in addition to cold and hyperosmotic stress. However, it remains to be determined whether Hik33 regulates the expression of different sets of genes in a stress-specific manner in each case.

The structure of Hik33 can be divided into an amino-terminal signal-input domain and a carboxy-terminal kinase domain. Between these two domains, there are a HAMP region (Aravind and Ponting 1999; also known as a type-P linker; Williams and Stewart 1999), a leucine zipper domain, and a PAS domain (Fig. 2). The HAMP region, which is located immediately downstream of the second transmembrane region, consists of two helical subregions in tandem. It has been proposed that, in *E. coli* and *Salmonella enterica* (Park and Inouye 1997; Butler and Falke 1998; Appleman and Stewart 2003), the HAMP region transduces extracellular signals via intramolecular structural changes. Such changes might involve intermolecular dimerization via interactions between the two helical regions (Aravind and Ponting 1999; Williams and Stewart 1999). Thus, it is possible that a conformational change in the HAMP region, induced by cold or osmotic stress, might generate a dimeric form of Hik33, with resultant activation of the kinase domain (Fig. 2). However, both (i) the way in which Hik33 perceives different kinds of stress and responds by dimerization and (ii) the way in which Hik33 transmits the stress signal to the appropriate signal-transduction pathway for induction of expression of the appropriate set of genes remain to be determined.

4.5 Hik16, Hik33, and Hik34 as salt sensors

There is some confusion about the terms "salt stress" and "hyperosmotic stress", and salt-inducible genes have sometimes been reported to as osmostress-inducible genes in the literature. We have found that responses to salt stress and to osmostress are clearly different when whole-genome patterns of transcription and changes in cell volume are compared in *Synechocystis* under these types of stress (Kanesaki et al. 2002). Salt stress due to 0.5 M NaCl and hyperosmotic stress due to 0.5 M sorbitol induced the expression of a total of 156 and 257 genes, respectively (Kanesaki et al. 2002; Mikami et al. 2002). The genes whose expression is regulated by salt stress, by hyperosmotic stress, and by both kinds of stress comprise distinct and separate groups. It is likely that Na^+ and Cl^- ions penetrate the plasma membranes through K^+ (Na^+) channels and Cl^- channels to produce strong ionic effects, although salt stress also produces slight and transient hyperosmotic effects (Kanesaki et al. 2002). However, *Synechocystis* senses these two types of effect in different ways.

Hik16 (Slr1805), Hik33, and Hik34 (Slr1285) were identified as putative salt sensors and Hik41 (Sll1229) was identified as a putative signal transducer by screening of a library of cells with mutations in Hiks with DNA microarrays (Marin et al. 2003). These three salt sensors act separately to regulate the expression of different sets of genes. We were surprised to find that a mutation in either Hik16 or Hik41 eliminated the inducibility by salt of three genes, namely, *slr0967, sll0938*, and *sll0939*, which encode proteins of unknown function. Since Hik41 contains a receiver domain in its amino-terminal region, it is very likely that these two kinases constitute a signal-transduction pathway in which Hik16 is the sensor and Hik41 is the transducer of salt stress (Marin et al. 2003).

Genes whose expression is regulated by Hik33, Hik34, Hik16, or Hik41 actually correspond to only about one-fifth of all salt-inducible genes (Marin et al. 2003). The remaining salt-inducible genes are regulated by as yet unidentified salt sensors. Since such sensors were not identified during screening of our library of *hik* mutants (Marin et al. 2003), it is possible that these unidentified salt sensors differ from Hiks in *Synechocystis*.

4.6 Hik7 and Rre29 as the sensor and signal transducer of a phosphate deficit

Phosphate is very important in various cellular processes and, in particular, in the synthesis of nucleic acids and membrane lipids. Since phosphate forms insoluble salts with calcium, iron, and aluminium, the supply of free phosphate from the natural environment is very limited. Thus, cells have developed mechanisms for the incorporation of free phosphate from the environment into the cytoplasm for maintenance of an appropriate concentration of phosphate. Under phosphate-limiting conditions, alkaline phosphatase is synthesized and released outside cells to generate free phosphate in the cell's surroundings. In bacteria, this process involves the induction of genes that encode alkaline phosphatase and subunits of a phosphate-specific transporter (Pst) system (Torriani-Gorini et al. 1994). In *Synechocystis*, expression of the gene for alkaline phosphatase (Sll0654) and of two different operons that encode subunits of the Pst system, namely, a phosphate-binding protein, PstS, PstC, PstA, and PstB (Sll0679-Sll0683 and Slr1247-Slr1250), is induced under phosphate-limiting conditions (Hirani et al. 2001).

A search for homologies between Hiks of *Synechocystis* and two known phosphate sensors, SphS and PhoR, in *Synechococcus* sp. PCC 7942 and *Synechococcus* sp. WH7803, respectively (Aiba et al. 1993; Watson et al. 1996), identified Hik7 (Sll0337) as a candidate for a phosphate sensor in *Synechocystis.* A similar procedure identified Rre29 (Slr0081) as a candidate for a response regulator involved in phosphate signalling. In this case, a search was made for a protein similar to the response regulator SphR, which acts in concert with the phosphate sensor SphS in *Synechococcus* (Aiba et al. 1993).

In *Synechocystis*, mutants in which the *hik7* gene and/or the *rre29* gene had been inactivated by gene-targeted mutagenesis were defective in the production of alkaline phosphatase and in the uptake of phosphate under phosphate-limiting conditions, demonstrating that Hik7 is a positive regulator of genes whose expression is induced under phosphate-limited conditions (Hirani et al. 2001). Since SphR is a DNA-binding transcription factor (Aiba et al.1994; Nagaya et al. 1994), it is possible that Rre29 might also be a transcription factor that can bind to the promoter regions of genes that are induced by phosphate deficiency. DNA microarray analysis of the phosphate deficiency-induced expression of genes in cells with mutations in Hik7 and/or Rre29 should prove most informative.

4.7 Sensors of metal ions

Despite their toxicity at high levels, metal ions such as Ni^{2+}, Co^{2+}, Zn^{2+}, and Mn^{2+} are essential for the regulation of the activity of numerous enzymes. For instance, Ni^{2+} ions act as cofactors in reaction catalyzed by glyoxalase I, by peptide deformylases, by methyl-coenzyme M reductase, by urease, by superoxide dismutases and by hydrogenases (Ermler et al. 1998), whereas Mn^{2+} ions are required for formation of the catalytic centre of the oxygen-evolving machinery in the photosystem II complex (Yocum and Pecoraro 1999) and act as cofactors in reaction catalyzed by enzymes such as Mn superoxide dismutase and Mn catalase (Borgstahl et al. 2000; Barynin et al. 2001). Cells maintain appropriate internal concentrations of these metal ions by sensing extracellular or intracellular concentrations and by regulating the transport of these ions into or out of the cytoplasm. Recent data suggest the involvement of two-component systems in the regulation of expression of genes for the transporters of metal ions that control intracellular levels of metal ions in *Synechocystis*.

4.7.1 Hik27 and Rre16 as the sensor and signal transducer of manganese deficiency

When *Synechocystis* is grown at low concentration of Mn^{2+} ions, the expression of the *mntCAB* operon (*sll1598-1600*), which contains the genes for the manganese transporter (MntABC), is induced. The MntABC system takes up Mn^{2+} ions from the cell's environment and maintains an appropriate concentration of Mn^{2+} ions in the cytoplasm (Bartsevich and Pakrasi 1995, 1996).

Yamaguchi et al. (2002) screened a knockout library of Hiks using the DNA microarrays and found a mutant of Hik27 (Slr0640) in which the expression of the *mntCAB* operon was specifically enhanced under normal culture conditions. A response regulator, Rre16 (Slr1873), was also identified by screening of a library of *rre* mutants for cells in which the *mntCAB* operon was constitutively expressed (Yamaguchi et al. 2002). The level of expression of the *mntCAB* operon in cells with mutation in the *hik27* gene or in the *rre16* gene was unaffected by the absence of Mn^{2+} ions from the growth medium. Moreover, purified Rre16 bound specifically to the promoter region of the *mntCAB* operon. These findings indicated that the expression of the *mntCAB* operon was specifically induced by manganese deficiency and that reversal of the deficiency required only Hik27 and Rre16. Thus, Hik27 and Rre16 were identified as the sensor and signal transducer of manganese deficiency (Yamaguchi et al. 2002). Hik27 and Rre16 were renamed ManS (manganase sensor) and ManR (manganese regulator), respectively (Yamaguchi et al. 2002). ManS and ManR were also identified by monitoring the activation of the *mntCAB* promoter under normal culture conditions after random mutagenesis of the *Synechocystis* genome (Ogawa et al. 2002).

Intracellular levels of Mn^{2+} ions increased in cells with mutations in ManS and/or ManR and these mutations resulted in the inhibition of the growth of cells in the presence of high levels of Mn^{2+} ions in the growth medium (Yamaguchi et al. 2002). Since the repression of the *mntCAB* operon requires the activity of the

Fig. 3. Hypothetical models for the regulation of expression of genes in the *mntCAB* and *nrsBACD* operons and of the activity of their products. A, Regulation of the *mntCAB* operon by ManS (Hik27). ManS senses a manganese deficiency. It is active when concentration of Mn^{2+} ions in the cell is adequate and it acts negatively to regulate the expression of the *mntCAB* operon. When the cell lacks adequate Mn^{2+} ions, ManS is inactivated and releases the *mntCAB* operon from the negative regulation by ManR (Rre16). The ManABC complex transports Mn^{2+} ions from the periplasm to the cytoplasm. B, Regulation of the *nrsBACD* operon by NrsS (Hik30). NrsS senses an excess of Ni^{2+} ions in the cell. It is active and regulates, positively, the expression of the *nrsBACD* operon via NrsR (Rre33) when the concentration of Ni^{2+} ions in the cell is excessive. The NrsABCD complex pumps Ni^{2+} ions out of the cytoplasm. PM, Plasma membrane.

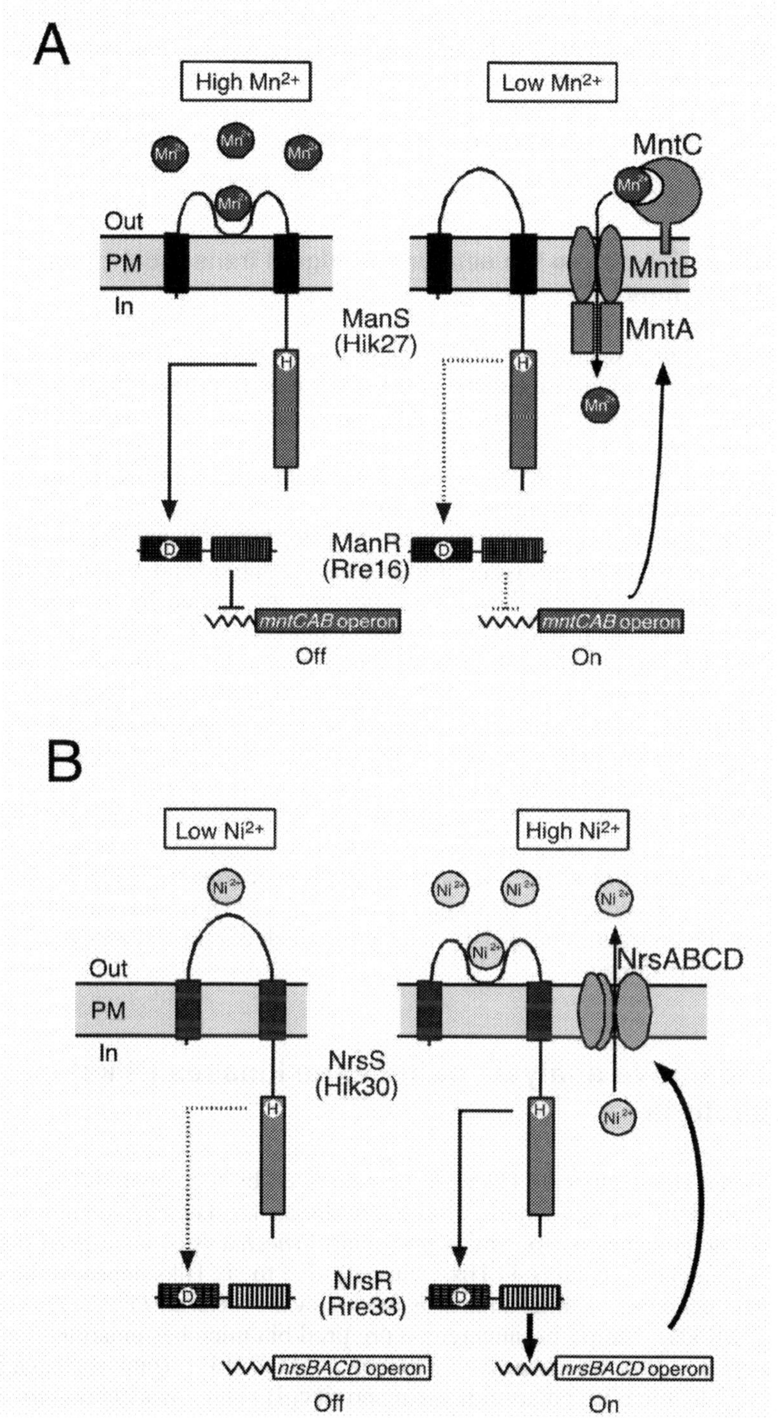
A
High Mn2+
Low Mn2+
MntC
Out
PM
In
MntB
MntA
ManS
(Hik27)
ManR
(Rre16)
mntCAB operon
Off
mntCAB operon
On
B
Low Ni2+
High Ni2+
NrsABCD
Out
PM
In
NrsS
(Hik30)
NrsR
(Rre33)
nrsBACD operon
Off
nrsBACD operon
On

ManSR two-component system under normal growth conditions, it is likely that the ManSR two-component system acts negatively to regulate the expression of the *mntCAB* operon in an Mn^{2+}-dependent manner (Fig. 3A). Thus, ManR is a repressor that loses its activity when ManS is inactivated by a manganese deficit. However, the way in which ManS (Hik27) senses the concentration of Mn^{2+} ions remains to be determined.

4.7.2 Hik30 and Rre33 as the sensor and signal transducer of an excess of Ni^{2+} ions

The nickel-resistance operon (*nrsBACD*; slr0793-slr0796) of *Synechocystis* contains genes that are required for the export of Ni^{2+} ions from the cytoplasm (García-Domínguez et al. 2000). However, it is likely that Ni^{2+} ions themselves are imported into cells by non-specific transporters of cations. Expression of the *nrsBACD* operon is induced by an increase in the concentration of Ni^{2+} ions in the growth medium (García-Domínguez et al. 2000).

An operon that contains genes for Rre33 (NrsR; Sll0797) and Hik30 (NrsS; Sll0798) is located 118 bp upstream of the *nrsBACD* operon and is transcribed in the opposite direction. When these two genes are inactivated by gene-targeted mutagenesis, the Ni^{2+}-dependent inducibility of the *nrsBACD* operon disappears and cells become very sensitive to high concentrations of Ni^{2+} ions (López-Maury et al. 2002). Moreover, purified NrsR binds to the promoter region of the *nrsBACD* operon (López-Maury et al. 2002). Therefore, it appears that the NrsSR two-component system acts positively to regulate the Ni^{2+}-dependent expression of the nickel-resistance operon (Fig. 3B).

A periplasmic region of NrsS includes amino acid sequences that are highly homologous to the Ni^{2+}-binding sites of methyl-coenzyme M reductases from other bacteria (Ermler et al. 1997), but there is no experimental evidence for the binding of Ni^{2+} ions to the periplasmic region of NrsS. Thus, the way in which NrsS senses the concentration of Ni^{2+} ions and transduces the excess-Ni^{2+} signal remains to be determined.

4.8 Comparative analysis of histidine kinases (Hiks) in cyanobacteria

The functional significance of Hiks can be addressed by a comparative study of the Hiks found in various strains of cyanobacteria. To date, the complete sequences of the genomes of six strains, including *Synechocystis,* have been determined. *Synechocystis* contains 44 Hiks, but only five to six Hiks appear to be encoded in the genomes of three strains of marine cyanobacteria, *Prochlorococcus marinus* MED4 (http://bahama.jgi-psf.org/prod/bin/microbes/pmarmed/home.pmarmed.cgi), *Prochlorococcus marinus* MIT9313 (http://bahama.jgi-psf.org/prod/bin/microbes/pmarmit/home.pmarmit.cgi) and *Synechococcus* sp.

strain WH8102 (http://bahama.jgi-psf.org/prod/bin/microbes/syn/home.syn.cgi). The genome of the thermophilic cyanobacterium *Thermosynechococcus elongates* BP-1 encodes 19 Hiks (Nakamura et al. 2002) and that of the filamentous nitrogen-fixing microorganism *Anabaena* sp. PCC 7120 encodes 129 Hiks (Kaneko et al. 2001). The number of genes for Hiks appears to be related to the size of genome. For example, marine cyanobacteria, which are major photosynthetic prokaryotes in the world's oceans (Partensky et al. 1999), have small genomes (approximately 1.7 - 2.4 Mbp) and the smallest numbers (5 or 6) of Hiks. The nutritional environment in the ocean may be poor but it is more stable than that on land or in fresh water. This stability might explain why these organisms have such small genomes and such small numbers of Hiks.

We found that five Hiks are conserved among the six cyanobacteral strains whose genomes have been sequenced. They are Hik33, Hik34, Hik7, Hik2, (Slr1147) and Hik8 (Sll0750). As mentioned above, Hik33 senses clod stress and hyperosmotic stress (Suzuki et al. 2000; Mikami et al. 2002); Hik34 is a salt sensor (Marin et al. 2003) and Hik7 senses phosphate deficiency (Hirani et al. 2001). A homolog of Hik33, Ycf26, has also been found to be encoded in the genomes of the plastids in the red algae *Porphyra purpurea* and *Cyanidium caldarium*, but the functions of these proteins in these plastids remain to be characterized (Glöckner et al. 2000).

Hik8 is a homolog of SasA, which is associated with a circadian clock-related protein KaiC in *Synechococcus* sp. PCC 7942 (Iwasaki et al. 2000; Kageyama et al. 2003). This observation suggests that Hik8 might be involved in the regulation of circadian rhythm in *Synechocystis*. In addition, our DNA microarray indicates that Hik2 might be the second cold sensor that we mentioned above (our unpublished results). The conservation of Hik8 and Hik2 in all six strains of cyanobacteria examined suggests that these two Hiks might also play essential roles in cyanobacteria.

4.9 Future perspectives

In *Synechocystis*, seven Hiks, namely, Hik33, Hik34, Hik16, Hik41, Hik7, Hik27, and Hik30 have been identified as sensors of abiotic stress and Rre29, Rre16, and Rre33 have been identified as transducers of the signals perceived by Hik7, Hik27, and Hik30, respectively, in the respective two-component systems. However, the functions of 37 Hiks and 39 Rres are unknown. It is difficult to predict the functions of these Hiks and Rres from their structural features and their homology to proteins in other organisms. Since we have already generated libraries of *hik* mutants and *rre* mutants, DNA microarray analysis using these mutants should help us to determine the functions of at least some of these Hiks and Rres.

Cyanobacteria have large numbers of signalling proteins that resemble eukaryotic enzymes, such as receptor serine/threonine (S/T) kinases and S/T phosphatases (Zang et al. 1998; Wang et al. 2002). In plants and other eukaryotes, S/T kinases are often found as components of signal transduction pathways, for example, the membrane-bound brassinosteroid receptor BRI1, in *A. thaliana* (Li and Chory, 1997), and a phosphorylation cascade that includes Hog1 mitogen-activated protein kinase in yeast, the activity of which is regulated by the Sln1-Ypd1-Ssk1 phospho-relay system for osmostress signalling (Maeda et al. 1994, 1995). Since *Synechocystis* cells contain a total of 13 membrane-bound or soluble S/T kinases (Zang et al. 1998), it is possible that this microorganism exploits S/T kinases as sensors or signal transducers. However, little is known about the signal transduction pathways in which the individual S/T kinases might be involved and the nature of the genes that are regulated by the transduced signals.

Synechocystis cells contain nine different sigma factors (Goto-Seki et al. 1999; Imamura et al. 2003) that direct the selection of promoters by RNA polymerases under various stress conditions (Ishihama 1993). It has been suggested that each individual sigma factor might be involved in the regulation of the expression of specific stress-inducible genes. SigE does, indeed, regulate the expression of a gene for type III glutamine synthetase under nitrogen-starvation conditions (Muro-Pastor et al. 2001). Moreover, since the expression of the genes for SigD and SigB is induced by strong light and by heat, respectively (Imamura et al. 2003), it seems likely that these sigma factors might regulate the expression of strong light-inducible or heat-inducible genes. However, the stress-inducible genes regulated by each sigma factor remain to be identified.

Construction of libraries of mutants with mutations in the genes for all S/T kinases and sigma factors and the analysis, using DNA microarrays, of the stress-inducible genes whose expression is affected by mutations of the various signal-transducing components should provide important information. Such analysis will help us to understand further details of the mechanisms responsible for the perception and signal transduction of abiotic stresses in *Synechocystis* in terms of stress-specific signal-transduction pathways and their associated sensors, transducers and sigma factors, as well as the cross-talk among the various pathways.

Acknowledgements

This work was supported in part by a Grant-in-Aid for Scientific Research (S) (no. 13854002) and by a Grant-in-Aid for Scientific Research on Priority Areas (2) (no. 14086207) to N.M.; by a Grant-in-Aid for Scientific Research (C) (no. 14540606) to K.M.; and by a Grant-in-Aid for Scientific Research on Priority Areas (C) ("Genome Biology"; no. 13206081) and a Grant-in-Aid for Exploratory Research (no. 14654169) to I.S. All the cited grants were awarded by the Ministry of Education, Culture, Sports, Science and Technology of Japan.

References

Aiba H, Nagaya M, Mizuno T (1993) Sensor and regulator proteins from the cyanobacterium *Synechococcus* species PCC7942 that belong to the bacterial signal-transduction protein families: implication in the adaptive response to phosphate limitation. Mol Microbiol 8:81-91

Aiba H, Mizuno T (1994) A novel gene whose expression is regulated by the response-regulator, SphR, in response to phosphate limitation in *Synechococcus* species PCC7942. Mol Microbiol 13:25-34

Appleman JA, Stewart V (2003) Mutational analysis of a conserved signal-transducing element: the HAMP linker of the *Escherichia coli* nitrate sensor NarX. J Bacteriol 185:89-97

Aravind L, Ponting CP (1999) The cytoplasmic helical linker domain of receptor histidine kinase and methyl-accepting proteins is common to many prokaryotic signalling proteins. FEMS Microbiol Lett 176:111-116

Bartsevich VV, Pakrasi HB (1995) Molecular identification of an ABC transporter complex for manganese: analysis of a cyanobacterial mutant strain impaired in the photosynthetic oxygen evolution process. EMBO J 14:1845-1853

Bartsevich VV, Pakrasi HB (1996) Manganese transport in the cyanobacterium *Synechocystis* sp. PCC 6803. J Biol Chem 271:26057-26061

Bartsevich VV, Shestakov SV (1995) The *dspA* gene product of the cyanobacterium *Synechocystis* sp. strain PCC 6803 influences sensitivity to chemically different growth inhibitors and has amino acid similarity to histidine protein kinases. Microbiology 141:2915-2920

Barynin VV, Whittaker MM, Antonyuk SV, Lamzin VS, Harrison PM, Artymiuk PJ, Whittaker JW (2001) Crystal structure of manganese catalase from *Lactobacillus plantarum*. Structure (Cambridge) 9:725-738

Borgstahl GE, Pokross M, Chehab R, Sekher A, Snell EH (2000) Cryo-trapping the six-coordinate, distorted-octahedral active site of manganese superoxide dismutase. J Mol Biol 296:951-959

Butler SL, Falke JJ (1998) Cysteine and disulfide scanning reveals two amphiphilic helices in the linker region of the aspartate chemoreceptor. Biochemistry 37:10746-10756

Chang C, Kwok SF, Bleecker AB, Meyerowitz EM (1993) *Arabidopsis* ethylene-response gene *ETR1*: Similarity of product to two-component regulators. Science 262:539-544

Ermler U, Grabarse W, Shima S, Goubeaud M, Thauer RK (1998) Active sites of transition-metal enzymes with a focus on nickel. Curr Opin Struct Biol 8:749-758

García-Domínguez M, Lopez-Maury L, Florencio FJ, Reyes JC (2000) A gene cluster involved in metal homeostasis in the cyanobacterium *Synechocystis* sp. strain PCC 6803. J Bacteriol 182:1507-1514

Glöckner G, Rosenthal A, Valentin K (2000) The structure and gene repertoire of an ancient red algal plastid genome. J Mol Evol 51:382-390

Goto-Seki A, Shirokane M, Masuda S, Tanaka K, Takahashi H (1999) Specificity crosstalk among group 1 and group 2 sigma factors in the cyanobacterium *Synechococcus* sp. PCC7942: in vitro specificity and a phylogenetic analysis. Mol Microbiol 34:473-484

Hirani TA, Suzuki I, Murata N, Hayashi H, Eaton-Rye JJ (2001) Characterization of a two-component signal transduction system involved in the induction of alkaline phosphatase under phosphate-limiting conditions in *Synechocystis* sp. PCC 6803. Plant Mol Biol 45:133-144

Hua J, Chang C, Sun Q, Meyerowitz EM (1995) Ethylene insensitivity conferred by *Arabidopsis ERS* gene. Science 269:1712-1714

Hwang I, Chen HC, Sheen J (2002) Two-component signal transduction pathways in *Arabidopsis*. Plant Physiol 129:500-515

Imamura S, Yoshihara S, Nakano S, Shiozaki N, Yamada A, Tanaka K, Takahashi H, Asayama M, Shirai M (2003) Purification, characterization, and gene expression of all sigma factors of RNA polymerase in a cyanobacterium. J Mol Biol 325:857-872

Inaba M, Suzuki I, Szalontai B, Kanesaki Y, Los DA, Hayashi H, Murata N (2003) Gene-engineered rigidification of membrane lipids enhances the cold inducibility of gene expression in *Synechocystis*. J Biol Chem 278:12191-12198

Ishihama A (1993) Protein-protein communication within the transcription apparatus. J Bacteriol 175:2483-2489

Iwasaki H, Williams SB, Kitayama Y, Ishiura M, Golden SS, Kondo T (2000) A KaiC-interacting sensory histidine kinase, SasA, necessary to sustain robust circadian oscillation in cyanobacteria. Cell 101:223-233

Kageyama H, Kondo T, Iwasaki H (2003) Circadian formation of clock protein complexes by KaiA, KaiB, KaiC, and SasA in cyanobacteria. J Biol Chem 278:2388-2395

Kanesaki Y, Suzuki I, Allakhverdiev SI, Mikami K, Murata N (2002) Salt stress and hyperosmotic stress regulate the expression of different sets of genes in *Synechocystis* sp. PCC 6803. Biochem Biophys Res Commun 290:339-348

Kaneko T, Sato S, Kotani H, Tanaka A, Asamizu E, Nakamura Y, Miyajima N, Hirosawa M, Sugiura M, Sasamoto S, Kimura T, Hosouchi T, Matsuno A, Muraki A, Nakazaki N, Naruo K, Okumura S, Shimpo S, Takeuchi C, Wada T, Watanabe A, Yamada M, Yasuda M, Tabata S (1996) Sequence analysis of the genome of the unicellular cyanobacterium *Synechocystis* sp. strain PCC6803. II. Sequence determination of the entire genome and assignment of potential protein-coding regions. DNA Res 3:109-136

Kaneko T, Nakamura Y, Wolk CP, Kuritz T, Sasamoto S, Watanabe A, Iriguchi M, Ishikawa A, Kawashima K, Kimura T, Kishida Y, Kohara M, Matsumoto M, Matsuno A, Muraki A, Nakazaki N, Shimpo S, Sugimoto M, Takazawa M, Yamada M, Yasuda M, Tabata S (2001) Complete genomic sequence of the filamentous nitrogen-fixing cyanobacterium *Anabaena* sp. strain PCC 7120. DNA Res 8:205-213

Kimura Y, Nakano H, Terasaka H, Takegawa K (2001) *Myxococcus xanthus mckA* encodes a histidine kinase-response regulator hybrid sensor required for development and osmotic tolerance. J Bacteriol 183:1140-1146

Knight H, Knight MR (2001) Abiotic stress signalling pathways: specificity and cross-talk. Trends Plant Sci 6:262-267

Laroche C, Beney L, Marechal PA, Gervais P (2001) The effect of osmotic pressure on the membrane fluidity of *Saccharomyces cerevisiae* at different physiological temperatures. Appl Microbiol Biotechnol 56:249-254

Li J, Chory J (1997) A putative leucine-rich repeat receptor kinase involved in brassinosteroid signal transduction. Cell 90:929-938

Lopez-Maury L, García-Domínguez M, Florencio FJ, Reyes JC (2002) A two-component signal transduction system involved in nickel sensing in the cyanobacterium *Synechocystis* sp. PCC 6803. Mol Microbiol 43:247-256

Los DA, Ray MK, Murata N (1997) Differences in the control of the temperature-dependent expression of four genes for desaturases in *Synechocystis* sp. PCC 6803. Mol Microbiol 25:1167-1175

Los DA, Murata N (1998) Structure and expression of fatty acid desaturases. Biochim Biophys Acta 1394:3-15

Los DA, Murata N (2000) Regulation of enzymatic activity and gene expression by membrane fluidity. Science's STKE http://stke.sciencemag.org/cgi/content/full/oc_sigtrans;2000/62/pe1

Lu Q, Park H, Egger LA, Inouye M (1996) Nucleoside-diphosphate kinase-mediated signal transduction via histidyl-aspartyl phosphorelay systems in *Escherichia coli*. J Biol Chem 271:32886-32893

Maeda T, Wurgler-Murphy SM, Saito H (1994) A two-component system that regulates an osmosensing MAP kinase cascade in yeast. Nature 369:242-245

Maeda T, Takekawa M, Saito H (1995) Activation of yeast PBS2 MAPKK by MAPKKKs or by binding of an SH3-containing osmosensor. Science 269:554-558

Marin K, Suzuki I, Yamaguchi K, Ribbeck K, Yamamoto H, Kanesaki Y, Hagemann M, Murata N (2003) Identification of histidine kinases that act as sensors in the perception of salt stress in *Synechocystis* sp. PCC 6803. Proc Natl Acad Sci USA 100:9061-9066

Mikami K, Kanesaki Y, Suzuki I, Murata N (2002) The histidine kinase Hik33 perceives osmotic stress and cold stress in *Synechocystis* sp. PCC 6803. Mol Microbiol 46:905-915

Mizuno T, Wurtzel ET, Inouye M (1982) Cloning of the regulatory genes (*ompR* and *envZ*) for the matrix proteins of the *Escherichia coli* outer membrane. J Bacteriol 150:1462-1466

Mizuno T, Kaneko T, Tabata S (1996) Compilation of all genes encoding bacterial two-component signal transducers in the genome of the cyanobacterium, *Synechocystis* sp. strain PCC 6803. DNA Res 3:407-414

Mount SM, Chang C (2002) Evidence for a plastid origin of plant ethylene receptor genes. Plant Physiol 130:10-14

Murata N, Wada H (1995) Acyl-lipid desaturases and their importance in the tolerance and acclimation to cold of cyanobacteria. Biochem J 308:1-8

Murata N, Los LA (1997) Membrane fluidity and temperature perception. Plant Physiol 115:875-879

Muro-Pastor AM, Herrero A, Flores E (2001) Nitrogen-regulated group 2 sigma factor from *Synechocystis* sp. strain PCC 6803 involved in survival under nitrogen stress. J Bacteriol 183:1090-1095

Nagaya M, Aiba H, Mizuno T (1994) The *sphR* product, a two-component system response regulator protein, regulates phosphate assimilation in *Synechococcus* sp. strain PCC 7942 by binding to two sites upstream from the *phoA* promoter. J Bacteriol 176:2210-2215

Nakamura Y, Kaneko T, Sato S, Ikeuchi M, Katoh H, Sasamoto S, Watanabe A, Iriguchi M, Kawashima K, Kimura T, Kishida Y, Kiyokawa C, Kohara M, Matsumoto M, Matsuno A, Nakazaki N, Shimpo S, Sugimoto M, Takeuchi C, Yamada M, Tabata S (2002) Complete genome structure of the thermophilic cyanobacterium *Thermosynechococcus elongatus* BP-1. DNA Res 9:123-130

Ogawa T, Bao DH, Katoh H, Shibata M, Pakrasi HB, Bhattacharyya-Pakrasi M (2002) A two-component signal transduction pathway regulates manganese homeostasis in *Synechocystis* 6803, a photosynthetic organism. J Biol Chem 277:28981-28986

Park H, Inouye M (1997) Mutational analysis of the linker region of EnvZ, an osmosensor in *Escherichia coli*. J Bacteriol 179:4382-4390

Partensky F, Hess WR, Vaulot D (1999) *Prochlorococcus*, a marine photosynthetic prokaryote of global significance. Microbiol Mol Biol Rev 63:106-127

Rampersaud A, Inouye M (1991) Procaine, a local anesthetic, signals through the EnvZ receptor to change the DNA binding affinity of the transcriptional activator protein OmpR. J Bacteriol. 173:6882-6888

Sakamoto T, Murata N (2002) Regulation of the desaturation of fatty acids and its role in tolerance to cold and salt stress. Curr Opin Microbiol 5:206-210

Schuster SS, Noegel AA, Oehme F, Gerisch G, Simon MI (1996) The hybrid histidine kinase DokA is part of the osmotic response system of *Dictyostelium*. EMBO J 15:3880-3889

Shinozaki K, Yamaguchi-Shinozaki K (2000) Molecular responses to dehydration and low temperature: differences and cross-talk between two stress signaling pathways. Curr Opin Plant Biol 3:217-223

Sugiura A, Hirokawa K, Nakashima K, Mizuno T (1994) Signal-sensing mechanisms of the putative osmosensor KdpD in *Escherichia coli*. Mol Microbiol 14:929-938

Suzuki I, Los DA, Kanesaki Y, Mikami K, Murata N (2000) The pathway for perception and transduction of low-temperature signals in *Synechocystis*. EMBO J 19:1327-1334

Suzuki I, Kanesaki Y, Mikami K, Kanehisa M, Murata N (2001) Cold-regulated genes under control of the cold sensor Hik33 in *Synechocystis*. Mol Microbiol 40:235-244

Szalontai B, Nishiyama Y, Gombos Z, Murata N (2000) Membrane dynamics as seen by Fourier transform infrared spectroscopy in a cyanobacterium, *Synechocystis* PCC 6803. The effects of lipid unsaturation and the protein-to-lipid ratio. Biochim Biophys Acta 1509:409-419

Tasaka Y, Gombos Z, Nishiyama Y, Mohanty P, Ohba T, Ohki K, Murata N (1996) Targeted mutagenesis of acyl-lipid desaturases in *Synechocystis*: evidence for the important roles of polyunsaturated membrane lipids in growth, respiration and photosynthesis. EMBO J 15:6416-6425

The Arabidopsis Genome Initiative (2000) Analysis of the genome sequence of the flowering plant *Arabidopsis thaliana*. Nature 408:796-815

Torriani-Gorini A, Yagil E, Silver S (1994) Phosphate in microorganisms: cellular and molecular biology. ASM Press, Washington DC

Urao T, Yakubov B, Satoh R, Yamaguchi-Shinozaki K, Seki M, Hirayama T, Shinozaki K (1999) A transmembrane hybrid-type histidine kinase in *Arabidopsis* functions as an osmosensor. Plant Cell 11:1743-1754

van Waasbergen LG, Dolganov N, Grossman AR (2002) *nblS*, a gene involved in controlling photosynthesis-related gene expression during high light and nutrient stress in *Synechococcus elongatus* PCC 7942. J Bacteriol 184:2481-2490

Vigh L, Maresca B, Harwood JL (1998) Does the membrane's physical state control the expression of heat shock and other genes? Trends Biochem Sci 23:369-374

Voelkner P, Puppe W, Altendorf K (1993) Characterization of the KdpD protein, the sensor kinase of the K^+-translocating Kdp system of *Escherichia coli*. Eur J Biochem 217:1019-1026

Wang L, Sun YP, Chen WL, Li JH, Zhang CC (2002) Genomic analysis of protein kinases, protein phosphatases and two-component regulatory systems of the cyanobacterium *Anabaena* sp. strain PCC 7120. FEMS Microbiol Lett 217:155-165

Watson GM, Scanlan DJ, Mann NH (1996) Characterization of the genes encoding a phosphate-regulated two-component sensory system in the marine cyanobacterium *Synechococcus* sp. WH7803. FEMS Microbiol Lett 142:105-109

Williams SB, Stewart V (1999) Functional similarities among two-component sensors and methyl-accepting chemotaxis proteins suggest a role for linker region amphipathic helices in transmembrane signal transduction. Mol Microbiol 33:1093-1102

Xiong L, Schumaker KS, Zhu JK (2002) Cell signaling during cold, drought, and salt stress. Plant Cell 14 Suppl: S165-S183

Yamaguchi K, Suzuki I, Yamamoto H, Lyukevich A, Bodrova I, Los DA, Piven I, Zinchenko V, Kanehisa M, Murata N (2002) A two-component Mn^{2+}-sensing system negatively regulates expression of the *mntCAB* operon in *Synechocystis*. Plant Cell 14:2901-2913

Yamazaki M, Ohnishi S, Ito T (1989) Osmoelastic coupling in biological structures: decrease in membrane fluidity and osmophobic association of phospholipid vesicles in response to osmotic stress. Biochemistry 28:3710-3715

Yocum CF, Pecoraro VL (1999) Recent advances in the understanding of the biological chemistry of manganese. Curr Opin Chem Biol 3:182-187

Zhang CC, Gonzalez L, Phalip V (1998) Survey, analysis and genetic organization of genes encoding eukaryotic-like signaling proteins on a cyanobacterial genome. Nucleic Acids Res 26:3619-3625

5 Oxidative stress signalling

Radhika Desikan, John T. Hancock and Steven J. Neill

Abstract

Oxidative stress arises from an imbalance in the generation and removal of reactive oxygen species (ROS) within cells. ROS are produced during photosynthesis and respiration, as by-products of metabolism, or via dedicated enzymes. Cells are equipped with a range of efficient antioxidant mechanisms to remove ROS. Changes in the cellular redox balance result from exposure to various abiotic and biotic stresses, with induction of both ROS generation and removal mechanisms. Recent transcriptomic analyses indicate that the expression of many genes is regulated by ROS. These include genes encoding antioxidants, cell rescue/defence proteins, and signalling proteins. Genetic studies have begun to elucidate the biological roles of ROS. These include programmed cell death, stomatal closure, and gravitropism. Further work will no doubt reveal new functions for ROS as signalling molecules.

5.1 Introduction

For plants, as for all aerobic organisms, oxygen is a double-edged sword. It is absolutely required for normal growth and development, yet continuous exposure to oxygen can result in cellular damage and ultimately death. This is because molecular oxygen is continually reduced within cells to several forms of Reactive Oxygen Species (ROS; sometimes referred to as Active Oxygen Species, AOS), in particular the superoxide free radical anion ($O_2^{\cdot-}$) and hydrogen peroxide (H_2O_2), that react with various cellular components to bring about acute or chronic damage sufficient to result in cellular death (Finkel and Holbrook 2000; Scandalios 2002a). In plant cells, ROS are generated in high amounts by both constitutive and inducible routes, but under normal situations, the redox balance of the cell is maintained via the constitutive action of a wide range of antioxidant mechanisms that have evolved to remove ROS. Various environmental stresses and endogenous stimuli perturb this redox balance via increased ROS production or reduced antioxidant activity, such that oxidative stress ensues (Fig. 1). In response to increased ROS, the expression of genes encoding antioxidant proteins is induced, as well as that of genes encoding proteins involved in a wider range of cellular rescue processes. In addition, it is increasingly clear that ROS also have signalling functions outside of oxidative stress (Fig. 1). Here, we outline the mechanisms that regulate redox balance in plant cells, describe and discuss cellular responses to

Topics in Current Genetics, Vol. 4
H. Hirt, K. Shinozaki (Eds.) Plant Responses To Abiotic Stress

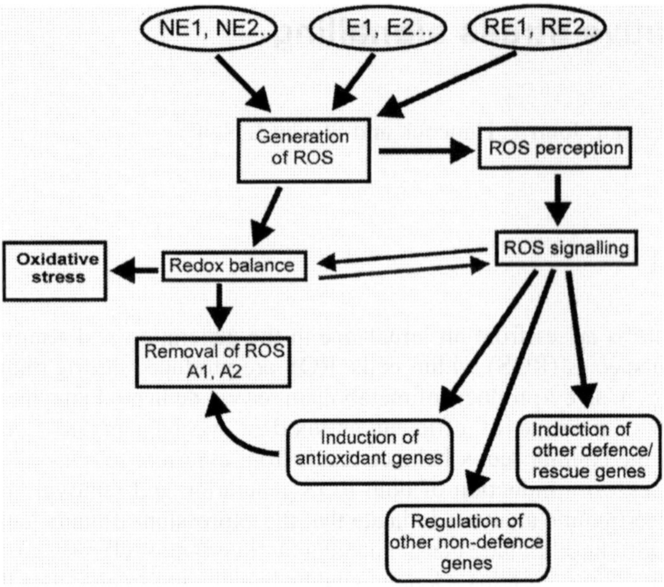

Fig. 1. Redox balance within a cell is determined by the relative rates of generation and removal of ROS. If generation exceeds removal (increased generation and/or decreased removal) then oxidative stress may result. ROS are perceived in cells via as yet-uncharacterised mechanisms. ROS perception and signalling induce the transcription of antioxidant genes, the products of which may restore redox balance and ameliorate the damaging effects of ROS. In addition, ROS also induce the expression of genes not obviously involved with oxidative stress but that may be induced by other stresses, as well as mediating other physiological and developmental processes. NE1, NE2: non-enzymatic sources of ROS; E1, E2: enzymes generating ROS as side-products. RE1, RE2: dedicated ROS-generating enzymes; A1, A2: non-enzymatic and enzymatic antioxidants.

ROS and the potential signalling mechanisms involved, and highlight some of the developmental and physiological processes in which ROS may participate.

5.2 Reactive oxygen species (ROS)

Oxygen is normally reduced by four electrons to produce water, a reaction catalysed by complex IV (cytochrome oxidase) of the mitochondrial electron transport chain. However, molecular oxygen is also converted to ROS by a single electron reduction to the superoxide anion (equation 1).

$$(1)\quad O_2 + e^- \rightarrow O_2^{\bullet -}$$

$$(2)\quad O_2^{\bullet -} + O_2^{\bullet -} + 2H^+ \rightarrow H_2O_2$$

$$(3)\quad O_2^{\bullet -} + H^+ \rightarrow HO_2^{\bullet}$$

(4) $H_2O_2 + O_2^{\bullet -} \rightarrow O_2 + OH^- + OH^\bullet$

(5) $O_2^{\bullet -} + NO^\bullet \rightarrow OONO^-$

As the extra electron is in an unpaired state in the outer orbital, superoxide is a free radical. It is relatively unstable, being either converted back to molecular oxygen, or to H_2O_2, either spontaneously, or in a reaction catalysed by the enzyme superoxide dismutase (SOD) (equation 2). H_2O_2 is a non-charged molecule, can diffuse through both aqueous and lipid environments and has a longer half-life than superoxide, and is thus a more likely signalling molecule than superoxide. As superoxide is a charged molecule, it is unlikely to permeate membranes. However, it can be protonated to form the non-charged perhydroxyl radical (equation 3).

Superoxide and hydrogen peroxide can together react to generate the hydroxyl radical, in the Haber-Weiss or Fenton reaction, catalysed by transition metal ions such as iron or copper (equation 4). The hydroxyl radical is highly reactive and thus less likely to act as a signalling molecule. Another relevant reaction is that between superoxide and nitric oxide, to produce peroxynitrite (equation 5). Peroxynitrite is very reactive, and undoubtedly causes cellular damage. It may have a direct signalling role, and an indirect one by virtue of removing nitric oxide (see section 5.5.1).

ROS and peroxynitrite react to varying degrees with biomolecules causing oxidative damage, with the hydroxyl radical being particularly harmful. Oxidation targets include proteins, DNA, and lipids. In addition, ROS, in particular H_2O_2, have come to function as signalling molecules, with regulated synthesis, specific effects and a range of removal mechanisms. The evolution of photosynthesis and aerobic metabolism inevitably led to the development of ROS-generating processes in chloroplasts, mitochondria, and peroxisomes. It seems likely that antioxidant mechanisms evolved to counter-act the damaging effects of these ROS (Scandalios, 2002a). As environmental stresses increase ROS generation, there would have been evolutionary pressure for selection of ROS signalling mechanisms that induce genes encoding antioxidant and cellular defence proteins. This 'defence' role for ROS and such proteins may be one reason why the induction of many cellular defence/rescue genes is a common response to several environmental stresses and oxidative stress, and help to explain acclimation and cross-tolerance, in which previous exposure to the same stress or a different stress induces tolerance to subsequent exposure (Bowler and Fluhr, 2000). Protective roles for ROS may also have been the driver for the evolution of enzymes such as NADPH oxidase for which the key reaction appears to be ROS generation *per se*, and for which enzyme activity can be regulated by environmental stresses. Thus, biotic and abiotic environmental stresses not only enhance ROS generation via non-specific mechanisms, but also trigger defence signalling mechanisms that, if successful, start with induction of ROS generation, continue with induction of defence responses, and end in removal of ROS to restore redox balance and cell survival. The ubiquity of ROS and the elaboration of ROS signalling mechanisms within cells may also have driven the adoption of ROS for broader signalling processes, such that controlled synthesis of ROS, and subsequent perception and signal transduction, came to form part of discrete signalling processes such as stomatal closure and root development.

5.3 Redox balance and the generation and removal of ROS

5.3.1 Redox balance

Several sources of ROS occur within plant cells, in different sub-cellular locations. Similarly, there also exist a number of antioxidant mechanisms, again found in various locations (Bray et al. 2000; Neill et al. 2002c). Under normal physiological conditions, cellular compartments may have a particular redox balance, determined by the relative rates of ROS generation and removal. Any stimulus that increases ROS and/or decreases antioxidant activity will disturb the redox balance and therefore induce oxidative stress (Fig. 1). In addition to damaging effects, oxidative stress may alter the cellular redox potential (recently termed the 'redox environment' (Schafer and Buettner 2001)). The intracellular environment is maintained within a range of voltages, usually lower than -200mV, a value complementing that of the reductants NADPH and NADH. The electronegativity of the cell is maintained by the millimolar concentrations of reduced glutathione (GSH). It is possible that signalling proteins have thiol groups, with mid-point potentials within the physiological range, that can exist in reduced or oxidised states, with the protein adopting a conformation commensurate with the state of reduction (see section 5.4.2.2). Altered conformations may modify protein function, for example, activating or inactivating the protein. Oxidative stress will cause the intracellular redox environment to become more electro-positive. This may induce a shift in the redox environment away from the physiological range for thiol groups, and thus potentially interfere with signalling pathways.

5.3.2. ROS generation

ROS are generated from both electron transport and enzymatic sources. ROS production is increased by stresses that include excessive light energy, wounding, ozone, drought, UV-irradiations, pathogen challenge, low and high temperatures, and heavy metals (Fig. 2; Dat et al. 2000; Bray et al. 2002; Neill et al. 2002c; Vranova et al. 2002b). It is difficult to discern the levels of ROS in control and stressed cells; a wide range of concentrations (μM to mM: see Neill et al. 2002c) has been estimated. There can be no doubt however, that various stresses do increase H_2O_2 generation substantially. Determination of the relative contributions of different cellular H_2O_2 sources and the impact of H_2O_2 on cell signalling will be greatly facilitated by the development of robust and quantitative means to monitor intracellular ROS concentrations.

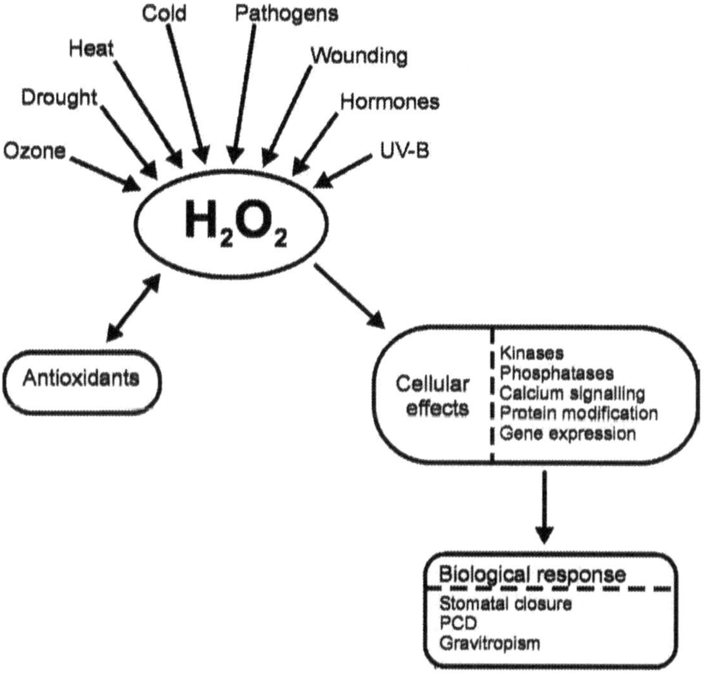

Fig. 2. Regulation, removal, and cellular effects of hydrogen peroxide (H_2O_2). Various abiotic and biotic stresses cause an increase in H_2O_2 within cells. Various antioxidants within the cell act as redox buffers to maintain the redox balance. Perception of H_2O_2 leads to activation of cellular responses such as reversible protein phosphorylation, release of calcium, direct modification of proteins on thiols and regulation of gene expression. These cellular changes result in biological responses, which include programmed cell death (PCD), stomatal closure and gravitropism.

ROS generation occurs via electron transport reactions in both chloroplasts and mitochondria. The Mehler reaction in chloroplasts generates superoxide that is readily converted to H_2O_2 (Polle 1996). Stresses such as high light intensity, drought stress, extreme temperatures, heavy metals, and UV radiations all enhance photosynthetic ROS generation (Dat et al. 2000). Superoxide also arises from electron leakage in mitochondria.

H_2O_2 is generated via several enzyme-mediated reactions in which it is likely that H_2O_2 is not the main 'raison d'etre'. These include glycollate oxidase, producing glyoxylate and H_2O_2 during photorespiration, and acyl CoA oxidase, producing H_2O_2 during the β oxidation of lipids, in peroxisomes (Wojtaszek 1997; Corpas et al. 2001). However, the effects of various stresses on H_2O_2 generation via these enzymatic routes are not yet clear.

H_2O_2 is also generated from dedicated enzymes, for which the key function appears to be H_2O_2 (or superoxide) synthesis. The best characterised of these is NADPH oxidase (sometimes referred to as *rboh* [for respiratory burst oxidase homologue]). In mammals, NADPH oxidase is a plasma membrane–located enzyme, initially isolated from phagocytic cells, and made up of several membrane and cytosolic sub-units that assemble at the phagocyte plasma membrane following cell stimulation. The key subunit is a large glycosylated flavin- and haem-containing protein (gp91) that transfers electrons from NADPH to molecular oxygen to generate $O_2^{\cdot -}$ (and subsequently H_2O_2) that, directly or indirectly, are microbicidal (Reeves et al. 2002). Associated with the $O_2^{\cdot -}$ generation is a respiratory burst (reflecting a hugely increased demand for oxygen). Homologues of the gp91 sub-unit (*rboh* genes and proteins) have been isolated from plants, but no genes encoding potential homologues of the NADPH oxidase cytosolic sub-units have been found. In fact, it may be that non-phagocytic animal cells also utilise the gp91 protein by itself to generate low levels of H_2O_2 for signalling purposes (Lambeth 2002).

The initial work on ROS generation in plants via NADPH oxidase-like enzymes started with plant-pathogen interactions, focussing on the "oxidative burst" (Doke 1983). The oxidative burst leading to the generation of H_2O_2 is typically induced in plants and cell cultures following challenge with either pathogens or elicitor molecules derived from them, and numerous studies have provided evidence for NADPH oxidase being the source of H_2O_2 (Lamb and Dixon 1997; Bolwell 1999). Recent data implicate NADPH oxidase as also being the source of H_2O_2 generated during drought and ozone exposure, or following ABA treatment, via inhibition of H_2O_2 generation by diphenylene iodonium (DPI) (Pei et al. 2000; Zhang et al. 2001c; Jiang and Zhang 2002; Wohlgemuth et al. 2002). It should be noted that DPI is not a specific inhibitor of NADPH oxidase, and may inhibit other flavin-containing enzymes (Bestwick et al. 1999).

There is now considerable molecular evidence for NADPH oxidase (*rboh*) genes in plants. *rboh* genes have been cloned from *Arabidopsis* (Desikan et al. 1998a; Keller et al. 1998; Torres et al. 1998); tomato (Amicucci et al. 1999), tobacco (Simon-Plas et al. 2002), and potato (Yoshioka et al. 2001). Moreover, six to eight *rboh* genes are present in *Arabidopsis*, with differential expression patterns (Torres et al. 1998; The Arabidopsis Genome Initiative, 2000), suggesting different biological functions. Some of the *rboh* genes are induced by H_2O_2 itself and by biotic stresses (Desikan et al. 1998a; Yoshioka et al. 2001; Simon-Plas et al. 2002).

*rboh*A is a 105 kDa plasma membrane protein (Keller et al. 1998), with six membrane –spanning domains (Fig. 3). NAD(P)H and FAD binding domains are conserved at the C-terminus and the haem binding domains are located between two histidine residues in the membrane spanning regions III and V (Keller et al. 1998; Torres et al. 1998). The plant protein does not seem to be heavily glycosylated (Keller et al. 1998), and contains an EF hand (a calcium binding domain) at the N-terminus (Desikan et al. 1998a; Keller et al. 1998; Torres et al. 1998), which has been shown to bind calcium *in vitro* (Keller et al. 1998). Direct calcium activation was demonstrated for tobacco and tomato plasma membrane NADPH oxidases, and the activity was increased by pathogen challenge (Sagi and Fluhr 2001). Expression of a calmodulin gene in tobacco resulted in elevated levels of

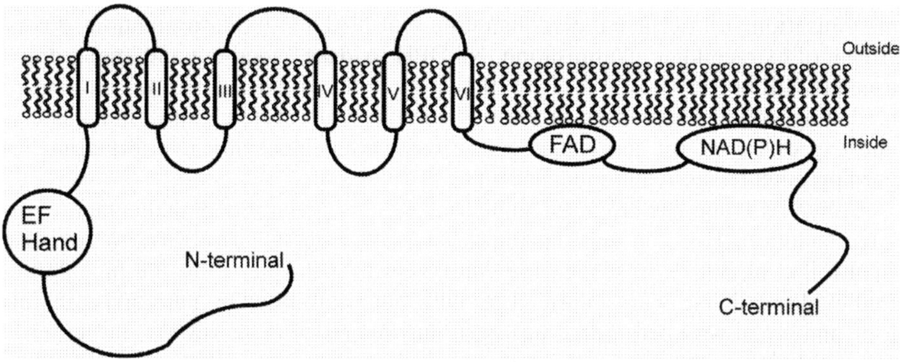

Fig. 3. Predicted structure of *Rboh* (NADPH oxidase) in *Arabidopsis* (derived from Torres et al. (1998)). The six transmembrane domains, position of EF hand in the N-terminal and FAD/NAD(P)H binding domains in the C-terminal regions are indicated.

NADPH and increased H_2O_2 generation following elicitor/pathogen challenge, potentially via activation of a calmodulin-dependent NAD kinase that affects NADPH availability and hence activity of NADPH oxidase (Harding et al. 1997). In *Arabidopsis*, intracellular calcium increased concomitant with H_2O_2 production during an incompatible plant-pathogen interaction (Grant M et al. 2000). However, DPI inhibited bacteria-induced ROS production but not calcium release, placing calcium upstream of H_2O_2 production. Thus, it seems likely that NADPH oxidase is regulated by calcium during biotic stresses. It is not yet known whether calcium also regulates NADPH oxidase activity increased by abiotic stresses (Rao et al. 1996; A-H- Mackerness et al. 1999b; Jiang and Zhang 2002; Wohlgemuth et al. 2002).

Pharmacological data implicate reversible protein phosphorylation and the action of G proteins in the regulation of NADPH oxidase activity during plant-pathogen responses (Lamb and Dixon 1997). Recent genetic and biochemical data have identified a protein kinase required for H_2O_2 synthesis following ABA treatment in *Arabidopsis* guard cells (Mustilli et al. 2002), and highlighted a role for Rac and Rho GTP-binding proteins in regulating NADPH oxidase activity during PCD and anoxia responses in rice and *Arabidopsis* respectively (Kawasaki et al. 1999; Baxter-Burrell et al. 2002).

An essential requirement for NADPH oxidase during pathogen-induced H_2O_2 production has been demonstrated recently by functional genomic experiments. Using a reverse genetic approach, *rboh* knock-out mutants were identified in an *Arabidopsis* T-DNA insertion population (Tissier et al. 1999). Characterisation of single and double mutants subsequently demonstrated that *AtRbohD* is largely responsible for H_2O_2 produced in response to bacterial challenge, whereas *AtrbohF* appears more important for fungal HR. Although it seems clear that a lack of *rboh* genes compromises H_2O_2 production, the effects on defence responses are less clear. Simon-Plas et al. (2002) generated antisense tobacco cells to reduce the expression of *NtRbohD*. These antisense cell lines had reduced H_2O_2 production after elicitation, indicating the requirement for NADPH oxidase. However, medium alkalinisation still occurred, demonstrating that NADPH oxidase functions downstream of extracellular alkalinisation. NADPH oxidase knock-out and knock-down mutants will be very useful tools to dissect the roles of individual *rboh* proteins in mediating cellular responses to various stimuli. Moreover, cloning, mutagenesis and expression of the individual genes may reveal biochemical information regarding potential activation mechanisms.

Bolwell's group has proposed an alternative, and perhaps complementary, regulated source of H_2O_2, at least during the oxidative burst – a cell wall peroxidase (Bolwell et al. 2002). In suspension cultures of French bean cells, the fungal elicited oxidative burst is not inhibited by DPI, but is inhibited by peroxidase inhibitors, and a cell wall peroxidase has been purified (Blee et al. 2001). Furthermore, this oxidative burst is entirely dependent on extracellular alkalinisation, as the oxidative burst does not occur when using a cell culture medium of pH 6. Moreover, simultaneous release of a reductant also occurs, the identity of which is not yet clear. Importantly, *Arabidopsis* plants transformed with an antisense construct to the bean peroxidase are hypersensitive to bacterial and fungal pathogens (Bolwell et al. 2002), indicating a biological role for this enzyme.

Another source of H_2O_2, oxalate oxidase, converts oxalate and oxygen to H_2O_2 and carbon dioxide. Germin-like oxalate oxidases in barley mediate H_2O_2 production in response to attack by a fungal pathogen (Zhang et al. 1995). Copper-containing amine oxidases located in the cell wall have been proposed as a source of ROS in elicitor-treated epidermal cells of tobacco (Allan and Fluhr 1997). Amine oxidases catalyse the oxidation of various amines to yield ammonia and H_2O_2 (Wojtaszek 1997). Further work is required to elucidate the role of oxalate oxidases and amine oxidases in ROS generation in response to different stresses.

5.3.3 Removal of ROS

Redox imbalance can also result from a reduction in antioxidant activity. Antioxidant defences are both constitutive and inducible by oxidative stress. Plant cells are particularly rich in antioxidants, and their activity and location will affect the concentration of H_2O_2 at any given time and place. Thus, high antioxidant levels might localise H_2O_2 within cellular microdomains, emanating from the point of origin or entry of H_2O_2 (Neill et al. 2002c).

ROS can be removed either via non-enzymatic or enzymatic mechanisms. Non-enzymatic antioxidants include vitamin C (ascorbate), vitamin E (tocopherol), glutathione, flavonoids, alkaloids, and carotenoids (Bray et al. 2002). Millimolar concentrations of ascorbate and glutathione are found in chloroplasts and other cellular compartments, as well as the apoplast (Noctor and Foyer 1998; Smirnoff, 2000), buffering cells against oxidative damage (Horemans et al. 2000). The *Arabidopsis vtc-1* mutant, deficient in ascorbate biosynthesis, has increased sensitivity to ozone, UV-B, and sulphur dioxide (Conklin et al. 1996). However, other ascorbate-deficient mutants are not hypersensitive to ozone (Conklin et al. 2000).

The tripeptide glutathione (γ-Glu-Cys-Gly, GSH) is a major redox buffer ubiquitous in aerobic cells (Foyer et al. 2001). Glutathione reacts with H_2O_2, being oxidised to GSSG, and functions in the ascorbate-glutathione cycle (see below). Increased levels of GSH have been found following chilling, heat shock, pathogen attack and drought stress (Noctor et al. 2002), and the activity of GSH biosynthetic enzymes is increased during environmental stress (Vanacker et al. 2000; Vernoux et al. 2002). Manipulation of GSH content via alteration of the activity of enzymes regulating GSH synthesis is a key research priority for enhancing oxidative stress tolerance.

Relatively little detail is known concerning flavonoids and carotenoids in ROS removal in plants. However, overexpression of β-carotene hydroxylase in *Arabidopsis*, leading to increased chloroplast xanthophyll content, resulted in enhanced tolerance towards oxidative stress induced in high light (Davison et al. 2002).

Enzymatic ROS scavenging mechanisms in plants include SOD (superoxide dismutase), present in many cellular compartments; catalase, located in peroxisomes; and the ubiquitous ascorbate-glutathione cycle. SOD catalyses the dismutation of superoxide to H_2O_2, and is thus one of the primary mediators of H_2O_2 production from intracellular sources of superoxide. Unlike most organisms, plants have multiple forms of the different types of SODs encoded by multiple genes (Scandalios 2002a). The various forms of SOD that occur in cells are categorised by the metal co-factor of the enzyme, e.g. Cu/Zn, Mn, Fe(III), or Ni (II/III). Catalase removes H_2O_2 by degrading it to water and oxygen. Catalase is located mainly in peroxisomes, where H_2O_2 synthesis occurs (see section 5.3.2). Catalase is thus well-positioned to remove excess H_2O_2 before it can leak out into other parts of the cell.

The ubiquity of the ascorbate-glutathione (Asc-GSH) cycle and the high concentrations of ascorbate and glutathione indicate the critical importance of this cycle as a regulator of cellular oxidative balance (Noctor and Foyer 1998). The Asc-GSH cycle involves metabolism of H_2O_2 via interactions between the antioxidant enzymes ascorbate peroxidase (APX), glutathione reductase (GR), and dehydroascorbate reductase (DHAR) (see Polle 1996). Dehydroascorbate is reduced by GSH to ascorbate, mediated by DHAR (and GSH becomes oxidised to GSSG). GSSG is then reduced to GSH requiring the action of GR and the consumption of NADPH, whilst ascorbate is oxidised to dehydroascorbate via APX, which also reduces H_2O_2 to water. Therefore, these enzymes are key targets for manipulating levels of ROS within cells.

Antioxidant enzyme activity can be modulated by different stimuli. Gibberellin decreases antioxidant activity (catalase, SOD and APX) in barley aleurone cells, thus increasing H_2O_2 levels and initiating cell death (Fath et al. 2001). On the other hand, antioxidant activity was increased by drought stress in several plant species (Zhu and Scandalios 1994; Gogorcena et al. 1995). Exposure of plants to high intensity light also led to an accumulation of transcripts encoding SOD (Tsang et al. 1991; Mishra et al. 1995) and APX (Karpinski et al. 1999), and heat tolerance was associated with increased SOD activity in tobacco (Tsang et al. 1991).

Specific roles for antioxidant enzymes have been explored via transgenic approaches. Over-expression of tobacco chloroplastic Cu/Zn SOD did not alter tolerance towards oxidative stress, suggesting that other antioxidant mechanisms might be limiting (Allen 1995). However, expression of a pea chloroplastic Cu/Zn SOD in tobacco did result in increased resistance to methyl-viologen-induced membrane damage (Allen 1995). It is possible that intracellular location is important for physiological effects. That SOD can protect against oxidative stress is apparent from the work of Zhu and Scandalios (1992). Yeast cells lacking MnSOD, and thus susceptible to oxidative stress, became more resistant to oxidative stress after transformation and expression of maize MnSOD, implying functional conservation between species.

Catalase was found to be indispensable for oxidative stress tolerance in transgenic tobacco. Willekens et al. (1997) showed that plants antisensed to catalase generated enhanced levels of H_2O_2 in response to both abiotic and biotic stresses. Reduced catalase activity resulted in the induction of other antioxidants (APX and GPX), suggesting a compensatory mechanism. Recently, Rizhsky et al. (2002) found that plants lacking both APX and catalase were less sensitive to oxidative stress, as assessed by paraquat-induced cell death, when compared to plants antisensed to catalase alone. Moreover, these double antisense plants were not compensated by an increase in the levels of other antioxidants. However, their photosynthetic activity was decreased, suggesting that reduction in the two key antioxidant enzymes result in the suppression of ROS production via chloroplasts (Rizhsky et al. 2002). Thus, suppression of photosynthetic activity during periods of environmental stress may offer a novel approach to oxidative stress tolerance.

The phenomena of acclimation and cross-tolerance have been linked to oxidative stress (Bowler and Fluhr, 2000). Increased ROS generation and induction of antioxidant genes during stress do indicate an involvement of oxidative stress, and it will be informative to determine whether acclimation still occurs in the absence of such oxidative stress, for example, in *rboh* knock-out mutants or in plants over-expressing antioxidant enzymes. The fact that some cellular defence genes are induced by several stresses, apparently independently of as well as via ROS, suggests that cellular responses are complex. Moreover, in field situations, plants are likely to be exposed to a number of stresses concurrently, e.g. cold, UV, heat/drought etc., and the interactions of such stresses with endogenous processes are yet to be unravelled.

5.4 Cellular responses

5.4.1 Effects on gene expression

Several studies have indicated a requirement for ROS signalling in the induction of genes induced by a range of stimuli such as pathogen challenge or exposure to UV or ozone (Neill et al. 2002b; Vranova et al. 2002b). Most data relate to H_2O_2, although there are some suggesting that $O_2^{\cdot-}$ is the key molecule (Jabs et al. 1997). These experiments, using treatments that inhibit H_2O_2 production or facilitate its removal with scavengers such as catalase, have identified genes encoding antioxidant enzymes such as APX as well as those encoding cellular defence proteins such as pathogenesis-related (PR) proteins, glutathione S-transferase (GST) and phenylalanine ammonia-lyase (PAL) (e.g. Levine et al. 1994; Desikan et al. 1998b; Karpinski et al. 1999). Alternatively, direct effects of H_2O_2 on gene expression have been determined. Clearly, H_2O_2 application may not induce the same responses as when generated internally – exogenous H_2O_2 is degraded very rapidly, H_2O_2 that enters the cell may not reach the same sub-cellular compartment(s) in which is it is generated or in which it accumulates following stimulation, and H_2O_2 challenge in the absence of other intracellular events that might normally be co-stimulated along with H_2O_2 production may not accurately reflect cellular responses during stress. Nevertheless, exposure to H_2O_2 has identified genes such as a receptor kinase (Desikan et al. 2000), annexin (Gidrol et al. 1996) as well as a peroxisome biogenesis gene (Lopez-Huertas et al. 2000) as being directly inducible by H_2O_2. Induction of peroxisome biogenesis suggests that peroxisome proliferation may be a key response to oxidative stress.

More recent experiments have used cDNA profiling and DNA microarray approaches to analyse large-scale gene expression in response to ROS. In order to clarify the role of ROS in pathogen-defence signalling, Durrant et al. (2000) performed a cDNA-AFLP analysis of tobacco genes that were induced by fungal challenge, in the presence or absence of DPI. Most of the pathogen-induced genes did not require ROS production, although direct induction of these genes by ROS was not assessed (Durrant et al. 2000). This implies that even during a defence response where ROS generation is a key step, cellular responses that ensue are not solely determined by ROS.

Using *Arabidopsis* suspension cultures, Desikan et al. (2001b) found that 1-2 % of the genes on the microarray utilised were altered in their expression following exposure to H_2O_2. Of the 175 genes that were regulated, 113 were induced and 62 repressed by H_2O_2. Genes encoding proteins with antioxidant properties were induced as well as a substantial number encoding proteins that had functions in other non-oxidative stress and cell defence processes. Signalling genes that were induced include those encoding a calmodulin, protein kinase, tyrosine phosphatase, histidine kinase, and small GTP binding protein. The expression of genes encoding transcription factors was also increased by H_2O_2, suggesting that downstream genes are also likely to be regulated. Expression analysis of a small sub-set of the H_2O_2-induced genes showed that some of them were also induced by other stimuli that involve oxidative stress, e.g. ABA, UV-B and elicitor treatments, revealing that expression of these genes occurred both via, and independently of, H_2O_2 (Desikan et al. 2001b).

Oxidative stress has also been shown to regulate the expression of some of the yeast genome (Gasch et al. 2000). Exposure of yeast cells to H_2O_2 induced the expression of genes involved in detoxification of ROS (such as catalase and SOD), as well as those involved in cellular redox reactions (such as thioredoxins and glutathione reductase). Interestingly, the use of yeast mutants deficient in transcription factors showed that most of the ROS-induced genes were dependent for their induction on the redox-active transcription factor, Yap1. Comparison of oxidative stress genes in yeast and plants may reveal common mechanisms.

Acclimation tolerance develops when a plant is exposed to sub-lethal doses of one stress that subsequently protects it from a further, normally lethal, dose of the same stress. It has been suggested that oxidative stress responses are essential to both abiotic and biotic stress tolerance in plants (Bowler and Fluhr 2000). However, the molecular mechanisms by which plants acclimate to oxidative stress are not very clear. Vranova et al. (2002a) pre-treated tobacco plants with sub-lethal doses of methyl viologen (MV, which generates superoxide), followed by a larger dose of MV, and followed global changes in gene expression using both mRNA differential display and microarray techniques. Approximately 2 % of the tobacco genes were altered in their expression in acclimated leaves, including several not previously associated with oxidative stress. Genes with predicted cytoprotective or detoxifying functions (such as an ABC transporter) and signal transduction (such as a calcium sensor-interacting protein kinase) were up-regulated in acclimated leaves, implying a variety of cellular responses during acclimation tolerance.

H_2O_2 has also been implicated as a signal mediating systemic resistance to both abiotic and biotic stresses. Karpinski et al. (1999) showed that exposure of plants harbouring a transgenic APX2-LUC gene to high intensity light caused an increase in APX2-LUC expression. Moreover, a systemic signal involving H_2O_2, was also generated, inducing an acclimatory response in untreated parts of the plant, a phenomenon termed "systemic acquired acclimation" (Karpinski et al. 1999). H_2O_2 generation also occurs both locally and systemically in response to wounding (Orozco-Cardenas and Ryan 1999), and recent work has shown that this requires H_2O_2 as a second messenger, mediating the expression of various defence-related genes in systemic parts of tomato plants (Orozco-Cardenas et al. 2001). Previously, Alvarez et al. (1998) had showed that the oxidative burst in pathogen challenged *Arabidopsis* leaves activates a secondary systemic burst in distal parts of the plant, leading to systemic immunity via the expression of defence-related genes. It is possible that H_2O_2 is not the primary signal that is transmitted, and interactions with other signalling intermediates such as salicylic acid (SA) could also be involved.

Heterologous systems have also been used to elucidate the function of oxidative stress tolerance in plants. The H_2O_2 inducible annexin-like protein identified in *Arabidopsis* rescued *E. coli oxyR* mutants from oxidative stress (Gidrol et al. 1996). OxyR is a transcriptional regulator of oxidative stress-induced defence genes in *E. coli*; thus, mutants lacking this gene are unable to grow at high H_2O_2 concentrations (Gidrol et al. 1996). Although the exact mechanisms by which annexin counteracts oxidative stress in plants are not known, the presence of domains in the protein also present in plant peroxidases suggested some form of antioxidative role (Gidrol et al. 1996). In other work, Belles-Boix et al. (2000) identified the *Arabidopsis* CEO1 protein, which protected yeast against oxidative damage. *CEO1* appears to be part of a gene family unique to plants and was not induced by oxidative stress. Whether the annexin and CEO1 proteins have similar physiological functions in plants remains to be seen.

Direct effects of H_2O_2 on the proteome have been demonstrated. Godon et al. (1999) identified the "H_2O_2 stimulon" in yeast, which included several heat shock proteins and antioxidant enzymes; as well as enzymes involved in protein degradation pathways. The effects of oxidative stress on the *Arabidopsis* mitochondrial proteome have been reported by Sweetlove et al. (2002). Sub-sets of proteins were identified that were up-regulated, down-regulated, or degraded following exposure of *Arabidopsis* cells to H_2O_2. Two classes of antioxidant defence proteins, peroxiredoxins and protein disulphide isomerase, were found to be increased by oxidative stress. Proteins associated with the TCA cycle were down-regulated, probably reflecting down-regulation of ATP synthesis. This is the first study to characterise the effects of oxidative stress on the protein profile of a sub-cellular organelle. It will be important to repeat this for other organelles as well as the whole cell, in order to compare and contrast the global effects of oxidative stress on cellular function.

Potential new insights into ROS signalling may be inferred from a recent microarray study by Moseyko et al. (2002). The expression of many genes was altered in *Arabidopsis* roots in response to gravity, and over 20% associated with oxidative stress and plant defence. Research elsewhere has implicated H_2O_2 as a mediator of gravitropic responses (Joo et al. 2001); thus, these data may reflect endogenous H_2O_2 synthesis and action. In other work, Swidzinski et al. (2002) identified several oxidative stress-related genes that were up-regulated following heat-treatment of *Arabidopsis* cells, potentially reflecting the involvement of ROS in heat responses. It is possible that further transcriptomic studies will identify novel genes whose functions were not previously associated with oxidative stress.

An alternative approach to study the effects of oxidative stress on the transriptome is to induce oxidative stress by a reduction of antioxidant activity. Gene expression profiles were monitored in tobacco plants antisensed to either catalase or ascorbate peroxidase (AS-CAT, AS-APX) or double antisense plants (dAS; Rizhsky et al. 2002). Both the single antisense plants had elevated expression of Cu/Zn SOD and glutathione reductase (GR), whereas the dAS plants had elevated expression of monodehydroascorbate reductase (MDAR), an enzyme involved in the regeneration of ascorbate. An increase in expression of enzymes involved in removing ROS could reflect a compensatory mechanism to cope with the oxidative stress. In similar work, tobacco plants deficient in catalase, that had elevated ROS levels in high light, accumulated genes encoding pathogen-responsive proteins, resulting in enhanced disease resistance (Chamnongpol et al. 1998).

These antioxidant-deficient plants offer an excellent tool to study the effects of oxidative stress on transcription and other cellular and whole plant responses. Even so, increased ROS due to depletion of one scavenging mechanism, e.g. peroxisomal catalase, may not have exactly the same effects as ROS induced at another cellular location, e.g. by plasma membrane NADPH oxidase. It is clear from these few large-scale gene expression experiments carried out so far that oxidative stress/ROS do alter the expression of much larger number of genes than had hitherto been identified. These genes include those encoding antioxidant enzymes, indicating a protective cellular response. Genes associated with other stress responses are also induced, suggesting a co-ordinating and complementary role for H_2O_2 in stress responses. Finally, signalling and other genes are also induced, perhaps reflecting a broader signalling role for H_2O_2.

Gene expression in response to oxidative stress may be co-ordinated via the interaction of transcription factors (TFs) with *cis*-elements common to the regulatory regions of these genes. There is some evidence for oxidative stress-responsive *cis*-elements in plants. The microarray analysis of H_2O_2-induced gene expression in *Arabidopsis* indicated potential H_2O_2-responsive *cis*-elements in genes regulated by H_2O_2 (Desikan et al. 2001b). One of these elements, the *as-1* promoter element, has high homology with the redox-sensitive AP-1 box (a *cis*-element) in mammals (Karin et al. 1997), and has also been found in other H_2O_2-inducible genes in plants (Desikan et al. 2001b), although recent experiments using transgenic plants over-expressing the *as-1* element indicate that oxidative species other than H_2O_2 activate this promoter (Garreton et al. 2002). Identification of H_2O_2-specific *cis*-elements in genes in plants is a research priority. Further bioinformatic analyses of all the H_2O_2 -responsive genes identified via transcriptomic analysis may indicate such regulatory sequences, and functional studies will be required to confirm their H_2O_2-responsiveness *in vivo*.

5.4.2 Signalling

5.4.2.1 Transcription

Although H_2O_2 is a signal molecule capable of effecting large changes in the transcriptome, it is not known whether it is actually the signal *per se,* or whether oxidation of molecular substrates by H_2O_2 (or other ROS) is required to generate an intracellular signal. Certainly, increased ROS in cellular compartments such as the mitochondrion or chloroplast results in new transcriptional profiles, so there must be signalling between these organelles and the nucleus. This might involve direct effects of H_2O_2 on TFs, activation by H_2O_2 of signalling pathways that result in altered activity of specific TFs and/or generation of secondary (and further) signalling molecules that in turn affect signalling pathways that then alter the activity or formation of TFs (Fig. 4).

Redox modulation of TF activity could potentially involve modifications of thiol residues altering protein conformation and therefore activity (see section 5.4.2.2). Such thiol modifications by H_2O_2 have been demonstrated *in vitro* for the yeast TF YAP-1 (Delauney et al. 2000); the situation *in vivo* is not yet known. TFs could also be activated by H_2O_2 via the activation of signalling proteins, such as protein kinases. A well-known signalling cascade in which signal perception leads to the activation of TF and thus alteration in gene expression is that involving mitogen activated protein kinases (MAPKs). Various groups have shown that H_2O_2 activates specific MAPKs in *Arabidopsis* and other species (Desikan et al. 1999; 2001a; Grant JJ et al. 2000; Kovtun et al. 2000; Samuel et al. 2000). However, neither the mechanism of activation nor the downstream targets of these MAPKs are yet known. Nevertheless, it seems likely that H_2O_2 activation of MAPKs is a central phenomenon mediating cellular responses to multiple stresses. Indeed, Kovtun et al. (2000) have shown that this can be the case. H_2O_2 activates the MAPKs AtMPK3 and AtMPK6 via the MAPK kinase kinase (MAPKKK) enzyme

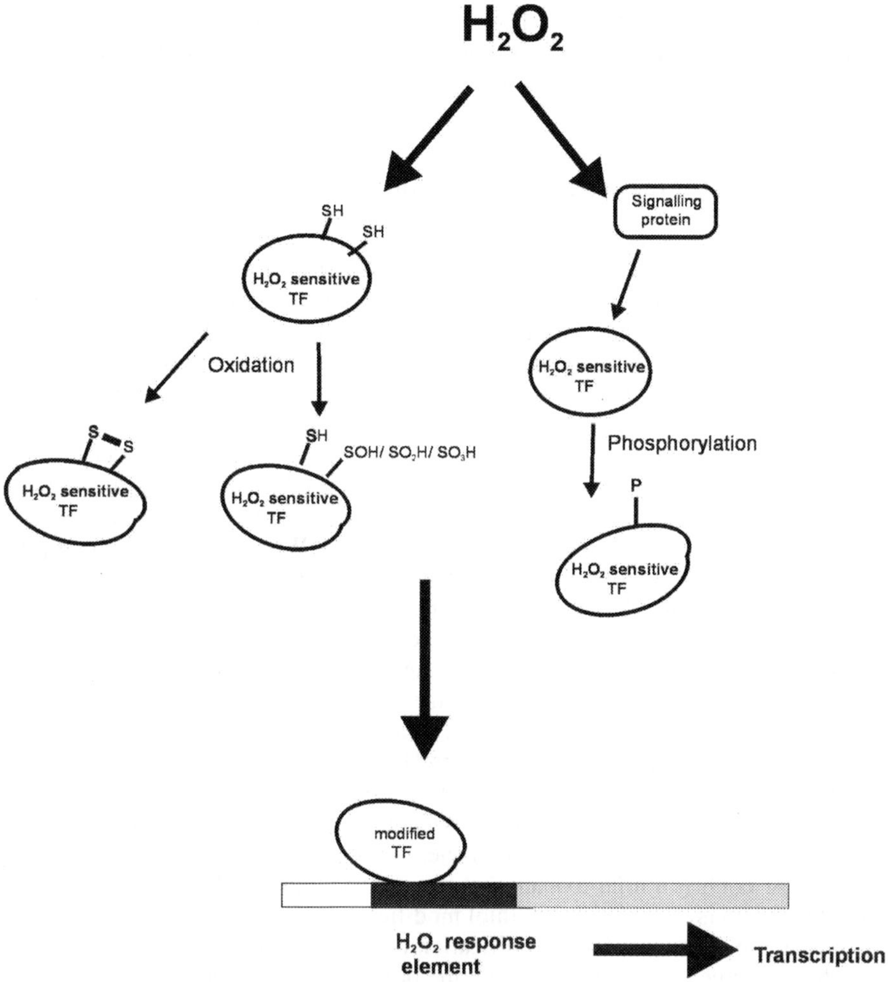

Fig. 4. Regulation of gene expression by H_2O_2. H_2O_2 can activate transcription by oxidising H_2O_2-responsive transcription factors (TFs), either via oxidation of individual cysteine thiols to yield thiol derivatives, or via oxidation of two adjacent thiols to form a disulfide bridge (Cooper et al. 2002). H_2O_2 can also activate a signalling protein such as a protein kinase that then phosphorylates a TF. The modified TF subsequently interacts with a "H_2O_2 response element" leading to regulation of gene expression.

ANP1, and moreover, plants over-expressing ANP1 were tolerant to heat shock, freezing and salt stress. In related work, Moon et al. (2002) have shown recently that H_2O_2 increased expression of the *Arabidopsis* NDP kinase 2, a kinase found to interact with the H_2O_2- activated MAPKs AtMPK3 and AtMPK6. Over-expression of AtNDPK2 down-regulated the accumulation of H_2O_2, which in turn enhanced tolerance to multiple stresses including cold, salt, and oxidative stress. The authors suggested that AtNDPK2 activated antioxidant genes that in turn mediated multiple stress tolerance. Together, these data suggest a scenario in which various stresses induce H_2O_2 generation, that in turn activates a MAPK signalling cascade that subsequently induces expression of antioxidant genes, thereby reducing H_2O_2 levels and restoring cellular homeostasis.

5.4.2.2 Effects of H_2O_2 on cell signalling

Whilst the above studies do indicate that oxidative stress has a significant impact on the genome, there is relatively little known about how the signal is transduced to alter gene expression. Recent work with yeast has indicated that a two component histidine kinase module can function as a peroxisensor (Singh 2000). Yeast mutants lacking the Sln1 histidine kinase gene were highly susceptible to H_2O_2 (Singh 2000). Similar two-component systems are already known as redox sensors in bacteria (Vranova et al. 2002b). Plants contain a range of histidine kinases and hybrid histidine kinases similar to Sln1 (which contains both a histidine kinase transmitter domain and a receiver domain). In *Arabidopsis*, some of these have been assigned functions as sensors and receptors for stimuli such as osmotic stress, ethylene, and cytokinin (Hwang et al. 2002). Whether some of these proteins can function as peroxisensors is currently under investigation.

Genetic approaches based on H_2O_2 sensitivity screens may help to find H_2O_2 sensors. An *Arabidopsis* mutant has been identified recently that has reduced expression of a H_2O_2-responsive marker gene, an enhanced oxidative burst in response to pathogen challenge, is hypersensitive to low titres of avirulent bacteria, and lacks a stomatal response to H_2O_2 (L Mur et al. pers. comm.). Using a genetic screen for altered root growth responses to H_2O_2, He et al. (2002) have also identified an *Arabidopsis* mutant lacking in H_2O_2 responsiveness. Preliminary data indicate that this mutant has increased antioxidant activity, suggesting that the lack of responses may reflect enhanced H_2O_2 generation.

Because H_2O_2 is a mild oxidant that can oxidise thiol (-SH) residues, it is possible that H_2O_2 is sensed via such thiol modification, potentially in proteins with a wide range of functions. Such modification depends on the pK_a of the thiol group and the molecular environment around the protein (Finkel 2000; Rhee et al. 2000; Danon 2002). Protein thiol groups that can be reversibly modified by ROS and reactive nitrogen species have recently been described as "nanotransducers" (Cooper et al. 2002), and potentially represent a very important cell signalling mechanism. Recent work has identified the *Arabidopsis* protein phosphatase 2C enzymes ABI1 and ABI2 as targets for H_2O_2 modification of cysteine residues *in vitro* (Meinhard et al. 2001; 2002). Identification of other H_2O_2-reactive proteins, perhaps using thiol-specific fluorescent dyes (Wu et al. 1998), followed by purification and mass spectrometric analysis, must be on the research agenda.

H_2O_2 interaction with the ubiquitous signalling messenger calcium is likely. Regulation of H_2O_2 homeostasis by calcium is complex, as H_2O_2 is positioned both upstream and downstream of calcium. In an *in vitro* study, Yang and Poovaiah (2002) have shown that calcium/calmodulin binds to and activates catalase. As calmodulin is present in peroxisomes, the cellular location of catalase, it is possible that calmodulin binding *in vivo* enhances degradation of H_2O_2 by activating calcium. In some signalling pathways such as in the guard cell response to ABA (see section 5.5.2), H_2O_2 activates calcium channels, thereby positioning calcium downstream of H_2O_2 (Pei et al. 2000). The activity of K^+ channels and H^+ATPases is also affected by H_2O_2 (Zhang et al. 2001a, b), but the mechanisms are not known.

Although to date most studies of H_2O_2 signalling, like those generally, have adopted a reductionist approach, it is well-recognised, that cell signalling is complex, with many parallel and interconnecting pathways. Indeed, it is already clear that H_2O_2 interacts closely with nitric oxide (see section 5.5), and probably with salicylic acid (SA) and jasmonic acid (JA) (A-H-Mackerness et al. 1999a).

5.5 H_2O_2 biology

5.5.1 Oxidative burst and PCD

The oxidative burst, in which large amounts of H_2O_2 are generated, is induced by pathogen challenge (Lamb and Dixon 1997) as well as by abiotic stresses such as ozone and UV (Rao et al. 1996; Wohlgemuth et al. 2002). During plant-pathogen interactions, the oxidative burst is part of a concerted series of events that involves induction of several defence responses including phytoalexin production and the Hypersensitive Response (HR; Lamb and Dixon 1997). Cell death occurring during the HR, potentially limiting the spread of disease from the infection point, is a genetically defined, programmed process (Programmed Cell Death, PCD; Greenberg 1997). During incompatible reactions (which result in HR), biphasic H_2O_2 production is observed, with one very rapid, and one prolonged burst of H_2O_2. However, during compatible interactions only the first burst of H_2O_2 occurs (Baker and Orlandi 1995). It is not yet known whether two different sources of H_2O_2 mediate these two distinct bursts, or the contribution of each burst to downstream events. The NADPH oxidase and peroxidase mutants will be helpful here. It is possible that pathogens themselves are capable of ROS generation and removal. For example, recent data indicate that the *slyA* gene is required for virulence and protection from oxidative stress in the phytopathogenic bacterium *Pseudomonas syringae* (Lindgren et al. 2002).

Although the oxidative burst is a primary response following pathogen challenge leading to cell death (Bolwell 1999), there is some data indicating that H_2O_2 is not required for PCD (Glazener et al. 1996; Ichinose et al. 2001). However, H_2O_2 does induce PCD in various systems (Levine et al. 1994; Desikan et al. 1998b; Solomon et al. 1999). A "presentation time" for exposure of cells to H_2O_2 is required, during which period transcription and translation are necessary (Desikan et al. 1998b; Solomon et al. 1999). Pharmacological data indicate that removal of H_2O_2 during pathogen or elicitor challenge reduces PCD (Levine et al. 1994; Desikan et al. 1998b). *Arabidopsis* knock-outs lacking functional *rboh* genes displayed reduced H_2O_2 generation and HR cell death following bacterial challenge (Torres et al. 2002). Tobacco plants that were antisense to either catalase or ascorbate peroxidase showed increased cell death to low doses of bacteria, compared to wild type plants (Mittler et al. 1999).

PCD occurs not only as a result of the oxidative burst following pathogen challenge, but also following exposure to abiotic stresses such as ozone. The ozone – induced oxidative burst results in a cell death process similar to the HR during plant-pathogen interactions. Ozone-induced cell death was inhibited by DPI in both *Arabidopsis* and tomato leaves (Wohlgemuth et al. 2002), suggesting a role for endogenous ROS. Interestingly, different *Arabidopsis* accessions appeared to generate either superoxide or H_2O_2. Developmentally-induced PCD has also been found to be driven by changes in redox balance. GA-induced PCD in barley aleurone was associated with increased ROS; however, this was due to a reduction in the antioxidant capacity rather than ROS generation (Fath et al. 2001). Thus, it is obvious that a close interplay between the oxidative and antioxidative capacity of the cell determines the cellular outcome of a physiological stimulus.

H_2O_2-induced PCD requires gene expression (Desikan et al. 1998b), but it is not yet known whether any PCD-specific genes exist that are regulated by H_2O_2. It remains to be seen whether any of the genes regulated by oxidative stress in *Arabidopsis* and tobacco are functionally involved in PCD (Desikan et al. 2001b; Vranova et al. 2002a). Identification and analysis of knock-out mutants in *Arabidopsis* insertion libraries should facilitate an analysis of the role of individual genes in PCD. Moreover, analysis of gene expression profiles in *rboh* mutants following exposure to pathogen challenge might identify those PCD-related requiring endogenous H_2O_2. Swidzinski et al. (2002) found oxidative stress-related genes up-regulated during heat-induced PCD in *Arabidopsis* cells, suggesting the involvement of ROS/ H_2O_2.

The parallels between animal and plant PCD are not clear. However, expression of animal cell death suppressor genes (Bcl-xl and Ced-9) in tobacco plants resulted in a suppression of oxidative stress-induced cell death (Mitsuhara et al. 1999). A role for mitochondria in H_2O_2-induced PCD is possible. Mitochondrial H_2O_2 production was increased by exposure of *Arabidopsis* cells to H_2O_2, resulting in altered mitochondrial function and PCD (Tiwari et al. 2002). Maxwell et al. (2002) reported that inhibition of mitochondrial electron transport or exposure to H_2O_2 induced intracellular H_2O_2 production and the expression of several PCD-associated genes. Expression of these genes was inhibited by an inhibitor of mitochondrial permeability pore formation, implying mitochondrion-nuclear signalling during H_2O_2-induced PCD. MAPK activation may also be linked to H_2O_2 generation and PCD. Over-expression of the MAPK kinases AtMEK4 and AtMEK5 in transgenic *Arabidopsis* plants induced HR-like cell death, preceded by the activation of endogenous MAPKs and ROS generation (Ren et al. 2002).

PCD regulation by H_2O_2 is likely to be complex, with interaction with other signalling intermediates and redox-active molecules such as nitric oxide (NO). Delledonne et al. (1998) showed that NO generation also occurs during the HR, with synergistic effects on cell death. Further work indicated that a critical ratio of H_2O_2 to NO is essential for PCD to occur in soybean cells. Reaction of $O_2^{\cdot-}$ with NO giving rise to peroxynitrite prevented PCD, and the rate of conversion of $O_2^{\cdot-}$ to either peroxynitrite or H_2O_2 determined the extent to which PCD occurred

(Delledonne et al. 2000). Bacterial challenge also elicited NO and H_2O_2 production in *Arabidopsis* cells. Here, however, cell death in the presence of H_2O_2 and NO was additive (Clarke et al. 2000), possibly reflecting differences in antioxidant capacity. For example, *Arabidopsis* protoplasts are more sensitive to H_2O_2 than are cells, reflecting their antioxidant status (Neill et al. 2002b).

5.5.2 H_2O_2 and stomata

Recent work has shown that H_2O_2 is an essential signal mediating stomatal closure induced by ABA. ABA is an endogenous anti-transpirant, synthesised in response to drought stress and inducing a range of survival responses including stomatal closure. Earlier work had shown that H_2O_2 induces stomatal closure (McAinsh et al. 1996) and that guard cells synthesise H_2O_2 in response to elicitor challenge (Allan and Fluhr 1997; Lee et al. 1999). The data of Pei et al. (2000) demonstrated that H_2O_2 is an endogenous component of ABA signalling in *Arabidopsis* guard cells. ABA increased H_2O_2 synthesis (via a putative NADPH oxidase, as observed by DPI inhibition of stomatal closure and requirement for NAD(P)H [Murata et al. 2001]), which induced stomatal closure, probably via activation of plasma membrane calcium channels (Pei et al. 2000). ABA-induced H_2O_2 production in guard cells has also been demonstrated for other species. Zhang et al. (2001c) showed that ABA-induced H_2O_2 synthesis occurs in *Vicia faba*, and suggested two sources of H_2O_2 – one located in the plasma membrane and another in the chloroplast. Pea guard cells also generate H_2O_2 in response to ABA, and ABA-induced stomatal closure is inhibited by removal of H_2O_2 (via catalase) or inhibition of synthesis (via DPI); NADPH oxidase-like genes are expressed in guard cells, and dark-induced closure also requires H_2O_2 synthesis (Desikan et al. unpublished). Identification and manipulation of guard cell sources of H_2O_2 is clearly important.

Various *Arabidopsis* mutants have been used to dissect ABA and H_2O_2 signalling in guard cells. In the *gca2* mutant, ABA increased H_2O_2 synthesis, but H_2O_2-induced calcium channel activation and stomatal closure were lacking (Pei et al. 2000), suggesting that the GCA2 protein is involved in H_2O_2 signalling. The two *Arabidopsis* H_2O_2 signalling mutants described in section 5.4.2.2 are deficient in guard cell H_2O_2 responses and will no doubt prove to be useful research tools. Reversible protein phosphorylation is central to guard cell signalling. Murata et al. (2001) used the ABA-insensitive *abi1* and *abi2* mutants, mutated in the ABI1 and ABI2 protein phosphatase 2C enzymes, to dissect H_2O_2 signalling in *Arabidopsis*. ABA-induced H_2O_2 generation was deficient in *abi1*, whereas *abi2* mutants synthesised H_2O_2 but could not respond to it, placing ABI1 upstream and ABI2 downstream of H_2O_2 synthesis. As mentioned earlier, the ABI1 and ABI2 proteins can be oxidised *in vitro* by H_2O_2, but whether this happens in guard cells is not yet known.

The ABA, H_2O_2 and guard cell story has recently expanded to include a protein kinase between ABA perception and H_2O_2 synthesis. Mustilli et al. (2002) identified an ABA responsive mutant and isolated the gene, *OST1*, by positional cloning. The gene encodes a protein kinase that is activated by ABA in both roots and guard cell protoplasts from wild type but not *ost1* plants. ABA-induced H_2O_2 synthesis was absent in *ost1* plants, although *ost1* stomata still closed in response to H_2O_2. It will be interesting to determine whether OST1 actually interacts with NADPH oxidase, leading to generation of H_2O_2 in guard cells.

As with other signalling systems, H_2O_2 is likely to interact with various signalling intermediates in guard cells. The recent findings that NO is a novel signal mediating ABA-induced stomatal closure (Neill et al. 2002a) indicate that, as with PCD, both H_2O_2 and NO appear to be made and to act in tandem.

5.5.3 H_2O_2 and roots

A new role for H_2O_2 in auxin signalling and gravitropism in maize roots was revealed recently by Joo et al. (2001). Gravity and asymmetric auxin application induced H_2O_2 generation, and moreover, asymmetric application of H_2O_2 promoted gravitropism. An intracellular source of H_2O_2 was indicated, as catalase application had no effect on gravitropism. The identification of *Arabidopsis* gravitropism-induced genes related to oxidative stress (Moseyko et al. 2002) may be indicative of a wider role for H_2O_2 in gravistimulation. NO has also been implicated in auxin effects on root growth (Pagnussat et al. 2002), suggesting, yet again, cross-talk between H_2O_2 and NO.

5.5.4 Anoxia and H_2O_2

Plants are commonly exposed to anoxic and hypoxic conditions due to flooding and poor soil drainage. Recent work by Baxter-Burrell et al. (2002) shows that regulation of H_2O_2 content by Rops (Rho-like small G proteins) is critical for oxygen deprivation tolerance in *Arabidopsis* seedlings. Rops have previously been shown to regulate various signalling processes in plants, including H_2O_2 generation (Yang 2002). The data by Baxter-Burrell et al. (2002) suggest a model in which oxygen deprivation activates Rop signalling to activate NADPH oxidase and hence H_2O_2 synthesis, resulting in the expression of oxygen deprivation-tolerance genes such as alcohol dehydrogenase. H_2O_2 also induces the expression of a gene encoding RopGAP, leading to the deactivation of Rop and subsequent reduction in H_2O_2. In previous work, Amor et al. (2000) had shown that pre-exposure of soybean cells to anoxic conditions protect against subsequent H_2O_2-induced cell death, via activation of peroxidases and alternate oxidase. It is possible that in soybean cells anoxia induces H_2O_2 synthesis that in turn induced peroxidases that were protective against subsequent H_2O_2 exposure.

5.6 Conclusions

Oxidative stress occurs as a consequence of several environmental factors that perturb the redox balance of plant cells. At its worst, oxidative stress causes cellular damage to biomolecules that may induce cell death. On the other hand, oxidative stress induces a range of cellular defence responses that are protective against various stresses, such that these responses may be key components of acclimation and cross-tolerance. ROS such as H_2O_2, and perhaps oxidative stress *per se*, due to the alterations in cellular redox environment, function as signals in plant cells, modulating cellular processes that may or may not directly operate as part of a stress response. Thus, oxidative stress activates cell signalling pathways that alter protein activities and transcription profiles. Developments in post-genomics technologies will drive the identification of more ROS-sensitive proteins and genes, and functional genomics approaches will facilitate analyses of the roles of these proteins and their regulation in cellular functions.

References

A-H-Mackerness S, Surplus SL, Blake P, John CF, Buchanan-Wollaston V, Jordan BR, Thomas B (1999a) Ultraviolet-B-induced stress and changes in gene expression in *Arabidopsis thaliana*:role of signalling pathways controlled by jasmonic acid, ethylene and reactive oxygen species. Plant Cell Environ 22:1413-1423

A-H-Mackerness S, Thomas B (1999b) Effects of UV-B radiation on plants:gene expression and signal transduction pathways. In Plant Responses to Environmental Stress. Smallwood MF, Calvert CM, Bowles DJ (eds), Bios Scientific Publishers, UK.

Allan CA, Fluhr R (1997) Two distinct sources of elicited reactive oxygen speices in tobacco epidermal cells. Plant Cell 9:1559-1572

Allen RD (1995) Dissection of oxidative stress tolerance using transgenic plants. Plant Physiol 107:1049-1054

Alvarez ME, Lamb C (1997) Oxidative burst-mediated defense responses in plant disease resistance. In:Oxidative Stress and the Molecular Biology of Antioxidant Defenses (ed:Scandalios JG), Cold Spring Harbor Laboratory Press, Cold Spring Harbor, NY, pp 815-839

Alvarez ME, Pennell RI, Meijer P-J, Ishikawa A, Dixon RA, Lamb C (1998) Reactive oxygen intermediates mediate a systemic signal network in the establishment of plant immunity. Cell 92:773-784

Amicucci E, Gaschler K, Ward JM (1999) NADPH oxidase genes from tomato (*Lycopersicon esculentum*) and curly-leaf pondweed (*Potamogeton crispus*). Plant Biol 1:524-528

Amor Y, Chevion M, Levine A (2000) Anoxia pretreatment protects soybean cells against H_2O_2-induced cell death:possible involvement of peroxidases and of alternative oxidase. FEBS Lett 477:175-180

Baker CJ, Orlandi EW (1995) Active oxygen in plant pathogenesis. Annu Rev Phytopathol 33:299-321

Baxter-Burrell A, Yang Z, Springer PS, Bailey-Serres J (2002) RopGAP4-dependent Rop GTPase rheostat control of *Arabidopsis* oxygen deprivation tolerance. Science 296:2026-2028

Belles-Boix E, Babiychuk E, Van Montagu M, Inze D, Kushnir S (2000) CEO1, a new protein from *Arabidopsis thaliana*, protects yeast against oxidative damage. FEBS Lett 482:19-24

Bestwick C, Bolwell P, Mansfield J, Nicole M, Wojtaszek P (1999) Generation of the oxidative burst – scavenging for the truth. Trends Plant Sci 4:88-89

Blee KA, Jupe SC, Richard G, Zimmerlin A, Davies DR, Bolwell GP (2001) Molecular identification and expression of the peroxidase responsible for the oxidative burst in French bean (*Phaseolus vulgaris* L.) and related members of the gene family. Plant Mol Biol 47:607-620

Bolwell GP (1999) Role of active oxygen species and NO in plant defence responses. Curr Opin Plant Biol 2:287-294

Bolwell GP, Bindschedler LV, Blee KA, Butt VS, Davies DR, Gardner SL, Gerrish C, Minibayeva F (2002) The apoplastic oxidative burst in response to biotic stress in plants:a three-component system. J Exp Bot 53:1367-1376

Bowler C, Fluhr R (2000) The role of calcium and activated oxygens as signals for controlling cross-tolerance. Trends Plant Sci 5:241-245

Bray EA, Bailey-Serres J, Weretilnyk E (2000) Responses to Abiotic Stress. In Biochemistry and Molecular Biology of Plants (eds:Buchanan B, Gruissem W, Jones, R) American Society of Plant Biologists, Rockville, USA, pp 1158-1203

Chamnongpol S, Willekens H, Moeder W, Langebartels C, Sandermann H Jr, Van Montagu M, Inze D, Van Camp W (1998) Defense activation and enhanced pathogen tolerance induced by H_2O_2 in transgenic tobacco. Proc Natl Acad Sci USA 95:5818-5823

Clarke A, Desikan R, Hurst RD, Hancock JT, Neill SJ (2000) NO way back:nitric oxide and programmed cell death in *Arabidopsis thaliana* suspension cultures. Plant J 24:667-677

Conklin PL, Williams EH, Last RL (1996) Environmental stress sensitivity of an ascorbic acid-deficient *Arabidopsis* mutant. Proc Natl Acad Sci USA 93:9970-9974

Conklin PL, Saracco SA, Norris SR, Last RL (2000) Identification of ascorbic acid-deficient *Arabidopsis thaliana* mutants. Genetics 154:847-856

Cooper CE, Patel RP, Brookes PS, Darley-Usmar VM (2002) Nanotransducers in cellular redox signalling:modification of thiols by reactive oxygen and nitrogen species. Trends Biochem Sci 27:489-492

Corpas FJ, Barroso JB, del Rio LA (2001) Peroxisomes as a source of reactive oxygen species and nitric oxide signal molecules in plant cells. Trends Plant Sci 6:145-150.

Czernic P, Visser B, Sun W, Savoure A, Deslandes L, Marco Y, Van Montagu M, Verbruggen N (1999) Characterisation of an *Arabidopsis thaliana* receptor-like protein kinase gene activated by oxidative stress and pathogen attack. Plant J 18:321-327

Danon A (2002) Redox reactions of regulatory proteins:do kinetics promote specificity? Trends Biochem Sci 27:197-203

Dat J, Vandenabeele S, Vranova E, Van Montagu M, Inze D, Van Breusegem FV (2000) Dual action of the active oxygen species during plant stress responses. CMLS Cell Mol Life Sci 57:779-795

Davison PA, Hunter CN, Horton P (2002) Overexpression of β carotene hydroxylase enhances stress tolerance in *Arabidopsis*. Nature 418:203-206

Delaunay A, Isnard A-D, Toledano MB (2000) H_2O_2 sensing through oxidation of the Yap1 transcription factor. EMBO J 19:5157-5166

Delledonne M, Xia Y, Dixon RA, Lamb C (1998) Nitric oxide functions as a signal in plant disease resistance. Nature 394:585-588

Delledonne M, Zeier J, Marocco A, Lamb C (2001) Signal interactions between nitric oxide and reactive oxygen intermediates in the plant hypersensitive disease resistance response. Proc Natl Acad Sci USA 98:13454-13459

Desikan R, Burnett E, Hancock JT, Neill SJ (1998a) Harpin and hydrogen peroxide induce the expression of a homologue of gp91-*phox* in *Arabidopsis thaliana* suspension cultures. J Exp Bot 49:1767-1771

Desikan R, Reynolds A, Hancock JT, Neill SJ (1998b) Harpin and hydrogen peroxide both initiate programmed cell death but have differential effects on gene expression in *Arabidopsis* suspension cultures. Biochem J 330:115-120

Desikan R, Clarke A, Hancock JT, Neill SJ (1999) H_2O_2 activates a MAP kinase-like enzyme in *Arabidopsis thaliana* suspension cultures. J Exp Bot 50:1863-1866

Desikan R, Neill SJ, Hancock JT (2000) Hydrogen peroxide-induced gene expression in *Arabidopsis thaliana.* Free Rad Biol Med 28:773-778

Desikan R, Hancock JT, Ichimura K, Shinozaki K, Neill SJ (2001a) Harpin induces activation of the Arabidopsis mitogen-activated protein kinases AtMPK4 and AtMPK6. Plant Physiol 126:1579-1587

Desikan R, A-H Mackerness S, Hancock JT, Neill SJ (2001b) Regulation of the *Arabidopsis* transcriptome by oxidative stress. Plant Physiol 127:159-172

Doke N (1983) Involvement of superoxide anion generation in the hypersensitive response of potato tuber tissues to infection with an incompatible race of *Phytophthora infestans* and to the hyphal wall components. Physiol Plant Pathol 23:345-357

Durrant WE, Rowland P, Piedras P, Hammond-Kosack KE, Jones JDG (2000) cDNA-AFLP reveals a striking overlap in race-specific resistance and wound-response gene expression profiles. Plant Cell 12:963-977

Fath A, Behke PC, Jones RL (2001) Enzymes that scavenge reactive oxygen species are down-regulated prior to gibberellic acid-induced programmed cell death in barley aleurone. Plant Physiol 126:156-166

Finkel T (2000) Redox-dependent signal transduction. FEBS Letts 476:52-54

Finkel T, Holbrook NJ (2000) Oxidants, oxidative stress and the biology of ageing. Nature 408:239-247

Foyer CH, Theodoulou FL, Delrot S (2001) The functions of inter- and intracellular glutathione transport systems in plants. Trends Plant Sci 6:486-492

Garreton V, Carpinelli J, Jordana X, Holuigue L (2002) The as-1 promoter element is an oxidative stress-responsive element and salicylic acid activates it via oxidative species. Plant Physiol 130:1516-1526

Gasch A, Spellman P, Kao C, Harel O, Eisen M, Storz G, Botstein D, Brown P (2000) Genomic expression programs in the response of yeast cells to environmental changes. Mol Biol Cell 11:4241-4257

Gidrol X, Sabelli PA, Fern YS, Kush AK (1996) Annexin-like protein from *Arabidopsis thaliana* rescues *oxyR* mutant of *Escherichia coli* from H_2O_2 stress. Proc Natl Acad Sci USA 93:11268-11273

Glazener JA, Orlandi EW, Baker CJ (1996) The active oxygen response of cell suspensions to incompatible bacteria is not sufficient to cause hypersensitive cell death. Plant Physiol 110:759-763

Godon C, Lagniel G, Lee J, Buhler J-M, Kieffer S, Perrot M, Boucherie H, Toledano MB, Labarre J (1998) The H_2O_2 stimulon in *Saccharomyces cerevisiae*. J Biol Chem 34; 22480-22489

Gogorcena Y, Iturber-Ormaetxe I, Escuredo PR, Becana M (1995) Antioxidant defenses against activated oxygen in pea nodules subjected to water stress. Plant Physiol 108:753-759

Grant JJ, Yun B-W, Loake GJ (2000) Oxidative burst and cognate redox signalling reported by luciferase imaging:identification of a signal network that functions independently of ethylene, SA and Me-JA but is dependent on MAPKK activity. Plant J 24:569-582

Grant M, Brown I, Adams S, Knight M, Ainslie A, Mansfield J (2000) The *RPM1* plant disease resistance gene facilitates a rapid and sustained increase in cytosolic calcium that is necessary for the oxidative burst and hypersensitive cell death. Plant J 23:441-450

Greenberg JT(1997) Programmed cell death in plant-pathogen interactions. Annu Rev Plant Physiol Plant Mol Biol 48:525-545

Guan LM, Zhao J, Scandalios JG (2000) *Cis*-elements and trans-factors that regulate expression of the maize Cat1 antioxidant gene in response to ABA and osmotic stress:H_2O_2 is the likely intermediary signaling molecule for the response. Plant J 22:87-95

Harding SA, Oh S-H, Roberts DM (1997) Transgenic tobacco expressing a foreign calmodulin gene shows an enhanced production of active oxygen species. EMBO J 16:1137-1144

He J-h, Dong F-C, An G-y, Song C-P (2002) Isolation and characterisation of reactive oxygen species-insensitive mutants in *Arabidopsis thaliana*. Xibei Zhiwu Xuebao 22:496-504

Hernandez JA, Corpas FJ, Gomez M, del Rio LA, Sevilla F (1993) Salt-induced oxidative stress in chloroplasts of pea plants. Plant Physiol 89:103-110

Horemans N, Foyer CH, Asard H (2000) Transport and action of ascorbate at the plant plasma membrane. Trends Plant Sci 5:263-267

Hwang I, Chen H-C, Sheen J (2002) Two-component signal transduction pathways in *Arabidopsis*. Plant Physiol 129:500-515

Ichinose Y, Andi S, Doi R, Tanaka R, Taguchi F, Sasabe M, Toyoda K, Shiraishi T, Yamada T (2001) Generation of hydrogen peroxide is not required for harpin-induced apoptotic cell death in tobacco BY-2 cell suspension cultures. Plant Physiol Biochem 39:771-776

Jabs T, Tschope M, Collinge C, Hahlbrock K, Scheel D (1997) Elicitor-stimulated ion fluxes and $O_2^{\cdot-}$ from the oxidative burst are essential components in triggering defense gene activation and phytoalexin synthesis in parsley. Proc Natl Acad Sci USA 94:4800-4805

Jiang M, Zhang J (2001) Effect of abscisic acid on active oxygen species, antioxidant defence system and oxidative damage in leaves of maize seedlings. Plant Cell Physiol 42:1265-1273

Jiang M, Zhang J (2002) Involvement of plasma membrane NADPH oxidase in abscisic acid- and water stress-induced antioxidant defense in leaves of maize seedlings. Planta 215:1022-1030

Joo JH, Bae YS, Lee JS (2001) Role of auxin-induced reactive oxygen species in root gravitropism. Plant Physiol 126:1055-1060

Karin M, Liu Z, Zandi E (1997) AP-1 function and regulation. Curr Opin Cell Biol 9:240-246

Karpinski S, Reynolds H, Karpinska B, Wingsle G, Creissen G, Mullineaux P (1999) Systemic signaling and acclimation in response to excess excitation energy in *Arabidopsis*. Science 284:654-657

Kawasaki T, Henmi K, Ono E, Hatakeyama S, Iwano M, Satoh H, Shimamoto K (1999) The small GTP-binding protein Rac is a regulator of cell death in plants. Proc Natl Acad Sci USA 96:10922-10926

Keller T, Damude HG, Werner D, Doerner P, Dixon RA, Lamb C (1998) A plant homolog of the neutrophil NADPH oxidase gp91-*phox* subunit gene encodes a plasma membrane protein with Ca^{2+} binding motifs. Plant Cell 10:255-266.

Kovtun Y, Chiu W-L, Tena G, Sheen J (2000) Functional analysis of oxidative stress-activated mitogen-activated protein kinase cascade in plants. Proc Natl Acad Sci USA 97:2940-2945

Lamb C, Dixon RA (1997) The oxidative burst in plant disease resistance. Annu Rev Plant Physiol Plant Mol Biol 48:251-275

Lambeth JD (2002) Nox/Duox family of nicotinamide adenine dinucleotide (phosphate) oxidases. Curr Opin Hematol 9:11-17

Lee S, Choi H, Suh S, Doo I-S, Oh K-Y, Choi EJ, Taylor ATS, Low PS, Lee Y (1999) Oligogalacturonic acid and chitosan reduce stomatal aperture by inducing the evolution of reactive oxygen species from guard cells of tomato and *Commelina communis*. Plant Physiol 121:147-152

Levine A, Tenhaken R, Dixon R, Lamb C (1994) H_2O_2 from the oxidative burst orchestrates the plant hypersensitive disease resistance response. Cell 79:583-593

Lindgren PB, Jaokbek JL, Boutt EA, Libby SJ (2002) The *slyA* gene of the bacterial plant pathogen *Pseudomonas syringae* is required for virulence. Abstracts of the American Society for Microbiology 102:66-67

Lopez-Huertas E, Charlton WL, Johnson B, Graham IA, Baker A (2000) Stress induces peroxisome biogenesis genes. EMBO J 19:6770-6777

Maxwell DP, Nickels R, McIntosh L (2002) Evidence of mitochondrial involvement in the transduction of signals required for the induction of genes associated with pathogen attack and senescence. Plant J 29:269-279

McAinsh MR, Clayton H, Mansfield TA, Hetherington AM (1996) Changes in stomatal behaviour and guard cell cytosolic free calcium in response to oxidative stress. Plant Physiol 111:1031-1042

Meinhard M, Grill E (2001) Hydrogen peroxide is a regulator of ABI1, a protein phosphatase 2C from *Arabidopsis*. FEBS Letts 508:443-446

Meinhard M, Rodriguez PL, Grill E (2002) The sensitivity of ABI2 to hydrogen peroxide links the abscisic acid-response regulator to redox signalling. Planta 214:775-782

Mishra NP, Fatma T, Singhal GS (1995) Development of antioxidative defense systems of wheat seedlings in response to high light. Physiol Plant 95:77-82

Mitsuhara I, Malik KA, Miura M, Ohashi Y (1999) Animal cell-death suppressors Bcl-$x_L$and Ced-9 inhibit cell death in tobacco plants. Curr Biol 9:775-778

Mittler R (2002) Oxidative stress, antioxidants and stress tolerance. Trends Plant Sci 7:405-410

Mittler R, Herr EH, Orvar BL, van Camp W, Willekens H, Inze D, Ellis BE (1999) Transgenic tobacco plants with reduced capability to detoxify reactive oxygen intermediates are hyperresponsive to pathogen infection. Proc Natl Acad Sci USA 96:14165-14170

Moon H, Lee B, Choi G, Shin D, Prasad T, Lee O, Kwak S-S, Kim DH, Nam J, Bahk J, Hong JC, Lee SY, Cho MJ, Lim CO, Yun D-J (2003) NDP kinase 2 interacts with two oxidative stress-activated MAPKs to regulate cellular redox state and enhances multiple stress tolerance in transgenic plants. Proc Natl Acad Sci USA 100:358-363

Murata Y, Pei Z-M, Mori IC, Schroeder J (2001) Abscisic acid activation of plasma membrane Ca^{2+} channels in guard cells requires cytosolic NAD(P)H and is differentially disrupted upstream and downstream of reactive oxygen species production in *abi1-1* and *abi2-1* protein phosphatase 2C mutants. Plant Cell 13:2513-2523

Mustilli A-C, Merlot S, Vavasseur A, Fenzi F, Giraudat J (2002) *Arabidopsis* OST1 protein kinase mediates the regulation of stomatal aperture by abscisic acid and acts upstream of reactive oxygen species production. Plant Cell 14:3089-3099

Neill SJ, Desikan R, Clarke A, Hancock JT (2002a) Nitric oxide is a novel component of abscisic acid signalling in stomatal guard cells. Plant Physiol 128:13-16

Neill SJ, Desikan R, Clarke A, Hurst RD, Hancock JT (2002b) Hydrogen peroxide and nitric oxide as signalling molecules in plants. J Exp Bot 53:1237-1247

Neill S, Desikan R, Hancock J (2002c) Hydrogen peroxide signalling. Curr Opin Plant Biol 5:388-395

Noctor G, Foyer CH (1998) Ascorbate and glutathione:keeping active oxygen under control. Annu Rev Plant Physiol Plant Mol Biol 49:249-279

Noctor G, Gomez L,Vanacker H, Foyer CH (2002) Interactions between biosynthesis, compartmentation and transport in the control of glutathione homeostasis and signalling. J Exp Bot 53:1283-1304

Orozco-Cardenas ML, Ryan C (1999) Hydrogen peroxide is generated systemically in plant leaves by wounding and systemin via the octadecanoid pathway. Proc Natl Acad Sci USA. 96:6553-6557

Orozco-Cardenas ML, Narvaez-Vasquez J, Ryan CA (2001) Hydrogen peroxide acts as a second messenger for the induction of defense genes in tomato plants in response to wounding, systemin, and methyl jasmonate. Plant Cell 13:179-191

Pagnussat GC, Simontacchi M, Puntarulo S, Lamattina L (2002) Nitric oxide is required for root organogenesis. Plant Physiol 129:954-956

Pei Z-M, Murata Y, Benning G, Thomine S, Klusener B, Allen GJ, Grill E, Schroeder JI (2000) Calcium channels activated by hydrogen peroxide mediate abscisic signalling in guard cells. Nature 406:731-734

Polle A (1996) Mehler reaction:friend or foe in photosynthesis? Botanica Acta 109:84-89

Rao MV, Paliyath G, Ormrod DP (1996) Ultraviolet-B- and ozone-induced biochemical changes in antioxidant enzymes *Arabidopsis thaliana.* Plant Physiol 110:125-136

Reeves EP, Lu H, Lortat H, Jacobs L, Messina CGM, Bolsover S, Gabella G, Potma EO, Warley A, Roes J, Segal AW (2002) Killing activity of neutrophils is mediated through activation of proteases by K+ flux. Nature 416:291-297

Ren D, Yang H, Zhang S (2002) Cell death mediated by MAPK is associated with hydrogen peroxide production in *Arabidopsis.* J Biol Chem 277:559-565

Rhee SG, Bae YS, Lee S-R, Kwon J (2000) Hydrogen peroxide:a key messenger that modulates protein phosphorylation through cysteine oxidation. Science's STKE http://stke.sciencemag.org/cgi/content/full/OC_sigtrans;2000/53/pe1

Rizhsky L, Hallak-Herr E, Van Breusegem F, Rachmilevitch S, Barr JE, Rodermel S, Inze D, Mittler R (2002) Double antisense plants lacking ascorbate peroxidase and catalase are less sensitive to oxidative stress than single antisense plants lacking ascorbate peroxidase or catalase. Plant J 32:329-342

Sagi M, Fluhr R (2001) Superoxide production by plant homologues of the gp91^{phox} NADPH oxidase. Modulation of activity by calcium and by tobacco mosaic virus infection. Plant Physiol 126:1281-1290

Samuel MA, Miles GP, Ellis BE (2000) Ozone treatment rapidly activates MAP kinase signalling in plants. Plant J 22:367-376

Scandalios JG (2002a) The rise of ROS. Trends Biochem Sci 27:483-486

Scandalios JG (2002b) Oxidative stress responses – what have genome-scale studies taught us? Genome Biol 3:1019.1-1019.6

Schafer FQ, Buettner GR (2001) Redox environment of the cell as viewed through the redox state of the glutathione disulfide/glutathione couple. Free Rad Biol Med 11:1191-1212

Schroeder JI, Allen GJ, Hugouvieux V, Kwak JM, Waner D (2001) Guard cell signal transduction. Annu Rev Plant Physiol Plant Mol Biol 52:627-658

Simon-Plas F, Elmayan T, Blein J-P (2002) The plasma membrane oxidase Nt*rboh*D is responsible for AOS production in elicited tobacco cells. Plant J 31:137-147

Singh KK (2000) The *Saccharomyces cerevisiae* SLN1P-SSK1P two-component system mediates response to oxidative stress and in an oxidant-specific fashion. Free Rad Biol Med 29:1043-1050

Smirnoff N (2000) Ascorbic acid:metabolism and functions of a multi-facetted molecule. Curr Opin Plant Biol 3:229-235

Solomon M, Belenghi B, Delledonne M, Menachem E, Levine A (1999) The involvement of cysteine proteases and protease inhibitor genes in the regulation of programmed cell death in plants. Plant Cell 11:431-443

Sweetlove LJ, Heazlewood JL, Herald V, Holtzapffel R, Day DA, Leaver CJ, Millar AH (2002) The impact of oxidative stress on *Arabidopsis* mitochondria. Plant J 32:891-904

Swidzinski JA, Sweetlove LJ, Leaver CJ (2002) A custom microarray analysis of gene expression during programmed cell death in *Arabidopsis thaliana.* Plant J 30:431-446

The Arabidopsis Genome Initiative (2000) Analysis of the genome of the flowering plant *Arabidopsis thaliana*. Nature 408:796-815

Tissier AF, Marillonnet S, Klimyuk V, Patel K, Torres MA, Muphy G, Jones JDG (1999) Multiple independent defective *suppressor-mutator* transposon insertions in Arabidopsis:a tool for functional genomics. Plant Cell 11:1841-1852

Tiwari BS, Belenghi B, Levine A (2002) Oxidative stress increased respiration and generation of reactive oxygen species, resulting in ATP depletion, opening of mitochondrial permeability transition, and programmed cell death. Plant Physiol 128:1271-1281

Torres MA, Onouchi H, Hamada S, Machida C, Hammond-Kossack KE, Jones JDG (1998) Six *Arabidopsis thaliana* homologues of the human respiratory burst oxidase (gp91-*phox*). Plant J 14:365-370

Torres MA, Dangl JL, Jones JDG (2002) *Arabidopsis* gp91^{phox} homologues *AtrbohD* and *AtrbohF* are required for accumulation of reactive oxygen intermediates in the plant defense response. Proc Natl Acad Sci USA 99:517-522

Tsang EWT, Bowler C, Herouart D, Van Camp W, Villarroel R, Genetello C et al (1991) Differential regulation of superoxide dismutases in plants exposed to environmental stress. Plant Cell 3:783-792

Vanacker H, Carver TLW, Foyer CH (2000) Early H_2O_2 accumulation in mesophyll cells leads to induction of glutathione during the hypersensitive response in the barley-powdery mildew interaction. Plant Physiol 123:1289-1300

Vernoux T, Sanchez-Fernandez R, May M (2002) Glutathione biosynthesis in plants. In:Inze D, Montagu MV (eds) Oxidative stress in plants. Taylor and Francis, London, pp 297-311

Vranova E, Atichartpongkul S, Villarroel R, Van Montagu M, Inze D, Van Camp W (2002a) Comprehensive analysis of gene expression in *Nicotiana tabacum* leaves acclimated to oxidative stress. Proc Natl Acad Sci USA 99:10870-10875

Vranova E, Inze D, Van Breusegem F (2002b) Signal transduction during oxidative stress. J Exp Bot 53:1227-1236

Willekens H, Chamnongpol S, Davey M, Schraudner M, Langebartels C, Van Montagu M, Inze D, Van Camp W (1997) Catalase is a sink for H_2O_2 and is indispensable for stress defence in C-3 plants. EMBO J 16:4806-4816

Wohlgemuth H, Mittelstrass K, Kschieschan S, Bender J, Weigel H-J, Overmyer K, Kangasarvi J, Sandermann H, Langebartels C (2002) Activation of an oxidative burst is a general feature of sensitive plants exposed to the air pollutant ozone. Plant Cell Env 25:717-726

Wojtaszek P (1997) Oxidative burst:an early plant response to pathogen infection. Biochem J 322:681-692

Wu Y, Kwon K-S, Rhee SG (1998) Probing cellular protein targets of H_2O_2 with fluorescein-conjugated iodoacetamide and antibodies to fluorescein. FEBS Lett 440:111-115

Yang Z (2002) Small GTPases:versatile signaling switches in plants. Plant Cell S375-S388

Yang T, Poovaiah BW (2002) Hydrogen peroxide homeostasis:Activation of plant catalase by calcium/calmodulin. Proc Natl Acad Sci USA 99:4097-4102

Yoshioka H, Sugie K, Park HJ, Maeda H, Tsuda N, Kawakita K, Doke N (2001) Induction of plant gp91*phox* homologue by fungal cell wall arachidonic acid, and salicylic acid in potato. Mol Plant-Microbe Interact 14:725-736

Zhang X, Dong FC, Cao JF, Song CP (2001a) Hydrogen peroxide-induced changes in intracellular pH of guard cells precede stomatal closure. Cell Res 11:37-43

Zhang X, Miao YC, An GY, Zhou Y, Shangguan ZP, Gao JF, Song CP (2001b) K^+ channels inhibited by hydrogen peroxide mediate abscisic acid signalling in *Vicia* guard cells. Cell Res 11:195-202

Zhang X, Zhang L, Dong F, Gao J, Galbraith DW, Song C-P (2001c) Hydrogen peroxide is involved in abscisic acid-induced stomatal closure in *Vicia faba.* Plant Physiol 126:1438-1448

Zhang Z, Collinge DB, Thordal-Christensen H (1995) Germin-like oxalate oxidase, a H_2O_2-producing enzyme, accumulates in barley attacked by the powdery mildew fungus. Plant J 8:139-145

Zhu D, Scandalios JG (1992) Expression of the maize MnSOD (Sod3) gene in MnSOD-deficient yeast rescues the mutant yeast under oxidative stress. Genetics 131:803-809

Zhu D, Scandalios JG (1994) Differential accumulation of manganese-superoxide dismutase transcripts in maize in response to abscisic acid and high osmoticum. Plant Physiol 106:173-178

Abbreviations

ABA: abscisic acid

AFLP: amplified fragment length polymorphism
AOS: active oxygen species
APX: ascorbate peroxidase
DHAR: dehydroascorbate reductase
DPI: diphenylene iodonium
GR: glutathione reductase
GST: glutathione S-transferase
HR: hypersensitive response
LUC: luciferase
MAPK: mitogen activated protein kinase
MDAR: monodehydroascorbate reductase
PAL: phenylalanine ammonia-lyase
PCD: programmed cell death
rboh: respiratory burst oxidase homologue
ROS: reactive oxygen species
SA: salicylic acid
SOD: superoxide dismutase

6 Signal transduction in plant cold acclimation

Pekka Heino and E. Tapio Palva

Abstract

Temperate plants respond to low temperature by activating a cold acclimation program leading to enhanced tolerance to freezing temperatures. This acclimation process is accompanied by altered expression of a number of stress response genes controlling production of proteins and metabolites that protect cellular structures and functions from the adverse effects of freezing and freeze-induced cellular dehydration. The changes in cold responsive gene expression are controlled by a set of dedicated transcription factors responding to the low temperature stimulus. We review the complex signal network that is required for sensing and transduction of the low temperature signal to altered gene expression and discuss the interactions of the signal pathways involved.

6.1 Introduction

6.1.1 Low temperature stress

Plants, due to their sessile and poikilothermic nature, are constantly exposed to a variety of biotic and abiotic stresses. This has led to evolution of adaptive mechanisms that enable plant cells to sense the environmental changes and activate responses that increase their tolerance to subsequent stresses. One of the most severe environmental challenges to plants is low temperature, which not only affects the growth and distribution of plants but also causes serious damage to a number of crops. Different plant species vary widely in their ability to tolerate low temperature stress (Levitt 1980; Sakai and Larcher 1987). Chilling-sensitive tropical species can be irreparably damaged even at temperatures significantly higher than the freezing temperature of the tissues. Injuries are caused by impairment of metabolic processes, by alterations in membrane properties, changes in structure of proteins and interactions between macromolecules as well as inhibition of enzymatic reactions. Chilling tolerant but freezing sensitive plants are able to survive temperatures slightly below zero, but are severely damaged upon ice formation in the tissues. On the other hand, frost tolerant plants are able to survive variable levels of freezing temperatures, the actual degree of tolerance being dependent on the species, developmental stage, and duration of the stress.

Exposure of plants to subzero temperatures results in extracellular freezing of tissues, due to the higher freezing point and presence of more active ice nucleators

Topics in Current Genetics, Vol. 4
H. Hirt, K. Shinozaki (Eds.) Plant Responses To Abiotic Stress

in the extracellular solution compared to the cell sap. Extracellular ice formation reduces the water potential outside the cell leading to efflux of water from the symplast and cellular dehydration. Therefore, on the cellular level, freezing stress is accompanied by dehydration stress and consequently, freezing tolerance is strongly correlated with tolerance to dehydration (caused by e.g. drought or high salinity). Freeze-induced dehydration can cause various perturbations in the membrane structures, including membrane fusions and lamellar to hexagonal II phase transitions (Steponkus and Webb 1992). Indeed, such membrane lesions appear to be the main cause of freezing damage (Levitt 1980, Steponkus 1984, Steponkus and Webb 1992, Steponkus et al. 1993). Although freeze-induced cellular dehydration is a central cause of freezing damage, additional factors contribute to freezing injury. Growing ice crystals can cause mechanical damage to cells and tissues. Furthermore, freezing temperatures per se or freeze-induced dehydration can have direct effects on cellular processes due to e.g. denaturation of proteins and disruption of macromolecular complexes.

A common denominator in several stresses, including low temperature is the production of reactive oxygen species (ROS), which can generate damage to different macromolecules in the cells (McKersie and Bowley 1998). Low temperatures, especially in combination with high light can cause excessive production of ROS and hence tolerance to freezing also correlates with effective scavenging systems for ROS to cope with this oxidative stress (Inzé and Van Montagu 1995)

6.1.2 Cold acclimation

The plant species native to temperate and boreal regions are regularly encountering and need to survive subzero temperatures. These species often employ environmental cues, mainly low temperature, as signals to increase their freezing tolerance. This adaptive process known as cold acclimation has been the focus for intensive studies since the beginning of this century, but only recently has the knowledge about the molecular details underlying the acclimation capacity started to accumulate. Plants need to adjust to both daily and seasonal fluctuations in temperature, seasonal acclimation being typical for overwintering herbaceous and woody plants. In overwintering woody plants, acclimation is normally a two-step process. Initially, the shortening of the photoperiod below a critical value causes growth cessation, development of dormancy and leads to a moderate increase in freezing tolerance. The second phase of acclimation is triggered by subsequent exposure to low temperature and is required for development of full frost hardiness (Weiser 1970, Welling et al. 1997). Although independent exposure to short photoperiod or low temperature can trigger some development of freezing tolerance (Welling et al. 2002) the full cold, acclimation response requires synergistic action of both factors (Puhakainen, Boije-Malm, Li, Heino and Palva, personal communication). The perception of the photoperiod, presumably involving phytochrome A (PhyA) (Olsen et al. 1999), is the critical component of this type of acclimation as demonstrated by the altered timing of acclimation in different latitudinal ecotypes of the same plant species (Junttila 1980, Li et al. 2002, 2003).

In annual and many overwintering herbaceous plants, low temperature alone is able to trigger full acclimation, regardless of the photoperiod. However, recent studies have implicated that phytochrome mediated processes can have a role in the acclimation process also in herbaceous plants (Tepperman et al. 2001, Kim et al. 2002). Furthermore, controlled photosynthesis is necessary for acclimation, partly because the acclimation process requires energy provided by photosynthesis (Wanner and Junttila 1999) and partly to prevent formation of excess exitation energy, which would otherwise lead to photoinhibition and formation of reactive oxygen species (Foyer et al. 1994).

The close association of freezing to other stresses resulting in water deficit, such as drought or high salinity suggests that adaptations to these stresses could share common components. Increase in the level of the phytohormone abscisic acid (ABA) is one of the early responses to water deficit stress (Ingram and Bartels 1996). Indeed, there is also a transient increase in the level of ABA during cold acclimation (Chen et al. 1983, Lalk and Dörffling 1985, Ryu and Li 1994, Lång et al. 1994). Even if the direct involvement of ABA in cold acclimation is not unequivocally demonstrated, other lines of evidence support a role for this hormone in the acclimation process: (i) Increased freezing tolerance can be achieved by application of ABA to the plants at normal growth temperatures (Chen et al. 1983, Lång et al. 1989). (ii) Both ABA-deficient and ABA-insensitive mutant of *Arabidopsis* (Koornneef et al. 1982, 1984) appear to be impaired in their ability to cold acclimate (Heino et al. 1990, Gilmour and Thomashow 1991, Mäntylä et al. 1995). Recently, Llorente et al. (2000) and Xiong et al. (2001a) isolated *Arabidopsis* mutants deficient in acclimation and expression of low temperature responsive genes. The mutations, *frs1* and *los5*, respectively were found to be allelic to each other and to the *aba3* mutation, and lead to ABA deficiency. FRS1/LOS5/ABA3 has a central role in ABA biosynthesis; it is involved in the generation of the sulphurated form of a molybdenum cofactor needed for the function of the aldehyde oxidase catalyzing the last step in ABA biosynthesis. However, the loss of low temperature responsiveness in *los5* seems not to be due to ABA deficiency, because ABA addition did not restore the cold inducibility of gene expression (Xiong et al. 2001a). It appears that in addition to its role in ABA biosynthesis, LOS5 has another, currently unknown function in cold acclimation.

The ability to cold acclimate is a polygenic trait, controlled by a number of genes, each likely having a small but additive effect on freezing tolerance development. Environmentally controlled expression of these genes is in turn leading to a number of physiological, cellular and molecular alterations, including changes in membrane lipid composition, accumulation of compatible solutes, changes in phytohormone and antioxidant levels and synthesis of novel proteins (Fig. 1) (Graham and Patterson 1982, Guy 1990, Thomashow 1999, Xin and Browse 2000). For the most part, the alterations are derived from altered gene expression. Moreover, recent studies have started to define the molecular basis of these changes and this has led to the characterization of a large number of genes induced by low temperature (Thomashow 1998, 1999). A recent estimate is that close to 25 % of the transcriptome of *Arabidopsis* is affected in low non-freezing temperature (Kreps et al. 2002). The current challenge in cold acclimation research is to define, which of

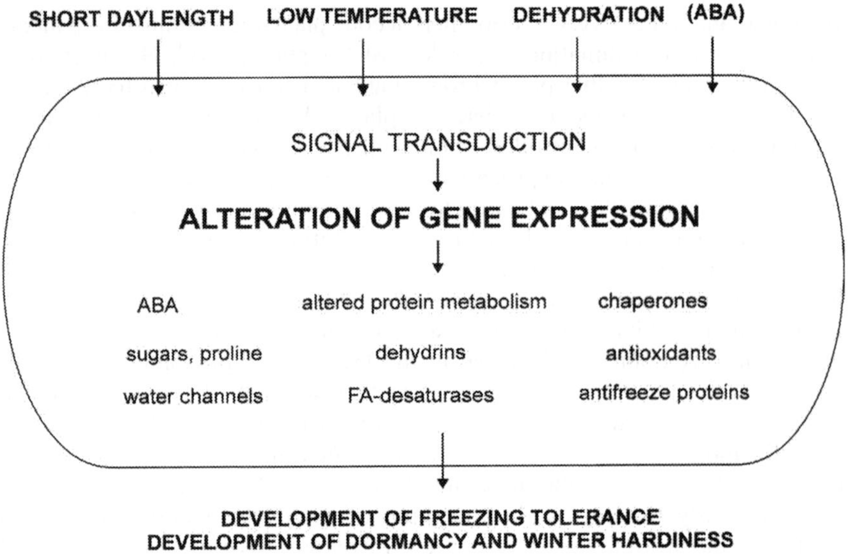

Fig. 1. Several environmental cues may trigger expression of cold acclimation related genes. The acclimation process is accompanied by alterations in protein and metabolite profiles, including changes in components involved in protection against low temperature per se or against freeze induced dehydration as well as components allowing growth at a lower temperature regime

these genes are actually involved in development of freezing tolerance. Cold responsive genes identified so far fall into two distinct main categories. Firstly, genes encoding enzymes or structural components of the cells, that are believed to participate in direct protection the cells against freezing damage (Thomashow 1998, 1999, Palva and Heino 1997, Hiilovaara-Teijo and Palva 1998). Secondly, genes encoding transcription factors and other regulatory proteins, that are believed to regulate the low temperature responses either transcriptionally or post-transcriptionally (Thomashow 1998, 1999, 2001, Nuotio et al. 2001, Viswanathan and Zhu 2002). Despite recent progress, we have only started to understand the molecular details of the regulatory mechanisms controlling low temperature responses. The complexity of the cold responsive transcriptome and the multitude of stimuli triggering acclimation suggest that plants have several parallel and interacting pathways that can lead to enhanced freezing tolerance. The presence of these multiple pathways was already suggested by the fact that several low temperature responsive genes, whose expression correlates with increased freezing tolerance, are also responsive to exogenous ABA (Palva 1994). Early gene expression studies by utilizing the *aba1* and *abi1* mutants of *Arabidopsis* demonstrated that the expression of a subset of the low temperature and drought responsive genes is ABA dependent, whereas some of them are activated through both ABA-independent and ABA-mediated pathways (Nordin et al. 1991, Gilmour and Thomashow 1991, Lång and Palva 1992, Palva 1994). Current studies in several laboratories have led to identification of components of such signal pathways (recently reviewed by Viswanathan and Zhu 2002) and this has started to clarify how plants perceive the low temperature signal and transduce this information into the nucleus to activate specific gene expression leading to enhanced freezing tolerance.

6.1.3 Molecular dissection of cold acclimation

A combination of several types of approaches has been instrumental in identifying components and elucidating molecular mechanisms of cold acclimation. These include pharmacological dissection of signal pathways, molecular analysis of cold responsive genes and their expression (reviewed by Viswanathan and Zhu 2002) and more recently expression profiling of the cold responsive transcriptome (Seki et al. 2001, Fowler and Thomashow 2002, Kreps et al. 2002). One of the most powerful approaches to elucidate the cold acclimation process in the model plant *Arabidopsis* has been isolation of mutants. Several mutational screens have been employed to explore the mechanisms of freezing tolerance and signal transduction pathways leading to low temperature responsive gene expression. Warren and colleagues isolated *Arabidopsis* mutants that fail to develop full freezing tolerance after cold acclimation and thus should be very informative for understanding the cold acclimation response (McKown et al. 1996, Warren et al. 1996). Most of the seven isolated *sfr* (sensitive to freezing) mutants did not compromise induction of previously characterized low temperature responsive genes, even if they are deficient in their cold acclimation capacity. These mutants may thus define novel components involved in freezing tolerance development. However, one of the mutants, *sfr6*, was shown to be deficient in induction of a subset of cold regulated genes. SFR6 appears to specifically affect the cold induction of genes whose activation is mediated by the DRE/CRT-element in their promoter. This element is the binding site for the CBF/DREB1 family of transcription factors (see 6.5.2) and consequently *sfr6* appears to be deficient in CBF-mediated target gene activation (Knight et al. 1999). Interestingly SFR6 appears also to be involved in ABA regulation of gene expression mediated by the abscisic acid response element (ABRE), because ABA induction of the *kin1* gene (Kurkela and Borg-Franck 1992) was also abolished in this mutant (Knight et al. 1999). Positional cloning of the *sfr* genes should enlighten their role in cold acclimation.

In a related screen, Xin and Browse have isolated *Arabidopsis* mutants that are constitutively freezing tolerant (Xin and Browse 1998). One of the mutants, *eskimo1* (*esk1*) exhibited constitutive freezing tolerance of about 80% of the tolerance level of fully acclimated wild type plants, and enhanced tolerance of acclimated mutant plants. However, this mutant did not show a general constitutive expression of cold induced genes. Only the transcript of *RAB18* (Lång and Palva 1992) was 2-3 fold elevated both in non-acclimated as well as in acclimated plants (Xin and Browse 1998). Instead, *esk1* had elevated levels of sugars and constitutively high proline content, which was also correlating with highly increased expression of the gene encoding Δ^1-pyrroline-5-carboxylate synthase, catalyzing the first committing step in proline biosynthesis (Xin and Browse 1998). It is possible that the increased proline content is directly contributing to the freezing tolerance of the *esk1*.

Zhu and colleagues have developed an efficient and sophisticated screen to isolate mutants in stress signal transduction (Ishitani et al. 1997). They generated a transgenic *Arabidopsis* line harbouring a chimeric gene construct, containing the firefly luciferase gene connected to the DRE/CRT and ABRE containing promoter of the cold, drought and ABA responsive *RD29A/LTI78* gene. The transgenic plants exhibit stress and ABA responsive bioluminescence and signalling mutants can be isolated by screening for alterations in bioluminescence. Seeds of transgenic plants were mutagenized with ethyl methanesulfonate (EMS) and more recently by T-DNA mutagenesis, and the M_2 seedlings screened for altered bioluminescence. The screens resulted in isolation of several mutants, which could be divided in three categories; *cos* for constitutive expression of osmotically responsive genes, *los* for low expression of osmotically responsive genes and hos for high expression of osmotically responsive genes (Ishitani et al. 1997). The cloning and analysis of the mutated genes is providing a wealth of information regarding regulation of gene expression in response to abiotic stresses.

The rapid progress in cold acclimation research has recently been described in several excellent reviews (Thomashow 1998, 1999, Shinozaki and Yamaguchi-Shinozaki 2000, Nuotio et al. 2001, Viswanathan and Zhu 2002). The current review is focused on discussing the recent progress in understanding the mechanisms of cellular signalling leading to activation of low temperature responsive genes and development of freezing tolerance.

6.2 Signal perception and low temperature sensing

6.2.1 Perception of cold

Theoretically, changes in temperature could be recognized in any part of the cell but the cellular components that are most likely directly affected by fluctuations in temperature are membranes and proteins. Thus, a temperature induced alteration in e.g. structure, folding, or compartmentalization of a protein could either initiate or modulate a signal transduction cascade activating the expression of cold responsive target genes. In plants, one of the transcriptional regulators for low temperature responsive genes, CBF1, is exhibiting cold-dependent denaturation on regions outside its DNA binding domain (Kanaya et al 1999). This structural alteration has been proposed to alter the function of the protein by facilitating its interaction with other components of the transcriptional activation complex (Kanaya et al. 1999).

Membrane receptors and ion channel proteins provide other potential targets that could be affected by temperature. Ding and Pickard (1993) have identified a mechanosensitive calcium channel exhibiting temperature dependent modulation and suggested that this protein could act as a low temperature sensor. A major class of membrane receptors is constituted of receptor protein kinases, in which ligand binding to the extracellular domain induces the kinase activity on the cytoplasmic side of the receptor. Theoretically, the lack of ligand would not support the existence of a receptor kinase in low temperature sensing. However, low temperature could cause an alteration of the structure of the sensory domain of the protein, either directly or through structural changes in the membrane. Such changes could either activate the kinase domain directly or allow protein-protein interactions needed for activation. Interestingly, receptor-like protein kinase genes have recently been demonstrated to be upregulated in response to low temperature in *Arabidopsis* (Hong et al. 1997, Kreps et al. 2002). Whether receptor-like kinases are involved in temperature sensing or mediating other, unknown, responses involved in low temperature, signalling remains to be seen.

Two component regulatory systems, composed of a membrane bound sensor histidine kinase and a corresponding response regulator protein, are central to sensing of environmental cues in prokaryotes. In *Synecocystis* PCC6803 a histidine kinase, Hik33, has been identified as a putative low temperature sensor (Suzuki et al. 2000). Hik33 autophosphorylation is induced by membrane rigidification caused by low temperature and this leads to activation of a subset of low temperature responsive genes, including genes for fatty acid desaturases (Suzuki et al. 2000, 2001). Interestingly, two component systems have been identified as components of ethylene and cytokinin signal transduction pathways in plants (Chang et al. 1993, Inoue et al. 2001, Urao et al. 2000) and recently a two component sensor kinase, AtHK1 from *Arabidopsis* was associated with osmoregulation (Urao et al.1999). AtHK1 was shown to complement the *sln* mutant and mediate osmosensing in yeast (Urao et al. 2000). Consequently, Urao et al. (2000) proposed that the AtHK1 kinase might act as an osmosensor in *Arabidopsis*. Interestingly, the corresponding gene is also responsive to low temperature.

6.2.2 Membrane rigidification

Membrane fluidity is directly affected by changes in temperature, and may hence be involved in low temperature sensing. Murata and Los (1997) suggested that a primary signal upon a change in temperature might be a change in membrane fluidity, which is one of the most rapid effects of temperature on the plasma membrane. Pd-catalyzed hydrogenation of the membrane lipids - a treatment expected to reduce membrane fluidity - rapidly induced expression of the *desA* gene encoding a fatty acid desaturase in *Synchocystis* PCC6803 (Vigh et al. 1993). The involvement of membrane rigidification in activation of low temperature responsive genes in *Synecocystis* PCC6803 was further confirmed by studies utilizing a double mutant *desA/desD*. This mutant only synthesizes a saturated C16 fatty acid and a mono-unsaturated C18 fatty acid and consequently the cells have more rigid membranes even at physiological temperatures (Inaba et al. 2003). This rigidification enhanced the cold induction of a set of low temperature responsive genes but had no effect on heat induction of gene expression (Inaba et al. 2003). The mechanism by which reduction in membrane fluidity leads to gene activation in *Syneco-*

cystis appears to be at least partly mediated by activation of the plasma membrane histidine kinase receptor Hik33 (Suzuki et al. 2000).

Örvar et al. (2000) have recently demonstrated that membrane rigidification, connected with structural changes in cytoskeleton could be involved in thermosensing also in higher plants. They demonstrated that treatment of alfalfa suspension cultures with either the membrane rigidifier DMSO or the actin microfilament destabilizer cytochalasin D (CD) resulted in cold acclimation and expression of the low temperature responsive gene *CAS30* even at normal growth temperature. Conversely, treatments with the membrane fluidizer benzyl alcohol (BA) or the actin filament stabilizer jasplakinolide (JK) were shown to prevent cold acclimation and induction of the *CAS30* gene in these cell cultures at 4°C (Örvar et al. 2000). One of the early events in cold acclimation is temperature dependent calcium influx to the cytosol (Knight et al. 1991, Monroy et al. 1993, Plieth et al. 1999). By analyzing the Ca^{2+}-influx into alfalfa cells after treatments with DMSO, CD, BA, or JK Örvar et al. also showed that DMSO and CD led to a Ca^{2+}-influx at 25°C, whereas BA and JK treatments inhibited the low temperature induced Ca^{2+}-influx at 4°C (Örvar et al. 2000). Recently, Sangwan et al. (2001) have confirmed the results using transgenic *Brassica napus* seedlings harbouring a GUS fusion to the promoter of the low temperature responsive gene *BN115.* They showed that GUS production was induced with DMSO, the microfilament destabilizer latrunculin B and microtubule destabilizers oryzalin and colchicine at 25°C, whereas BA, JK and the microtubule stabilizer taxol inhibited the activation of the *BN115* promoter at 4°C (Sangwan et al. 2001). Furthermore, they showed that treatment of plants with gadolinium, a mechanosensitive Ca^{2+}-channel blocker, prevents the induction of the *BN115* promoter after low temperature, DMSO, latrunculin B, oryzalin or cholchicine treatments (Sangwan et al. 2001). Consequently, they proposed that low temperature induced membrane rigidification, that might occur in distinct microdomains of the membrane (Murata and Los 1997), could lead to reorganization of the cytoskeleton and activation of mechanosensitive Ca^{2+}-channels. The resulting Ca^{2+}-influx could then trigger further events in signal transduction pathways leading to specific gene expression (Fig. 2) (Örvar et al. 2000, Sangwan et al 2001).

6.3 Role of Ca^{2+} in cold acclimation

Calcium is frequently involved as a second messenger in plant responses to external stimuli (Trewavas and Malhó 1997). Several lines of evidence suggest that calcium is also acting as a second messenger in low temperature signal transduction. A transient increase in cytosolic Ca^{2+}-levels has been demonstrated in response to cold shock (Knight et al. 1991, 1996, Polisensky and Braam 1996, Plieth et al. 1999). Monroy and Dhindsa (1995) have demonstrated that in alfalfa cells, low temperature triggers an influx of calcium into the cytosol. Treatment of cells with calcium chelators or Ca^{2+}-channel blockers prevented the calcium influx as well as the expression of low temperature responsive *cas15* gene and the devel-

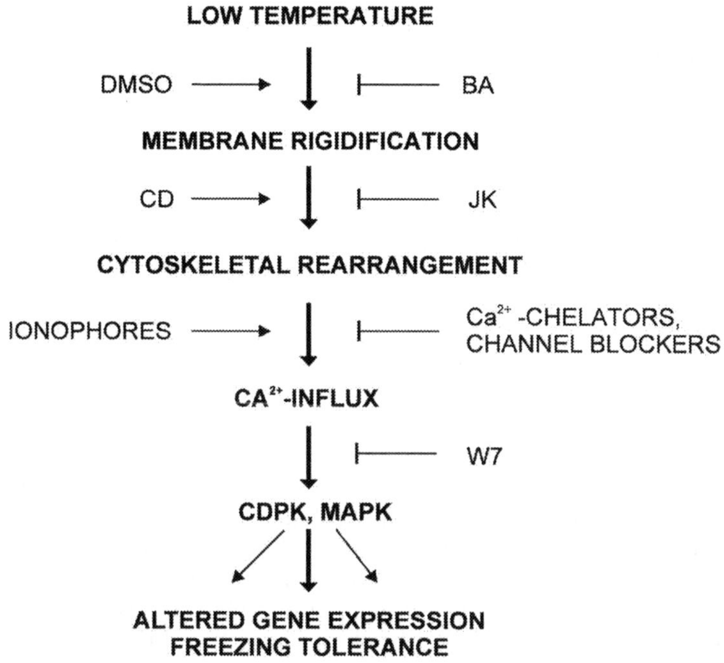

Fig. 2. A model showing initial events in cold signalling. Low temperature causes membrane rigidification, which leads to cytoskeletal rearrangements and subsequent Ca^{2+}-influx. Increased cytosolic Ca^{2+}-concentration is recognized by Ca^{2+}-binding proteins, including calmodulin, and leads to CDPK activation. See text for details.

opment of freezing tolerance (Monroy et al. 1993, Monroy and Dhindsa 1995, Sangwan et al. 2001). When Ca^{2+} influx in alfalfa cells or *Brassica napus* leaves was artificially increased by using a ionophore or a Ca^{2+}-channel agonist, cold acclimation specific genes were induced and freezing tolerance increased at 25°C (Monroy and Dhindsa 1995, Sangwan et al. 2001). In analogous studies, Ca^{2+}-channel blockers and a Ca^{2+} chelator were found to inhibit the low temperature activation of *kin* genes in *Arabidopsis* (Knight et al. 1996, Tähtiharju et al. 1997). However, in *Arabidopsis*, these treatments caused only a partial inhibition of cold induced Ca^{2+}-influx and low temperature responsive gene expression, suggesting that also an intracellular Ca^{2+} source might be involved.

Inositol trisphosphate (IP_3) and cyclic adenosine 5′-diphosphate ribose (cADPR) are able to release calcium from beet storage root vacuoles (Allen et al. 1995). Both IP_3 and cADPR have been implicated as regulators of Ca^{2+}-channels in response to low temperature (Knight et al. 1996, Sangwan et al. 2001, Xiong et al. 2001b). By using single cell based analysis in tomato Wu et al. (1997) demonstrated that cADPR can mediate activation of low temperature responsive genes, indicating that Ca^{2+}-release from intracellular stores is also involved in acclimation. They microinjected tomato hypocotyl cells with contructs, where promoters of two cold- and ABA-responsive Arabidopsis genes *LTI78/COR78/RD29A* (Nordin et al. 1991, Gilmour and Thomashow 1991, Yamaguchi-Shinozaki and Shinozaki 1993) or *KIN2* (Kurkela and Borg-Franck 1992) were coupled to the reporter gene *uidA* (GUS) and monitored their activation in response to ABA and specific pharmacological agents known to modulate Ca^{2+}-homeostasis in the cytocol. External application of ABA or coinjection with Ca^{2+} activated the two stress-responsive genes. Coinjection with cADPR or ADP-ribocyl cyclase was sufficient to activate the genes in the absence of ABA, whereas coinjection with 8-amino-cADPR, a competitive inhibitor of cADPR, prevented the activation of the genes, even when ABA was present (Wu et al.1997). Sangwan et al. (2001) have recently shown that cADPR treatment can induce cold acclimation and activate the low temperature responsive *BN115* gene in *Brassica napus* seedlings at 25°C, indicating that cADPR could indeed be involved in generation of the Ca^{2+}-influx during low temperature exposure.

Wu et al. (1997) also studied the effect of IP_3 in activation of the *RD29A/LTI78* and *KIN2* genes. Coinjection of the reporter constructs with IP_3 activated the genes and the activation was inhibited by heparin, a specific blocker of IP_3 receptors. However, heparin had no effect on ABA-induced expression of the reporter genes, indicating that IP_3 is not the primary mediator of intracellular Ca^{2+} release in ABA responses (Wu et al.1997). However, IP_3 has been suggested to have a role in low temperature signalling (Knight et al. 1996). IP_3 is produced by hydrolysis of phosphatidyl-inositol-4,5-bisphosphate (PIP_2). PIP_2 is synthesized by phosphatidylinositol 4-phosphate 5-kinase and an *Arabidopsis* gene encoding this enzyme has been shown to be induced by osmotic stress and ABA (Mikami et al. 1998). Hydrolysis of PIP_2 is mediated by an activated phosphoinoside-specific phpspholipase C (PI-PLC) (Trewavas and Malhó 1997). A stress and ABA responsive gene encoding a PI-PLC, whose activity is depending of Ca^{2+} has been isolated from *Arabidopsis* (Hirayama et al. 1995). Thus, regulation of both production and activity of PI-PLC, as well as the availability of its substrate PIP_2 during stress might control the IP_3-mediated signalling. Recently, a direct connection between phosphoinositide metabolism and stress signal transduction was shown by Xiong et al. (2001b). They mutagenized transgenic *Arabidopsis* plants harbouring a luciferase fusion to the promoter of the *RD29A/LTI78/COR78* gene and isolated a mutant, *fiery1* (*fry1*), that showed enhanced constitutive expression of low temperature induced genes and super-induction of them in response to cold, ABA, salt and osmotic stress (Xiong et al. 2001b). Interestingly, even if low temperature responsive gene expression is enhanced in the *fry1* mutant, the plants are unable to cold acclimate. Positional cloning of the *FRY1* revealed that it encodes an inositol polyphosphate 1-phosphatase, an enzyme that mediates the catabolism of IP_3. The *fry1* mutant plants were shown to contain significantly higher basal level of IP_3 compared to the wild type plants. In the wild type IP_3 level markedly increased after 1 min of ABA treatment and returned to the basal level after 10 min of treatment, whereas in the *fry1* mutant accumulation of IP_3 was detected after 30 min of treatment (Xiong et al. 2001b). These results demonstrate that IP_3 is involved in mediating ABA and stress signalling and indicates that a critical issue of tolerance development could be the ability to attenuate the IP_3 signal, which otherwise could lead to disturbances in Ca^{2+}-homeostasis (Xiong et al. 2001b).

One of the critical issues in Ca^{2+}-mediated processes in cells is the transient nature of Ca^{2+} increase. Ca^{2+}- channels in the plasma membrane or in the intracellular membranes are responsible for the Ca^{2+}-influx, whereas Ca^{2+}-ATPases and Ca^{2+}/H^{+} antiporters are mediating the Ca^{2+}-efflux from cytosol to maintain Ca^{2+} homeostasis. Genes encoding Ca^{2+}-ATPases have been cloned from *Arabidopsis* (Sanders et al. 1999), but their role in maintaining Ca^{2+}-homeostasis during stress is not clear. Recently, Puhakainen et al. (1999) have shown that low temperature treatment increases the activity of Ca^{2+}-ATPase activity in leaves of winter rye. Furthermore, overexpression of an *Arabidopsis* Ca^{2+}/H^{+} antiporter in tobacco resulted in sensitivity to cold shock, indicating that antiporter activity is needed for low temperature adaptation (Hirschi 1999).

Cytosolic Ca^{2+}-levels have been found to change in response to a variety of different stimuli in addition to cold, such as light, growth regulators, pathogen attack, wind, and touch (Gilroy and Trewavas 1994). A central question is where is the specificity in the signal? One answer could lie in information encoded in the amplitude, frequency, and spatial localization of the changes in Ca^{2+} concentration in the cell (Gilroy and Trewavas 1994, McAinsh and Hetherington 1998). Stress induced changes in cytosolic Ca^{2+} levels exhibit enormous variability in amplitude and temporal and spatial distribution. For example, touch, wind, and cold shock all cause sharp spikes in cytosolic calcium levels in tobacco seedlings within 15 seconds (Knight et al. 1991, 1996), whereas oxidative and salt stresses cause relatively low Ca^{2+} transients, lasting for several minutes (Price et al. 1994). These differences may allow plant cells to distinguish one kind of stress from another and to induce distinct gene expression required for adaptation to a particular stress. Ratio- and confocal imaging have indeed revealed spatially and temporally localized changes in calcium levels, implying that different parts of the cytoplasm may be regulated differently in response to a stimulus. Consequently, plant cells can distinguish between different stimulus-induced increases in cellular calcium. The experiments of Gong et al. (1998) with transgenic, aequorin-expressing tobacco seedlings have demonstrated that heat shock increases cytosolic Ca^{2+}. However, after the initial shock there was a refractory period in which additional heat shock signals failed to increase the Ca^{2+}-level. Throughout this refractory period, cells retained full responsiveness to other stimuli and for example responded to cold shock by a Ca^{2+} influx. Kinetics of the cytosolic Ca^{2+} increase after a cold shock was similar in both cold sensitive tobacco and cold tolerant *Arabidopsis*. However, tobacco was able to recover its ability to respond to cold shock 30 minutes after the initial shock, whereas *Arabidopsis* was not (Knight et al. 1996). The authors suggest that this altered response to repeated cold stimulation is important in the cold acclimation process.

How are the cold acclimation related Ca^{2+} signatures recognized? Signal transduction initiated by Ca^{2+}-influx is generally mediated through Ca^{2+}-binding proteins. Calmodulin is a highly conserved protein that has been considered as the primary sensor for changes in cytosolic Ca^{2+}-levels (Rudd and Frankling-Tong 1999). In *Arabidopsis*, environmental stimuli, including low temperature, trigger rapid activation of genes encoding CaM and CaM related proteins. This low temperature responsive expression of CaM genes is partially regulated by Ca^{2+}

(Polisensky and Braam 1996). Studies by Monroy et al. (1993) showing that treatment of alfalfa cells with a CaM antagonist prevented cold acclimation and reduced expression of cold regulated genes, indicate a role for CaM in low temperature signalling. On the other hand, Townley and Knight (2002) have shown that overexpression of a gene encoding CaM in *Arabidopsis* leads to reduced expression of cold responsive genes, suggesting that CaM might have a role as a negative regulator during cold acclimation. Recent studies by Zhu and colleagues have implicated another type of Ca^{2+}-sensor in stress signalling. Liu and Zhu (1997) first identified an *Arabidopsis* mutant, *sos3*, which was hypersensitive to Na^{+}. Cloning of the *SOS3* gene revealed that it encoded a protein highly similar to the regulatory B subunit of Ca^{2+}/calmodulin dependent phosphatase calcineurin in animals and yeast (Liu and Zhu 1998). This calcineurin B-like (CBL) protein was shown to mediate salt stress signalling in *Arabidopsis* by activating a specific protein kinase, SOS2, which then activates SOS1, a Na^{+}/H^{+} antiporter (Halfer et al. 2000, Liu et al. 2000, Shi et al. 2000, Zhu 2002). A gene family encoding CBLs in *Arabidopsis* was recently characterized and one of the genes, *AtCBL1*, shown to be responsive to low temperature, drought and wounding (Kudla et al. 1999). CBLs appear to be Ca^{2+}-sensors that activate a specific family of protein kinases called CIPKs (CBL-interacting protein kinases) and AtCBL1 has been shown to interact with the constitutively expressed CIPK1 in calcium dependent manner (Shi et al. 1999). Both CBLs and CIPKs are encoded by a multigene family in Arabidopsis, but the specific functions of the corresponding proteins are mostly unknown.

In conclusion, it seems clear that Ca^{2+}- influx to the cytosol is part of the initial response to low temperature. The Ca^{2+}-signal appears to be recognized by distinct Ca^{2+}-binding proteins, which then transmit the signal further by activating proteins, at least part of which appear to be protein kinases.

6.4 Protein phosphorylation

6.4.1 Protein kinases

It is now well established that protein phosphorylation/dephosphorylation is involved in signal transduction during cold acclimation. Monroy et al. (1993) originally demonstrated that in alfalfa cell suspension cultures changes in the phosphorylation pattern of pre-existing proteins are part of the low temperature response. W7, an antagonist of Ca^{2+}-dependent protein kinases (CDPKs), was shown to inhibit low temperature responsive gene expression and development of freezing tolerance in both *Arabidopsis* (Tähtiharju et al. 1997) and alfalfa (Monroy et al., 1993). Two CDPK encoding genes in alfalfa have been demonstrated to be responsive to low temperature, supporting a role in cold signalling (Monroy and Dhindsa 1995). Furthermore, Martin and Busconi (2001) have characterized a membrane bound CDPK, whose activity is enhanced by low temperature treatment. Recently, overexpression of a cold and salt stress inducible CDPK encoding gene, *OsCDPK*, has been shown to enhance low temperature tolerance of chilling sensitive rice plants (Saijo et al. 2000). Taken together, these studies indicate that CDPKs could play a central role in mediating Ca^{2+}-signals during acquisition of cold of chilling tolerance. As described above, CBLs are calcium-binding proteins that transmit Ca^{2+}-signals by activating CIPKs. Recently Kim et al. (2003) have shown that the gene encoding one of the members of the CIPK family, CIPK3, is responsive to low temperature, drought, salt, and ABA. By isolating and analyzing a T-DNA insertion mutant of *CIPK3*, they demonstrated that CIPK3 is regulating stress- and ABA responsive gene expression. Interestingly, CIPK3 appears only to regulate cold, salt, and ABA responses, because drought induction of genes was not affected in the *cipk3* mutant (Kim et al. 2003). The ABA induction of several different classes of ABA responsive genes was also inhibited in the *cipk3* mutant and consequently Kim et al. (2003) suggested that CIPK is acting downstream the Ca^{2+}-signal but upstream from transcription factors that regulate low temperature and ABA responsive promoters. Consequently, CIPK3 appears to define a component involved in cross-talk between cold and ABA signalling during acclimation (Kim et al. 2003).

Mitogen activated protein kinase (MAPKs) are mediators in several signal transduction pathways in eukaryotic cells, including responses to a variety of environmental stresses. A MAP kinase cascade involves three protein kinases. MAPKKKs are the primary signal receivers, which upon activation phosphorylate and activate MAPKKs. Active MAPKKs are dual specificity protein kinases, which phosphorylate MAPKs at both tyrosine and threonine residues in the conserved TXY motif. MAPKs in turn regulate transcription factors to generate specific responses. Components of several MAP kinase cascades have been isolated from plants (Mizoguchi et al.1997). Jonak et al. (1996) characterized components of a low temperature and drought regulated MAP kinase cascade in alfalfa. They isolated a cDNA corresponding to a gene encoding for a MAPK, MMK4. *MMK4* mRNA accumulated in response to low temperature and although the MMK4 protein levels were not affected by cold, the kinase activity of the protein was strongly enhanced after 10 min of low temperature treatment, reaching maximum activity after 60 min, and then returning to the basal level after 120 min (Jonak et al. 1996). Recently, it was shown that the cold activation of the MMK4 (also known as SAMK, for stress-activated MAP kinase) is mediated by membrane rigidification and cytoskeletal remodelling. SAMK activity was induced in alfalfa cell cultures at 25°C after treatment with a membrane rigidifier, DMSO, whereas a membrane fluidizer, BA, inhibited the cold responsive activation of the SAMK (Sangwan et al. 2002). Pre-treatment of cells with either the microfilament stabilizer jasplakinolide or the microtubule stabilizer taxol inhibited low temperature mediated activation of SAMK, whereas both microfilament and microtubul destabilizers, latrunculin B, and oryzelin, respectively, activated the SAMK at 25°C. Furthermore, cold- DMSO-, latrunculin B- and oryzalin-induced activation could be inhibited by Ca^{2+}-chelators EGTA and BAPTA and by Ca^{2+}-channel blockers lanthanum and gadolinium, demonstrating that Ca^{2+}-influx is needed for SAMK activation and that the Ca^{2+}-influx is downstream from the cytoskeleton remodelling (Sangwan et al. 2002).

Transcripts for a MAPK kinase kinase, ATMEKK1, and a MAPK, ATMPK3 have been shown to accumulate rapidly in *Arabidopsis* in response to low temperature (Mizoguchi et al. 1996). An H_2O_2 activated MAPKKK, ANP1, has been shown to be in the same kinase cascade as ATMPK3 and overexpression of *NPK1*, a tobacco ortholog of *ANP1*, rendered transgenic tobacco plants more cold tolerant (Kovtun et al. 2000). In *Arabidopsis*, the AtMPK4 and AtMPK6 are activated rapidly in response to low temperature (Ichimura et al. 2000). AtMPK4 has been placed in a cascade comprising ATMEKK1, MEK1/ATMKK2, and ATMPK4 (Mizoguchi et al. 1998). The molecular targets for these kinase cascades remain to be elucidated. Therefore, despite the obvious involvement of MAPK cascades in environmental signalling in plants, their exact role in cold acclimation is still unknown.

6.4.2 Protein phosphatases

Pharmacological studies were first used to demonstrate a role for protein phosphatases in cold signalling and cold acclimation. In alfalfa cells, protein phosphatase inhibitor okadaic acid induced the low temperature responsive *CAS15* gene at 25°C but had no effect on its expression at 4°C (Monroy et al. 1998). The protein kinase inhibitor staurosporine, on the other hand, had no effect on the non-induced level of *CAS15* transcripts at 25°C but prevented induction of the gene by low temperature. Similarly, treatments with genistein, H7, and wortmanin, inhibiting tyrosine kinases, protein kinase C and phosphoinositide kinase, respectively, were shown to prevent the activation of the *BN115* promoter and prevent development of freezing tolerance in *B. napus* leaves (Sangwan et al. 2001). Conversely, treatment of *B. napus* leaves with okadaic acid or calyculin A, inhibiting protein phosphatases 1 and 2A (PP1 and PP2A), respectively, activated the *BN115* promoter and conferred freezing tolerance even at 25°C (Sangwan et al. 2001). In a previous study, Wu et al. (1997) demonstrated that treatment with ocadaic acid was activating reporter constructs driven by the *RD29A* or *KIN2* promoters in microinjected tomato hypocotyls, even in the absence of ABA, whereas treatments with protein kinase inhibitors K252a and staurosporine inhibited the ABA-, cADPR-, and Ca^{2+}-mediated activation of *RD29A* or *KIN2* promoters (Wu et al 1997). On the other hand, PP2A activity has been shown to decrease dramatically at 4°C (Monroy et al. 1998). A protein that interacts with the catalytic subunit of an *Arabidopsis* PP2A was recently identified (Harris et al. 1999). This protein is a homolog of the yeast TAP42 protein, involved in the target-of-rapamycin (TOR) signalling pathway, presumably regulating protein synthesis. Interestingly, the gene for the PP2A interactor is induced with by temperature (Harris et al. 1999). However, the function of the target PP2A is currently not known and the significance of the interaction in cold signalling remains to be elucidated.

As discussed above, the ABA-insensitive mutant *abi1* of *Arabidopsis* exhibits delayed cold acclimation. In addition, the *abi* mutation prevents low temperature activation of the *RAB18* gene in *Arabidopsis* (Lång and Palva 1992, Mäntylä et al. 1995). The *ABI1* gene has been shown to encode a protein related to type 2C protein phosphatase (PP2C) (Leung et al. 1994, Meyer et al. 1994), which acts as negative regulator in ABA signalling (Gosti et al. 1999, Merlot et al. 2001). The gene encoding ABI1 has also been shown to be transiently induced by low temperature (Tähtiharju and Palva 2001) indicating that the ABI1 phosphatase could be involved in ABA signalling during cold acclimation. Tähtiharju and Palva (2001) characterized the role of a related PP2C, AtPP2CA, in *Arabidopsis*. The *AtPP2CA* gene was shown to be cold responsive, but while the induced expression of the *ABI1* was transient, the cold induced expression of *AtPP2C* remained on elevated level (Tähtiharju and Palva 2001). They also showed that transgenic plants expressing *AtPP2C* in antisense orientation exhibited superinduction of low temperature responsive genes during cold acclimation. In addition, cold acclimation was also accelerated in antisense plants. Therefore, AtPP2C appears to be a negative regulation of cold acclimation acting through an ABA-dependent pathway (Tähtiharju and Palva 2001).

It is evident that protein phosphorylation is involved in the signal transduction pathways leading to cold acclimation and several different types of kinases and phosphatases appear to be involved in the process. However, the exact position of the kinases in the signalling pathways and the substrates for the kinases/phosphatases are not known. One possibility is that the Ca^{2+}-signature generated in the early stages of signal transduction could be used to activate CDPKs, which then, directly or indirectly, could activate a MAP-kinase cascade. Active MAPK could then activate a transcription factor, which triggers altered gene expression.

6.5 Regulation of gene expression in response to low temperature

6.5.1 Gene expression in response to cold

Cold acclimation is accompanied by altered expression of a number of genes, some of which are likely to play a critical role in development of freezing tolerance. The cold responsive genes seem to exhibit a temporally complex pattern of expression involving both transcriptional and post-transcriptional controls (Hughes and Dunn 1996) The cold-responsive genes (*LTI/COR/CAS/KIN/ERD/RD*) are often also responsive to other stress-related stimuli, such as drought, salt, and ABA (Thomashow 1999, Shinozaki and Yamaguchi-Shinozaki 2000, Nuotio et al. 2001, Zhu 2001). Recent studies utilizing expression profiling have confirmed that a large amount of changes in gene expression is indeed involved in plant response to low temperature (Seki et al. 2001, Fowler and Thomashow 2002, Kreps et al. 2002). Kreps et al. (2002) have shown that out of the ~8000 *Arabidopsis* genes analyzed 2086 were responding to cold, 42% of those being induced, when the level of change in expression compared to the untreated control was at least 2-fold. This indicates that about 25% of the *Arabidopsis* transcriptome could be responding to low temperature.

Earlier expression studies with limited number of genes have already indicated clearly distinct temporal patterns of cold-induced gene expression. In *Arabidopsis,* many of the low temperature responsive transcripts are detectable after 1-2 hours of low temperature exposure and the transcript levels remain high as long as the plants are kept at low temperature, and rapidly return to low basal levels upon return to normal growth temperatures (Palva 1994, Thomashow 1999). However, many of the cold responsive genes are only transiently induced in the early and middle phases of cold acclimation. The recent expression profiling studies have expanded the previous work and underlined the complexity of the expression patterns. From the 2086 changes that Kreps et al. detected most were only transient. Similarly, Fowler and Thomashow (2002) have also profiled the expression of ~8000 *Arabidopsis* genes in response to low temperature. Out of the 218 genes that were found to be at least three-fold induced in response to low temperature, 156 showed only transient induction (Fowler and Thomashow 2002). The profiling studies clearly demonstrate that low temperature responses involve altered expression of a large set of genes and these genes differ in their temporal expression pattern. However, which of these genes are actually involved in cold acclimation and what is the role of the corresponding proteins in the acclimation process and in development of freezing tolerance is mostly unknown.

6.5.2 CRT/DRE/LTRE regulated gene expression

Understanding gene regulation during cold acclimation requires definition of the *cis*-elements that are required for the low temperature responsiveness of the genes and characterization of the *trans*-acting factors that bind to these sequences and mediate the response. The first suggestion for the identity of a low temperature responsive element (LTRE) was given for the *LTI78* promoter (Nordin et al. 1993). The role of this element in activation of genes during stress was subsequently demonstrated by Yamaguchi-Shinozaki and Shinozaki (1994). They performed deletion analysis of the *RD29A/LTI78/COR78* promoter and showed that the 9-bp element, TACCGACAT, with a core sequence of CCGAC, confers responsiveness to low temperature, drought, and high salinity, but not to ABA (Yamaguchi-Shinozaki and Shinozaki 1994). This low-temperature and dehydration-responsive element (DRE/LTRE) occurs also in several other promoters and was also characterized as the C-repeat (CRT) (Baker et al. 1994).

A transcription factor binding to the DRE/CRT element and activating cold induced gene expression was first identified by Stockinger et al. (1997). They employed yeast one-hybrid screening and were able to isolate an *Arabidopsis* gene encoding a DRE/CRT binding protein, CBF1 (C-repeat binding factor 1), belonging to the APETALA2/EREBP-family of transcription factors. Five additional genes encoding CBF1 homologs, called DREBs (DRE binding proteins) and two additional CBFs, CBF2, and CBF3 were subsequently cloned from *Arabidopsis* (Fig. 3) (Liu et al. 1998, Gilmour et al. 1998). The *DREB* genes could be divided into two distinct groups according to their responsiveness to low temperature and drought. The *DREB1* genes, *DREB1A, DREB1B*, and *DREB1* constitute a structur-

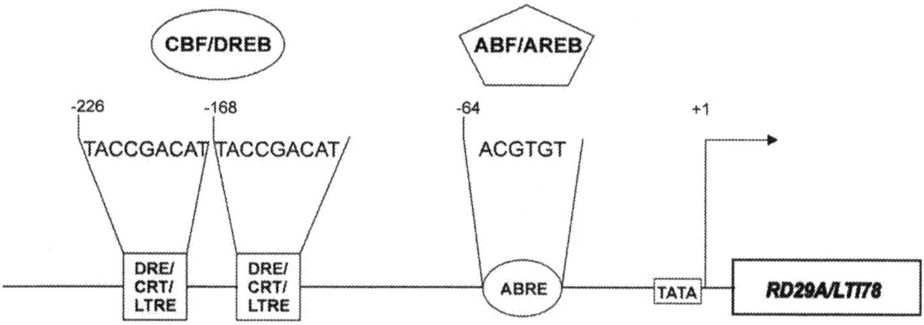

Fig. 3. Cold induced genes are members of more than one regulon. The promoter region of the *RD29A/LTI78/COR78* gene contains cis-elements recognized by different transcription factors, allowing the induction of the gene in response to different stimuli. DRE/LTRE/CRT elements are binding sites for CBF/DREB and ABRE binding site for ABF/AREB transcription factors, respectively.

ally homologous group of genes responsive to low temperature but not to drought, while the *DREB2* genes, *DREB2A* and *DREB2B*, were responsive to drought but not to low temperature (Liu et al. 1998). *CBF1* is identical with *DREB1B* and *CBF2* and *CBF3* are identical with *DREB1C* and *DREB1A*, respectively. Recent studies have resulted in isolation of an additional CBF homolog, CBF4 (Haake et al. 2002). Interestingly, the CBF4 gene is induced by drought but not by low temperature, indicating that also CBF4 might regulate drought responses in *Arabidopsis* (Haake et al. 2002). In addition to *Arabidopsis* CBF/DREB1, orthologs have been identified in other plant species, including *B. napus* (Jaglo et al. 2001, Gao et al. 2002), wheat, rye, tomato (Jaglo et al. 2001), barley, rice (Choi et al. 2002), and birch, *Betula pendula*, (Aalto, Ojala, Puhakainen, Heino and Palva, personal communication). Similar to *Arabidopsis*, the *CBF/DREB1* genes in other plant species also appear to consist of a small gene family. Jaglo et al. (2001) have isolated two *B. napus* and three rye cDNA clones encoding *CBF/DREB1* orthologs and shown that they are low temperature responsive. Recently Gao et al. (2002) isolated four *CBF/DREB1* genes from *B. napus* and demonstrated that the genes were responsive to low temperature and could activate reporter gene expression driven by LTRE/DRE/CRT elements in yeast. This demonstrates that the employment of CBF/DREB1 like transcription factors in activation of low temperature responsiveness of genes is highly conserved in the plant kingdom. The presence of low temperature responsive *CBF/DREB1* orthologs also in tomato indicate that the CBF/DREB1 pathway is not limited to plants that have the ability to cold acclimate (Jaglo et al. 2001).

Overexpression of *CBF1/DREB1B*, *CBF3/DREB1A*, or *CBF4* leads to constitutive expression of genes with promoters containing the DRE/CRT/LTRE element and to improved freezing and drought tolerance in non-acclimated plants (Jaglo-Ottosen et al. 1998, Kasuga et al. 1999, Haake et al. 2002). In addition, the overexpression of CBF3 leads to elevated levels of proline and sugars that are normally associated with cold acclimation (Gilmour et al. 2000). CBFs have been shown to activate reporter genes carrying CRT/DRE-elements in their promoters in yeast (Stockinger et al. 1997, Gilmour et al. 1998).

Recently, CBF1 activity in yeast was shown to be dependent on Gcn5, Ada2 and Ada3, which all are part of Ada and SAGA complexes in yeast and act as histone acetyltranserase (Gcn5) and transcriptional adapters (Ada2 and Ada3) (Stockinger et al. 2001). Cloning of *Arabidopsis* homologues of yeast *Gcn5* (*AtGCN5*) and Ada2 (*AtADA2a* and *AtADA2b*) revealed that AtGCN5 had indeed histone acetyltransferase activity and interacted with AtADA2a and AtADA2b. In addition, CBF1 was shown to interact with all three proteins (Stockinger et al. 2001). This indicates that CBF1 functions through Ada or SAGA like complexes in *Arabidopsis*.

CBF/DREB1 genes are themselves transiently regulated by low temperature (Gilmour et al. 1998, Liu et al. 1998, Medina et al.1999). The expression of all three *CBF* genes was detected after 30 min of low temperature exposure; the mRNA levels were highest after 1 hour and declined to the basal level by 6 hours (Medina et al. 1999). What are the factors and signal component controlling this cold induced expression of *CBF*s? Recently, an *Arabidopsis* mutant, *hos1*, which exhibited enhanced induction of the *CBF2* and *CBF3* genes, was isolated (Ishitani et al. 1998, Lee et al. 2001). In *hos1*, the levels of *CBF* mRNAs were significantly higher after 3 hours of low temperature exposure and they remained elevated up to 48 hours (Lee et al. 2001). Accordingly, target genes for CBFs were shown to be superinduced in the *hos1* mutant. Positional cloning of the *hos1* gene revealed that it encodes a novel RING finger protein that exhibited low temperature induced accumulation to the nucleus (Lee et al. 2001). Recently, RING finger proteins have been found to contain E3 ubiquitin ligase activity by which they transfer ubiquitin to proteins directing them to degradation (Yang et al. 2000). Because HOS1 is transferred to nucleus upon low temperature treatment and it can negatively regulate the expression of the *CBF* genes it is tempting to speculate that HOS1 is directly regulating the turnover of the hypothetical ICE protein (inducer of CBF expression), originally suggested by Thomashow to regulate CBF expression (Lee et al. 2001, Gilmour et al. 1998, Thomashow 2001). Interestingly, the *hos1* mRNA level was shown to decline transiently in response to low temperature, being almost undetectable after 30 min of treatment and raising to the basal level after 1 hour (Lee et al. 2001). If the HOS1 is indeed mediating degradation of the hypothetical ICE protein, then the low expression during the initial stages of cold treatment would allow ICE activation and subsequent transient induction of the *CBF* genes.

The *RD29A/LTI78* promoter seems to be the target of additional transcription factors. Sakamoto et al. (2000) have isolated three low temperature responsive genes encoding C_2H_2 type zinc finger proteins in *Arabidopsis*. Two of these, *AZF1* and *AZF3*, were novel genes, and the third *STZ/ZAT10* has been previously isolated and shown to rescue the salt sensitive phenotype of yeast calcineurin mutants and confer tolerance to elevated concentrations of Na^+ and Li+ in wild type yeast (Lippuner et al. 1996). Recently, The STZ/ZAT10 has been shown to repress the general activator mediated induction of genes and this repression was shown to be dependent on the EAR motif in the C-terminal part of the STZ/ZAT10 (Ohta et al.

2001). By using a *35S-STZ/ZAT10* effector and *RD29A-luc* reporter constructs in a transient expression assay, Lee et al. (2002) have shown that STZ/ZAT10 can repress the activity of the *RD29A* promoter.

By screening for alterations in the expression of the *RD29A-luc* transgene in response to low temperature Lee et al. (2002) isolated a mutant, *los2*, where the cold induction of the *RD29A* is highly reduced. While the endogenous low temperature responsive genes also exhibited reduced expression during cold, the *los2* mutation had no effect on expression of *CBF2* (Lee et al. 2002). The *LOS2* gene was cloned and it was shown to encode a protein having enolase activity. Interestingly, LOS2 was shown to be a bi-functional protein and to have analogously with previously characterized enolases from animals and fungi as well as DNA binding activity (Lee et al. 2002) and the ability to bind the promoter of the *STZ/ZAT10* gene. The expression of *STZ/ZAT10* in response to low temperature was enhanced and prolonged in the *hos2* mutant (Lee et al. 2002), suggesting that LOS2 is a negative regulator of *STZ/ZAT10*. Thus, the decreased expression of low temperature responsive genes in the *los2* mutant would be due to the increased expression of the negative regulator STZ/ZAT10 (Lee et al. 2002).

6.5.3 ABRE mediated gene expression

A set of characterized low temperature responsive genes do not carry the DRE/CRT/LTRE element in their promoters, e.g. the *Arabidopsis LTI6/RCI2A* gene (Capel et al. 1997, Nylander et al. 2001), indicating that also other elements confer cold responsiveness. Furthermore, most of the low temperature responsive genes are also induced by exogenous ABA. Consequently, their promoters should contain *cis*-elements mediating this response. Indeed, sequences closely resembling ABA response elements (ABREs) exist in the promoters of several low temperature responsive genes (Lång and Palva 1992, Baker et al. 1994, Yamaguchi-Shinozaki and Shinozaki 1994). The ABREs, cis-elements with the consensus sequence (C/T)ACGTGGC, have been shown to confer ABA-regulated expression of many genes when present in more that one copy (Guiltinan et al. 1990, Leung and Giraudat 1998).

Several regulatory proteins that can specifically bind to ABREs have been reported (e.g. Guiltinan et al. 1990). They contain the basic domain/leucine zipper (bZIP) motif found in many transcription factors. Two bZIP proteins that bind specifically to the ABRE elements mediating dehydration and ABA responsiveness of the *Arabidopsis RD29B/LTI65* gene were recently characterized (Uno et al. 2000). The genes encoding AREB1 and AREB2 (ABA-responsive element binding protein) are drought, salt and ABA induced but not responsive to low temperature. The AREB proteins can function as transcription factors, but need ABA for their activation. In another study, a family of four ABRE binding factors (ABFs), ABF1, 2, 3, and 4, was characterized also from *Arabidopsis* (Choi et al. 2000). One of these ABF proteins corresponds to AREB2 and the others are highly homologous to AREBs, indicating that at least five distinct transcription factors control stress-induced expression of ABRE containing genes (Fig. 3). The *ABF* genes respond differently to various environmental stresses, suggesting that they may act in different stress-response pathways. All *ABF*s are responsive to ABA but only the expression of the *ABF1* is also enhanced by low temperature (Choi et al. 2000). Recently, Kang et al. (2002) have shown that overexpression of ABF3 and ABF4 in transgenic plants leads to constitutive activation of ABA responsive genes and enhanced drought tolerance. However, freezing tolerance of the plants was not measured in this study and further studies are needed to elucidate the role of ABFs, especially ABF1, in activating ABA responsive genes during cold acclimation.

Kim et al. (2001a) have isolated a cold and ABA responsive gene that encodes a C_2H_2- type zinc finger protein SCOF1 in soybean. Overexpression of SCOF1 in transgenic *Arabidopsis* resulted in constitutive expression of the low temperature responsive genes *COR15A*, *COR47*, and *RD29B/LTI65*, and increased freezing tolerance in non-acclimated plants (Kim et al. 2001a). *SCOF1* is induced after 3 hours of low temperature treatment and the expression level is increasing at least up to 7 days. The temporal pattern of *SCOF1* expression and the constitutive expression of low temperature responsive genes in transgenic *Arabidopsis*, suggested that SCOF1 may act synergistically with CBF/DREB1 proteins and maintain the expression of CBF/DREB1 target genes after *CBF/DREB1* expression is decreased. However, the SCOF1 protein was not found to bind to either the DRE/CRT or ABRE sequences present in the promoters of *COR15a*, *COR47* (both elements), or *RD29B* (only ABRE) (Kim et al. 2001a). Interestingly, SCOF1 was shown to interact with the bZIP protein SGBF-1 in the yeast two-hybrid system (Kim et al. 2001). SGBF-1 in soybean has been shown to bind to the G-box, which shares the core sequence ACGT with the ABRE (Hong et al. 1995) and SCOF1 was shown to enhance the DNA binding activity of SGBF-1 to the ABRE sequence in a gel shift assay (Kim et al. 2001a). The fact that *SGBF-1* was also shown to be responsive to low temperature and ABA suggests that SCOF1 acts through interaction with SGBF-1 to regulate low temperature responsive genes through the *cis*-element ABRE (Kim et al. 2001a).

6.5.4 Regulation of transcription factors

The genes encoding transcription factors, such as CBF/DREB or ABF/AREB, shown to be involved in regulation of cold or ABA induced genes are themselves regulated by these stimuli. This raises the question how the initial stress signal is converted to altered gene expression. By definition, the primary signal acceptor in nucleus has to be present in non-stressed conditions being turned to an active form upon stress. Gilmour et al (1998) first hypothesized that an unknown transcription factor, designated ICE, would be activated in response to cold. ICE would then act on the *CBF/DREB1* promoters and activate the genes (see below). Recently, Guo et al. (2002) have demonstrated that all components needed for activation of *CBF/DRB1* genes are indeed present in non-acclimated plants. By screening a mutagenized population of *Arabidopsis* for reduced expression of the *RD29A-luc* fusion after low temperature treatment, they isolated a mutant, *los1-1*, in which the expression of the CBF/DREB1 target genes was highly reduced. Interestingly, the genes encoding all three *CBF*s were superinduced in response to low temperature (Guo et al. 2002). The cloning of the *LOS1* gene revealed that it encodes a translation elongation factor 2-like protein. In vivo protein labelling studies indicated that protein synthesis was specifically inhibited at low temperature. Thus, the *los1-1* mutant carries a low temperature sensitive allele of the *LOS1* gene (Guo et al. 2002). The fact that *CBF/DREB1* gene expression was not inhibited shows that protein synthesis is not needed for the low temperature activation of these genes. Interestingly, the CBF/DREB1 genes were clearly super-induced by cold. This indicates that either CBF/DREB1 proteins are feedback inhibiting their own expression or that one of their target genes is mediating the inhibition (Guo et al. 2002). Alternatively, protein synthesis might be needed to regulate the level of the hypothetical ICE protein. If e.g. the HOS1 is regulating the level of the ICE protein then inhibition of protein synthesis leads to reduced amount of HOS1 and subsequent increase in stability of ICE (Fig. 4).

The promoter regions of DREB1 genes contain sequences similar to the ABRE and MYB and MYC recognition motifs (Shinwari et al. 1998). However, because *DREB1* genes are not responsive to ABA, it appears that the ABRE motifs are not active in the context of *DREB1* promoters. MYC and MYB type of transcription factors, RD22BP1/AtMYC2 and AtMYB2, respectively, have been shown to activate ABA and drought stress responsive gene expression of the *RD22* gene (Abe et al. 1997). The transcripts for these factors are also induced by ABA and dehydration stress, but not by cold treatment (Urao et al. 1993, Abe et al. 1997). Recently, by using transgenic plants overexpressing AtMYC2 or AtMYB2 Abe et al. (2003) have shown that the expression of several ABA regulated genes is enhanced in these plants. Furthermore, a transposon insertion in the AtMYC2 gene reduced the ABA responsive expression of the *RD22* and *AtADH1* genes (Abe et al. 2003). This indicates that AtMYC2 and AtMYB2 are regulating genes in response to ABA. Recently, the first evidence of the involvement of MYC/MYB type of transcription factors in activation of gene expression in response to low temperature has been obtained. Zhu and colleagues have achieved identification of the gene encoding a putative ICE protein (Zhu, personal communication). By screening for mutations leading to altered expression of the *CBF1-luc* reporter gene in transgenic *Arabidopsis*, they isolated a mutant, where the low temperature induction of *CBF1* was highly reduced. The cloning of the corresponding gene revealed that it encodes a MYC type bHLH transcription factor that has affinity to the *CBF* promoters. As expected, the ICE gene is constitutively expressed (Fig. 4).

A novel putative negative regulator of CRT/DRE genes was recently identified (Xiong et al. 2002, Koiwa et al. 2002). By screening for altered responsiveness of the *RD29A-luc* reporter construct to stress in T-DNA mutagenized transgenic plants, Koiwa et al. (2002) isolated two mutants, *clp1* and *clp3*, where the luciferase activity was superinduced in response to cold ABA and salt (*clp1*) or only ABA (*clp3*). Slightly enhanced expression was also found for the endogenous *RD29A* gene and by nuclear run-on transcription, they showed that the increased expression was not due to more efficient initiation of transcription (Koiwa et al. 2002). The *CLP1* and *CLP3* genes are encoding proteins with high similarity to

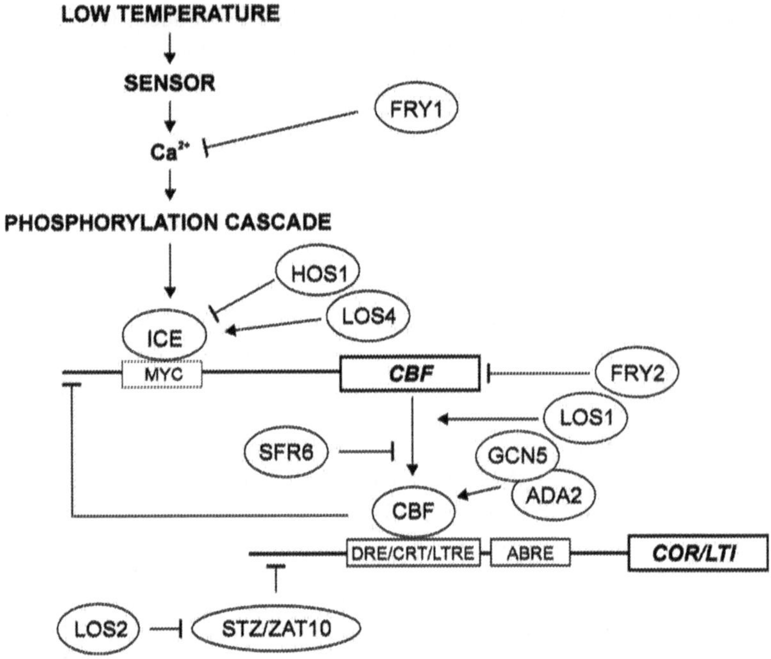

Fig. 4. A model for CBF-mediated expression of low temperature responsive genes. See text for details

CTD phosphatases, which dephosphorylate the conserved heptapeptide repeat of the carboxy-terminal domain (CTD) of RNA polymerase II. Phosphorylation status of CTD is known to regulate the promoter clearance and elongation stages of transcription; RNAP II with unphosphorylated CTD is entering the preinitiation complex, and subsequent phosphorylation of CTD is needed for turning the initiation complex into an elongation complex. While phosphorylated during the elongation phase, the CTD is again dephosphorylated at the end of transcription. Both AtCLP1 and AtCLP3 were shown to have phosphatase activity in vitro. In an analogous study using chemically mutagenized *RD29A-luc* plants, Xiong et al. (2002) isolated a mutant *fiery2* (*fry2*), which is allelic to the *clp1* mutant. The *fry2* mutant showed enhanced accumulation of several stress induced transcripts in response to low temperature, ABA and NaCl. The enhanced expression appeared to be restricted to genes that are regulated by CBFs/DREBs, as the mRNA levels for the *RD22* and *RD19* genes were not affected and the plants showed no apparent phenotype in non-stressed conditions (Xiong et al. 2002). The mRNAs for *DREB2A* and *CBF1*, *2*, and *3* also accumulated to higher levels in the mutant after NaCl and low temperature treatments, respectively. Xiong et al. (2002) have suggested that FRY2 acts as a negative regulator for the CBF/DREB regulated genes and the repression seems to take place at the level of *CBF/DREB* transcription (Fig. 4) (Xiong et al. 2002). Surprisingly, the *fry2* mutant had reduced cold acclimation capacity as compared with the wild type, even if the *CBF*s and their target genes were superinduced. This is in contrast with studies showing that constitutive overproduction of CBFs leads to enhanced expression of its target genes and increased freezing tolerance in transgenic plants (Jaglo-Ottosen et al. 1998, Liu et al. 1998, Kasuga et al. 1999). Interestingly, also the *fry1* mutant was shown to be deficient in acclimation, even if the expression of *CBF2* and several stress responsive genes was elevated (Xiong et al. 2001b) (see 6.3). This indicates that either *fry1* and *fry2* mutations have pleiotropic effects on processes involved in cold acclimation or that the downregulation of the *CBF*-genes is essential.

Recently, Gong et al. (2002) identified another positive regulator for the *CBF* genes. They isolated a mutant, *los4*, where the cold induction of the *RD29A-luc* reporter construct as well as endogenous low temperature responsive genes was reduced. The reduced expression of cold induced genes was reflecting the lower expression levels of the *CBF* genes (Gong et al. 2002). The *LOS4* gene encodes a DEAD-box RNA helicase, indicating that it functions in regulation of RNA metabolism. As expected, the *los4* mutant was deficient in cold acclimation. Interestingly *los4* was also found to be chilling sensitive, and the chilling sensitivity was greatly enhanced in darkness. Both chilling sensitivity and acclimation deficiency could be complemented by overexpression of *CBF3* (Gong et al. 2002) indicating that the CBF target genes are, in addition to development of freezing tolerance also required for chilling tolerance.

Light has previously been shown to be required for development of freezing tolerance, but not for expression of CBF-regulated target genes during cold acclimation (Wanner and Junttila 1999). However, by using transgenic plants harbouring a reporter construct where the *GUS*-gene was connected to four copies of *DRE/CRT* element, Kim et al. (2002) were able to demonstrate that the activation of gene expression through this element is requiring light and active PhyB. These results indicate that the expression of at least a subset of the CBF/DREB1 regulated genes in the absence of light is activated through a pathway not involving CBF/DREB1-factors. PhyA appears also to be needed for expression of the *CBF2* gene, at least under some conditions. Transient accumulation of the *CBF2* transcript has been shown in response to far red light, and this accumulation was found to be PhyA dependent (Tepperman et al. 2001). Crossatti et al. (1999) have also shown that the low temperature induction of the barley *COR14A* gene is requiring light. Taken together, it appears that light is a component regulating low temperature responsive gene expression and cold acclimation.

6.5.5 Post-transcriptional regulation of gene expression

Several lines of evidence indicate that some low temperature responsive genes appear to be regulated also at the post-transcriptional level. Crossatti et al. (1999) have shown that, in addition to be needed for full induction of gene expression, light also regulates the stability of the COR14b protein in barley. Results of Phillips et al. (1997) indicate that mRNA stability is modulated by a low-temperature-dependent protein factor. Interestingly, it has also been shown that the stability of the mRNA for the transcription factor SCOF-1 is regulated by low temperature

(Kim et al. 2001b). By constitutive overexpression of *SCOF-1* from 35S-promoter in transgenic tobacco, they showed that the amount of the *SCOF-1* transcript was elevated several fold when the plants were exposed to low temperature, even if the 35S-promoter itself is not responding to cold. Furthermore, they showed that degradation of the *SCOF-1* mRNA during deacclimation in soybean is depending on active gene expression, because treatment of cells with a transcription inhibitor, cordycepin, resulted in stabilization of the mRNA (Kim et al. 2001b).

An emerging theme in eukaryotic gene expression is the involvement of altered mRNA metabolism. Several RNA-binding proteins, which might stabilize or activate mRNA, have been found to be low temperature responsive e.g. in *Arabidopsis*, barley, leafy spurge, and potato (Carpenter et al. 1994, Dunn et al. 1996, Horwath and Olson 1998, Baudo et al. 1999). In prokaryotes, exposure to low temperature is causing a general repression of protein synthesis, but accumulation of a specific set of small proteins collectively known as cold shock proteins (CSPs). E.g. in *Escherichia coli*, the synthesis of the major CSP, CspA, can account up to 10% of total protein synthesis in early stages of low temperature exposure (Phadtare et al. 1999). CSPs are small polypeptides, consisting of two RNA binding domains RNP1 and RNP2 and they have likely functions as RNA chaperons and/or transcriptional antiterminators allowing translation and transcription of cold regulated genes (Phadtare et al. 1999, Jiang et al. 1997, Bae et al. 2000, Weber et al. 2002). A nucleic acid binding domain very similar to the CSPs has been identified in several eukaryotic proteins (Weber et al. 2002). This cold shock domain (CSD), together with auxiliary RNA binding domains, is regulating different processes by specific RNA/DNA binding (Graumann and Marahiel 1998). In plants, cold shock domain proteins have been identified in a variety of species, including lower plants, and both herbaceous and woody plants (Karlson and Imai (2003). In plants, the CSD-containing proteins can be divided in two classes: i) proteins with size and sequence highly similar to the prokaryotic CSPs, so far these have only been identified in barley and wheat and ii) proteins where the CSD is connected to auxiliary domains. The auxiliary domains appear to consist of a glycine rich region containing two or more copies of the CCHC-type of Zn^{2+}-fingers, originally identified as the RNA binding structures in retroviral capsid proteins (Karlson and Imai 2003). Recently Karlson et al. (2002) isolated a cDNA encoding a low temperature responsive CSD-protein in wheat. The WCSD1 protein contains a N-terminal CSD and a glycine rich domain containing three CCHC-type Zn^{2+}-fingers (Karlson et al. 2002). The *WCSD1* transcript was accumulating after 10 and 6 hours of low temperature treatment in shoots and roots of wheat, respectively and the induction was specific for low temperature, for no mRNA accumulation was seen after ABA, drought, or salt treatments (Karlson et al. 2002). By binding assays WCSD1 was shown to bind both ssDNA and dsDNA as well as RNA. Accordingly, Karlson et al. (2002) suggested that the WCSD1 could either participate in regulation of cold induced genes of be involved in recovery from translational arrest. *Arabidopsis* genome contains four genes encoding CSD-proteins all of them containing a glycine rich region and two or seven CCHC Zn^{2+}-fingers. The genes also appear to have altered expression pattern in response to low temperature (Karlson and Imai 2003, von Numers, Palva and

Heino, personal communication). The role of these CSD-proteins in cold acclimation remains to be elucidated.

6.6 Conclusions

The use of extensive mutant screens is providing a steadily increasing amount of information regarding the genes identifying factors that are involved in regulation of low temperature responsive genes. These approaches, combined with biochemical and molecular approaches will give us, in near future, a comprehensive picture of the molecular events leading to cold acclimation and development of freezing tolerance. It is now clearly demonstrated that the initial events in cold signalling include membrane rigidification and alterations in the cytoskeletal component of the cells, which is followed by Ca^{2+}-release from both extra- and intracellular sources and activation of protein kinase cascades, eventually leading to activation of transcription factors (Fig. 2). It is also evident that signal transduction in cold acclimation is following several independent, but interacting and converging pathways, which may involve cascades of transcription factors. The challenge for future research is to elucidate the molecular details of these pathways and analyze the cross-talk between them. This task will be greatly facilitated by analysis of alterations of transcriptomes and proteomes in response to stress. By using genome wide expression profiling or, in smaller scale, analyzing the expression patterns of selected set of genes by microarray-based techniques the global alterations in gene expression during acclimation can be analyzed and compared to the expression profiles in different mutants and transgenic plants. This will also aid in dissection of the different regulons involved in cold acclimation. Seki et al. (2001) and Fowler and Thomashow (2002) have already started to define the CBF/DREB1 regulon by performing expression profiling of 1300 and 8000 *Arabidopsis* genes, respectively, in transgenic plants overexpressing DREB1A (Seki et al. 2001) or CBF1, CBF2, or CBF3 (Fowler and Thomashow (2002). The analysis of transcriptome and proteome changes in response to low temperature will not only aid in determining the components of different regulons, thereby providing tools for genetic engineering, but also give valuable information of the target genes of these regulons, thereby giving an opportunity for metabolic profiling to predict and analyze the physiological changes that ultimately lead to increased freezing tolerance.

The elucidation of the signalling network leading to the activation of the different regulons involved in cold acclimation will also provide tools for regulon engineering. Low temperature stress is causing severe damage to a variety of crops each year and development of low temperature/freezing tolerant varieties would have a major impact in agriculture. Cold acclimation is a polygenic trait where a large number of gene products are contributing to achieve maximal freezing tolerance. Therefore, it is not likely that alteration of the expression pattern of one or few target genes would bring very significant changes in freezing tolerance. On the other hand, by genetic manipulation of signal transduction pathways or their end points, transcription factors, it is possible to simultaneously affect the expression of large amount of genes and consequently, obtain a major increase in tolerance. The power of regulon engineering has already been demonstrated in several studies. Transgenic *Arabidopsis* plants overexpressing *CBF/DREB1* genes have been shown to constitutively express *CBF/DREB1* target genes and exhibit constitutive freezing tolerance (Jaglo-Ottosen 1998, Liu et al. 1998, Kasuga et al. 1999, Gilmour et al. 2000). Jaglo et al. (2001) have also expressed *CBF1*, *CBF2*, and *CBF3* in *B. napus* and shown that this leads to enhanced expression of low temperature responsive *BN115* and *BN28* genes and increased freezing tolerance. In the future, the possible use of multiple transcription factors, activating several distinct regulons will further enhance the biotechnological applications of cold acclimation research.

Acknowledgements

The work in the authors' laboratory is supported by the Finnish Academy, Biocentrum Helsinki, and the National Technology Agency of Finland. We thank Dr. Jian-Kang Zhu for providing unpublished information and members of our lab for critical reading of the manuscript. We are grateful to MSc Elina Helenius for preparing the figures for the manuscript.

References

Abe H, Yamaguchi-Shinozaki K, Urao T, Iwasaki T, Hosakawa D and Shinozaki K (1997) Role of *Arabidopsis* MYC and MYB homologs in drought- and abscisic acid-regulated gene expression. Plant Cell 9:1859-1868

Abe H, Urao T, Ito T, Seki M, Shinozaki K and Yamaguchi-Shinozaki K (2003) *Arabidopsis* AtMYC2 (bHLH) and At MYB2 (MYB) Function as transcriptional activators in abscisic acid signaling. Plant Cell 15:63-78

Allen GJ, Muir SR and Sanders D (1995) Release of Ca^{2+} from individual plant vacuoles by both $InsP_3$ and cyclic ADP-ribose. Science 268:735-737

Bae W, Xia B, Inouye M and Severinov K (2000) *Escherichia coli* CspA-family RNA chaperones are transcription antiterminators. Proc Natl Acad Sci USA 97:7784-7789

Baker SS, Wilhelm, KS and Thomashow MF (1994) The 5'-region of *Arabidopsis thaliana cor15a* has *cis*-acting elements that confer cold-, drought- and ABA-regulated gene expression. Plant Mol. Biol 24:701-713

Baudo MM, Meza-Zepeda LA, Palva ET and Heino P (1999) Isolation of a cDNA corresponding to a low temperature- and ABA-responsive gene encoding a putative glycine rich RNA-binding protein in *Solanum commersonii*. J Exp Bot 50:1867-1868

Capel J, Jarillo JA, Salinas J and Martinez-Zapater JM (1997) Two homologous low-temperature-inducible genes from *Arabidopsis* encode highly hydrophobic proteins. Plant Physiol 115:569-576

Carpenter CD, Kreps JA and Simon AE (1994) Genes encoding glycine-rich *Arabidopsis thaliana* proteins with RNA-binding motifs are influenced by cold treatment and an endogenous circadian rhythm. Plant Physiol 104:1015-1025

Chang C, Kwok SF, Bleecker AB and Meyerowitz EM (1993) *Arabidopsis* ethylene-response gene *ETR1*: similarity of product to two-component regulators. Science 262:539-544

Chen HH, Li, PH and Brenner ML (1983) Involvement of abscisic acid in potato cold acclimation. Plant Physiol 71:362-365

Choi D-W, Rodriguez EM and Close TJ (2002) Barley *Cbf3* gene identification, expression pattern, and map location. Plant Physiol 129:1781-1787

Choi H, Hong J, Ha J, Kang J and Kim SY (2000) ABFs, a family of ABA-responsive element binding factors. J Biol Chem 275:1723-1730

Crosatti C, de Laureto PP, Bassi R and Cattivelli L (1999) The interaction between cold and light controls the expression of the cold-regulated barley gene *cor14b* and the accumulation of the corresponding protein. Plant Physiol 119:671-680

Ding JP and Pickard BG (1993) Modulation of mechanosensitive calcium-selective channels by temperature. Plant J 3:713-720

Dunn MA, Brown K, Lightowlers RL and Hughes MA (1996) A low-temperature- responsivegene from barley encodes a protein with single stranded nucleic acid binding activity which is phosphorylated *in vitro*. Plant Mol Biol 30:947-959

Fowler S and Thomashow MF (2002) *Arabidopsis* transcriptome profiling indicates that multiple regulatory pathways are activated during cold acclimation in additon to the CBF cold response pathway. Plant Cell 14:1675-1690

Foyer CH, Lelandais M and Kunert KJ (1994) Photooxidative stress in plants. Physiol Plant 92:696-717

Gao M-J, Allard G, Byass L, Flanagan AM and Singh J (2002) Regulation and characterization of four CBF transcription factors from *Brassica napus*. Plant Mol Biol 49:459-471

Gilmour S and Thomashow MF (1991) Cold acclimation and cold regulated gene expression in ABA mutants of *Arabidopsis thaliana*. Plant Mol Biol 17:1233-1240

Gilmour SJ, Zarka DG, Stockinger EJ, Salazar MP, Houghton JM and Thomashow MF (1998) Low temperature regulation of the *Arabidopsis CBF* family of AP2 transcriptional activators as an early step in cold-induced *COR* gene expression. Plant J 16:433-442

Gilmour SJ, Sebolt AM, Salazar MP, Everard JD and Thomashow MF (2000) Overexpression of the *Arabidopsis* CBF3 transcriptional activator mimics multiple biochemical changes associated with cold acclimation. Plant Physiol 124:1854-1865

Gilroy S and Trewavas A (1994) A decade of plant signals. BioEssays 16:677-682

Gong M, van der Luit AH, Knight MR and Trewavas AJ (1998) Heat-shock-induced changes in intracellular Ca^{2+} level in tobacco seedlings in relation to thermotolerance. Plant Physiol 116:429-437

Gong Z, Lee H, Xiong L, Jagendorf A, Stevenson B and Zhu J-K (2002) RNA helicase-like protein as an early regulator of transcription factors for plant chilling and freezing tolerance. Proc Natl Acad Sci USA 99:11507-11512

Gosti F, Beaudoin N, Serizet C, Webb AAR and Vartanian N (1999) ABI1 protein phosphatase 2C is a negative regulator of abscisic acid signaling. Plant Cell 11 1897-1909

Graham D and Patterson BD (1982) Responses of plants to low, nonfreezing temperatures: proteins, metabolism and acclimation. Annu Rev Plant Physiol 33:347-372

Graumann PL and Marahiel MA (1998) A superfamily of proteins that contain the cold-shock domain. Trends Biochem Sci 23:286-290

Guiltinan, M.J., Marcotte, W.R. and Quatrano, R.S. 1990. A plant leucine zipper protein that recognizes an abscisic acid response element. Science 250:267-271.

Guo Y, Xiong L, Ishitani M and Zhu J-K (2002) An *Arabidopsis* mutation in translation elongation factor causes superinduction of CBF/DREB1 transcription factor genes but blocks the induction of their downstream targets under low temperature. Proc Natl Acad Sci USA 99:7786-7791

Guy CL (1990) Cold acclimation and freezing stress tolerance: role of protein metabolism. Annu Rev Plant Physiol Plant Mol Biol 41:187-223

Haake V, Cook D, Riechmann JL, Pinede O, Thomashow MF and Zhang JZ (2002) Transcription factor CBF4 is a regulator of drought adaptation in *Arabidopsis*. Plant Physiol 130:639-648

Halfter U, Ishitani M and Zhu J-K (2000) The *Arabidopsis* SOS2 protein kinase physically interacts with and is activated by the calcium-binding protein SOS3. Proc Natl Acad Sci USA 97:3735-3740

Harris DM, Myrick TL and Rundle SJ (1999) The *Arabidopsis* homolog of yeast TAP42 and mammalian α4 binds to the catalytic subunit of protein phosphatase 2A and is induced by chilling. Plant Physiol 121:609-617

Heino P, Sandman G, Lång V, Nordin K and Palva ET (1990) Abscisic acid deficiency prevents development of freezing tolerance in *Arabidopsis thaliana* (L.) Heynh. Theor Appl Genet 79:801-809

Hirschi KD (1999) Expression of *Arabidopsis CAX1* in tobacco: altered calcium homeostasis and increased stress sensitivity. Plant Cell 11:2113-2122

Hiilovaara-Teijo M and Palva ET (1998) Molecular responses in cold adapted plants. In: Margesin R and Schinner F eds Cold-Adapted Organisms: Ecology, physiology, enzymology and molecular biology. Springer-Verlag, Heidelberg, pp 349-384

Hirayama, T, Ohto C, Mizoguchi T and Shinozaki K (1995) A gene encoding a phosphatidylinositol-specific phospholipase C is induced by dhydration and salt stress in *Arabidopsis thaliana*. Proc Natl Acad Sci USA 92:3903-3907

Hong JC, Cheong YH, Nagao RT, Bahk JD, Key JL and Cho MJ (1995) Isolation of two soybean G-box binding factors which interact with a G-box sequence of an auxin-responsive gene. Plant J 8:199-211

Hong SW, Jon JH, Kwak JM and Chua N-H (1997) Identification of a receptor-like protein kinase gene rapidly induced by abscisic acid, dehydration, high salt and cold treatments in *Arabidopsis thaliana*. Plant Physiol 113:1203-1212

Horwath DP and Olson PA (1998) Cloning and characterization of cold-regulated glycine-rich RNA-binding protein genes from leafy spurge (*Euphorbia esula* L.) and comparison to heterologous genomic clones. Plant Mol Biol 38:531-538

Hughes, MA and Dunn, MA (1996) The molecular biology of plant acclimation to low temperature. J Exp Bot 47:291-305

Ichimura K, Mizoguchi T, Yoshida R, Yuasa T and Shinozaki K (2000) Various abiotic stresses rapidly activate *Arabidopsis* MAP kinases ATMPK4 and ATMPK6. Plant J 24:655-665

Inaba M, Franchescelli S, Suzuki I, Szalontai B, Kanesaki Y, Los DA, Maresca B, Hayashi H and Murata N (2003) J Biol Chem (In Press) DOI 10.1074/jbc.M212204200

Ingram J and Bartels D (1996) The molecular basis of dehydration tolerance in plants. Annu Rev Plant Physiol Plant Mol Biol 47:377-403

Inoue T, Hihuchi M, Hashimoto Y, Seki M, Kobayashi M, Kato T, Tabata S, Shinozaki K and Kakimoto T (2001) Identification of CRE1 as a cytokinin receptor from *Arabidopsis*. Nature 409:1060-1063

Inzé D and Van Montagu M (1995) Oxidative stress in plants. Curr Opin Biotech 6:153-158

Ishitani M, Xiong L, Stevenson B and Zhu J-K (1997) Genetic analysis of osmotic and cold stress signal transduction in *Arabidopsis*: Interactions and convergence of abscisic acid-dependent and abscisic acid-independent pathways. Plant Cell 9:1935-1949

Ishitani M, Xiong L, Lee H, Stevenson B and Zhu J-K (1998) *HOS1*, a genetic locus involved in cold-responsive gene expression in *Arabidopsis*. Plant Cell 10:1151-1161

Jaglo KR, Kleff S, Amundsen KL, Zhang X, Haake V, Zhang JZ, Deits T and Thomashow MF (2001) Components of the *Arabidopsis* C-repeat/dehydration-responsive element binding factor cold-response pathway are conserved in *Brassica napus* and other plant species. Plant Physiol 127:910-917

Jaglo-Ottosen KR, Gilmour S, Zarka DG, Schabenberger O and Thomashow MF (1998) *Arabidopsis CBF1* overexpression induces *COR* genes and enhances freezing tolerance. Science 280:104-106

Jonak C, Kiegerl S, Ligternik W, Barkel PJ, Huskisson NS and Hirt H (1996) Stress signaling in plants: A mitogen-activated protein kinase pathway is activated by cold and drought. Proc Natl Acad Sci USA 93:11274-11279

Junttila O (1980) Effect of photoperiod and temperature on apical growth cessation in two ecotypes of *Salix* and *Betula*. Physiol Plant 48:347-352

Kanaya E, Nakajima N, Morikawa K, Okada K and Shimura Y (1999) Characterization of the transcriptional activator CBF1 from *Arabidopsis thaliana*. Evidence for cold denaturation in regions outside the DNA binding domain. J Biol Chem 274:16068-16076

Kang J-y, Choi H-i, Im M-y and Kim SY (2002) *Arabidopsis* basic leusine zipper proteins that mediate stress-responsive abscisic acid signaling. Plant Cell 14:343-357

Karlson D and Imai R (2003) Conservation of the cold shock domain protein family in plant. Plant Physiol 131:12-15

Karlson D, Nakaminami K, Toyomasu T and Imai R (2002) A cold regulated nucleic acid-binding protein of winter wheat shares a domain with bacterial cold shock proteins. J Biol Chem 277:35248-35256

Kasuga M, Liu Q, Miura S, Yamaguchi-Shinozaki K and Shinozaki K (1999) Improving plant drought, salt, and freezing tolerance by gene transfer of a single stress-inducible transcription factor. Nature Biotechnol 17:287-291

Kim JC, Lee SH, Cheong YH, Yoo C-M, Lee SI, Chun HJ, Yun D-J, Hong JC, Lee SY, Lim CO and Cho MJ (2001a) A novel cold-inducible zink finger protein from soybean, SCOF1, enhances cold tolerance in transgenic plants. Plant J 25:247-259

Kim JC, Jeong JC, Park HC, Yoo JH, Koo YD, Yoon HW, Koo SC, Lee S-H, Bahk JD and Cho MJ (2001b) Cold accumulation of SCOF-1 transcripts is associated with transcriptional activation and mRNA stability. Mol Cells 12:204-208

Kim H-J, Kim Y-K, Park J-Y and Kim J (2002) Light signaling mediated by phytochrome plays an important role in cold-induced gene expression through the C-repeat/dehydration responsive element (C/DRE) in *Arabidopsis thaliana*. Plant J 29:693-704

Kim K-N, Cheong YH, Grant JJ, Pandey GK and Luan S (2003) CIPK3, a calcium sensor-associated protein kinase that regulates abscisic acid and cold signal transduction in *Arabidopsis*. Plant Cell 15:411-423

Knight MR, Campbell AK, Smith SM and Trewavas AJ (1991) Transgenic plant aequorin reports the effects of touch and cold-shock and elicitors on cytoplasmic calcium. Nature 352:524-526

Knight H, Trwavas AJ and Knight MR (1996) Cold calcium signaling in *Arabidopsis* involves two cellular pools and a change in calcium signature after acclimation. Plant Cell 8:489-503

Knight H, Veale EL, Warren GJ and Knight MR (1999) The *sfr6* mutation in *Arabidopsis* suppresses low-temperature induction of genes dependent on the CRT/DRE sequence motif. Plant Cell 11:875-886

Koiwa H, Barb AW, Xiong L, Li F, McCully MG, Lee B-h, Sokolchik I, Zhu J, Gong Z, Reddy M, Sharkhuu A, Manabe Y, Yokoi S, Zhu J-K, Bressan RA and Hasegava PM (2002) C-terminal domain phosphatase-like family members (AtCPLs) differentially regulate *Arabidopsis thaliana* abiotic stress signaling, growth, and development. Proc Natl Acad Sci USA 99:10893-10898

Koornneef M, Jorna ML, Brinkhorst-van der Swan DLC and Karssen CM (1982) The isolation of abscisic acid (ABA) deficient mutants by selection of induced revertants in non-germinating gibberellin sensitive lines of *Arabidopsis thaliana* (L.) Heynh. Theor Appl Genet 61:385-393

Koornneef M, Reuling G and Karssen CM (1984) The isolation of abscisic acid-insensitive mutants of *Arabidopsis thaliana.* Physiol Plant 61:377-383

Kovtun Y, Chiu W-L, Tena G and Sheen J (2000) Functional analysis of oxidative-stress activated mitogen-activated protein kinase cascafe in plants. Proc Natl Acad Sci USA 97:2940-2945

Kreps JA, Wu Y, Chang H-S, Zhu T, Wang X and Harper JF (2002) Transcriptome changes for *Arabidopsis* in response to salt, osmotic and cold stress. Plant Physiol 130:2129-2141

Kudla J, Xu Q, Harter K, Gruissem W and Luan S (1999) Genes for calcineurin B-like proteins in *Arabidopsis* are differentially regulated by stress signals. Proc Natl Acad Sci USA 96:4718-4723

Kurkela S and Borg-Franck M (1992) Structure and expression of *kin2*, one of two cold- and ABA-induced genes in *Arabidopsis thaliana.* Plant Mol Biol 19:689-692

Lalk I and Dörffling K (1985) Hardening, abscisic acid, proline and freezing resistence in two winter wheat varieties. Physiol Plant 61:287-292

Lee H, Xiong L, Gong Z, Stevenson B and Zhu J-K (2001) The *Arabidopsis HOS1* gene negatively regulates cold signal transduction and encodes a RING finger protein that displayes cold-regulated nucleo-cytoplasmic partitioning. Genes Dev 15:912-924

Lee H, Guo Y, Ohta M, Xiong L, Stevenson B and Zhu J-K (2002) *LOS2*, a genetic locus required for cold responsive gene transcription encodes a bi-functional enolase. EMBO J 21:2692-2702

Leung J, Bourvier-Durant, M, Morris P-C, Guerrier D, Chefdor F and Giraudad J (1994) *Arabidopsis* ABA response gene *ABI1*: Features of a calcium-modulated protein phosphatase. Science 264:1448-1452

Leung, J. and Giraudat, J. 1998. Abscisic acid signal transduction. Annu. Rev. Plant. Physiol. Plant Mol. Biol. 49:199-222.

Levitt J (1980) Responses of Plants to Environmental Stresses: Chilling, Freezing and High Temperature Stresses. 2nd ed Academic Press New York

Li C, Puhakainen T, Welling A, Viherä-Aarnio A, Ernstsen A, Junttila O, Heino P and Palva ET (2002) Cold acclimation in silver birch (*Betula pendula*). Development of freezing tolerance in different tissues and climatic ecotypes. Physiol Plant 116:478-488

Li C, Viherä-Aarnio A, Puhakainen T, Junttila O, Heino P and Palva ET (2003) Ecotype dependent control of growth, dormancy and freezing tolerance under seasonal changes in *Betula pendula* Roth. Trees 17:127-132

Lippuner V, Cyert MS and Gasser CS (1996) Two classes of plant cDNA clones differentially complement yeast calcineurin mutants and increase salt tolerance of wild-type yeast. J Biol Chem 271:12859-12866

Liu J and Zhu J-K (1997) An *Arabidopsis* mutant that requires increased calcium for potassium nutrition and salt tolerance. Proc Natl Acad Sci USA 94:14960-14964

Liu J and Zhu J-K (1998) A calcium sensor homolog required for plant salt tolerance. Science 280:1943-1945

Liu J, Ishitani M, Halfter U, Kim CS and Zhu J-K (2000) The *Arabidopsis thaliana SOS2* gene encodes a protein kinase that is required for salt tolerance. Proc Natl Acad Sci USA 97:3730-3734

Liu Q, Kasuga M, Sakuma Y, Abe H, Miura S, Yamaguchi-Shinozaki K and Shinozaki K. (1998) Two transcription factors, DREB1 and DREB2, with an EREBP/AP2 DNA binding domain separate two cellular signal transduction pathways in drought- and low-temperature-responsive gene expression, respectively, in *Arabidopsis*. Plant Cell 10:1391-1406

Llorente F, Oliveros JC, Martinez-Zapater JM and Salinas J (2000) A freezing sensitive mutant of *Arabidopsis*, *frs1*, is a new *aba3* allele. Planta 211:648-655.

Lång V, Heino P and Palva ET (1989) Low temperature acclimation and treatment with exogenous abscisic acid induce common polypeptides in *Arabidopsis thaliana* (L.) Heynh. Theor Appl Genet 77:729-734

Lång V and Palva ET (1992) The expression of a *rab*-related gene, *rab18*, is induced by abscisic acid during the cold acclimation process of *Arabidopsis thaliana* (L.) Heynh. Plant Mol Biol 20: 951-962

Lång V, Mäntylä E, Welin B, Sundberg B and Palva ET (1994) Alterations in water status, endogenous abscisic acid content and expression of *rab18* gene during the development of freezing tolerance in *Arabidopsis thaliana.* Plant Physiol 104:1341-1349

Martin ML and Busconi L (2001) A rice membrane-bound calcium-dependent protein kinase is activated in response to low temperature. Plant Physiol 125:1442-1449

McKersie BD and Bowley SR (1998) Active oxygen and freezing tolerance in transgenic plants. In Plant Cold Hardiness (Li PH and Chen T eds Plenum Press New York pp 203-212

McAinsh MR and Hetherington AM (1998) Encoding specificity in Ca^{2+} signaling systems. Trends Plant Sci 3:32-35

McKown R, Kuroki G and Warren G (1996) Cold responses of *Arabidopsis* mutants impaired in freezing tolerance. J Exp Bot 47:1919-1925

Medina J, Bargues M, Terol J, Pérez-Alonso M and Salinas J (1999) The *Arabidopsis CBF* gene family is composed of three genes encoding AP domain-containing proteins whose expression is regulated by low temperature but not by abscisic acid or dehydration. Plant Physiol 119:463-469

Merlot S, Gosti F, Guerrier D, Vavasseur A and Giraudad J (2001) The ABI1 and ABI2 protein phosphatases 2C act in a negative feedback regulatory loop of the abscisic acid signaling pathway. Plant J 25:315-324

Meyer, K, Leube MP and Grill E (1994) A protein phosphatase 2C involved in ABA signal transduction in *Arabidopsis thaliana.* Science 264:1452-1455

Mikami K, Katagiri, T, Luchi S, Yamaguchi-Shinozaki K and Shinozaki K (1998) A gene encoding phosphatidylinositol 4-phosphate 5-kinase is induced by water stress and abscisic acid in *Arabidopsis thaliana.* Plant J 15:563-568

Mizoguchi T, Irie K, Hirayama T, Hayashida N, Yamaguchi-Shinozaki K, Matsumoto K and Shinozaki K (1996) A gene encoding a mitogen-activated protein kinase kinase kinase is induced simultaneously with genes for a mitogen-activated protein kinase and an S6 ribosomal protein kinase by touch, cold and water stress in *Arabidopsis thaliana.* Proc Natl Acad Sci USA 93:765-769

Mizoguchi T, Ichimura K and Shinozaki K (1997) Environmental stress response in plants: the role of mitogen-activated protein kinases. Trends Biotechnol 15:15-19

Mizoguchi T, Ichimura K, Irie K, Morris PC, Giraudad J, Matsumoto K and Shinozaki K (1998) Identification of a possible MAP kinase cascade in *Arabidopsis thaliana* based on pairwise yeast two-hybrid analysis and the functional complementation tests of the yeast mutants. FEBS Lett 437:56-60

Monroy AF and Dhindsa RS (1995) Low temperature signal tranduction: Induction of cold-acclimation specific genes of alfalfa by calcium at 25°C. Plant Cell 7:321-331

Monroy AF, Sarhan F and Dhindsa RS (1993) Cold induced changes in freezing tolerance, protein phosphorylation and gene expression: Evidence for a role of calcium. Plant Physiol 102:1227-1235

Monroy AF, Sangwan V and Dhindsa RS (1998) Low temperature signal transduction during cold acclimation: protein phosphatase 2A as an early target for cold inactivation. Plant J 13:653-660

Murata N and Los DA (1997) Membrane fluidity and temperature perception. Plant Physiol 115:875-879

Mäntylä E, Lång V and Palva ET (1995) Role of abscisic acid in drought-induced freezing tolerance, cold acclimation and accumulation of LTI78 and RAB18 proteins in *Arabidopsis thaliana.* Plant Physiol 107:141-148

Nordin K, Heino P and Palva ET (1991) Separate signal pathways regulate the expression of a low-temperature-induced gene in *Arabidopsis thaliana* (L.) Heynh. Plant Mol Biol 16:1061-1071

Nordin K, Vahala T and Palva ET (1993) Differential expression of two related, low-temperature-induced genes in *Arabidopsis thaliana* (L.) Heynh. Plant Mol Biol 21:641-653

Nuotio S, Heino P and Palva ET (2001) Signal transduction under low-temperature stress. In: Basra AB ed Crop Responses and Adaptations to Temperature Stress. Food Products Press, Binghamton, New York, pp151-176

Nylander M, Heino P, Helenius E, Palva ET, Ronne H and Welin BV (2001) The low temperature- and salt-induced *RCI2A* gene of *Arabidopsis* complements the sodium sensitivity caused by a deletion of the homologous yeast gene *SNA1.* Plant Mol Biol 45:341-352

Ohta M, Matsui K, Hiratsu K, Shinshi H and Ohme-Takagi M (2001) Repression domains of class II ERF transcriptional repressors share an essential motif for active repression. Plant Cell 13:1959-1968

Olsen JE, Junttila O, Nilsen J, Eriksson ME, Martinussen I, Olsson O, Sandberg G and Moriz T (1997) Ectopic expression of oat phytochrome A in hybrid aspen changes critical daylength for growth and prevents cold acclimatization. Plant J 12:1339-1350

Palva ET (1994) Gene expression under low temperature stress. In: Basra AS ed Stress Induced Gene Expression in Plants. Harwood Academic Publishers, New York, pp103-130

Palva ET and Heino P (1998) Molecular mechanism of plant cold acclimation and freezing tolerance. In: Li PH and Chen THH eds Plant Cold Hardiness. Plenum, New York, pp3-14

Phadtare S, Alsina J and Inouye M (1999) Cold-shock response and cold-shock proteins. Curr Opin Microbiol 2:175-180

Phillips JR, Dunn MA and Hughes MA (1997) mRNA stability and localization of the low-temperature-responsive barley gene family *blt14*. Plant Mol Biol 33:1013-1023

Plieth C, Hansen U-P, Knight H and Knight M (1999) Temperature sensing in plants: the primary characteristics of signal perception and calcium response. Plant J 18:491-497

Polisensky DH and Braam J (1996) Cold-shock regulation of the *Arabidopsis TCH* genes and the effects of modulating intracellular calcium levels. Plant Physiol 111:1271-1279

Price AH, Taylor A, Ripley SJ, Griffiths A, Trewavas AJ and Knight MR (1994) Oxidative signals in tobacco increase cytosolic calcium. Plant Cell 6:1301-1310

Puhakainen T, Pihakaski-Maunsbach K, Widell S and Sommarin M (1999) Cold acclimation enhances the activity of plasma membrane Ca^{2+} ATPase in winter rye leaves. Plant Physiol Biochem 37:231-239

Rudd JJ and Franklin-Tong VE (1999) Calcium signaling in plants. Cell Mol Life Sci 55:214-232

Ryu SB and Li PH (1994) Potato cold hardiness development and abscisic acid. II. *De novo* protein synthesis is required for the increase in free abscisic acid during potato (*Solanum commersonii*) cold acclimation. Physiol Plant 90:21-26

Saijo Y, Hata S, Kyozuka J, Shimamoto K and Izui K (2000) Over-expression of a single Ca^{2+}-dependent protein kinase confrs both cold and salt/drought tolerance on rice plants. Plant J 23:319-327

Sakai A and Larcher W (1987) Frost Survival of Plants: Responses and Adaptation to Freezing Stress. Springer-Verlag Berlin

Sakamoto H, Araki T, Meshi T and Iwabuchi M (2000) Expression of a subset of the *Arabidopsis* Cys_2/His_2-type zink-finger protein gene family under water stress. Gene 248:23-32

Sanders D, Brownlee C and Harper JF (1999) Communicating with calcium. Plant Cell 11:691-706

Sangwan W, Foulds, I, Singh J and Dhindsa RS (2001) Cold-activation of *Brassica napus BN115* promoter is mediated by structural changes in membranes and cytoskeleton, and requires Ca^{2+} influx. Plant J 27:1-12

Sangwan V, Örvar BL, Beyerly J, Hirt H and Dhindsa RS (2002) Opposite changes in membrane fluidity mimic cold and head stress activation of distinct plant MAP kinase pathways. Plant J 31:629-638

Seki M, Narusaka M, Abe H, Kasuga M, Yamaguchi-Shinozaki K and Shinozaki K (2001) Monitoring the expression pattern of 1300 *Arabidopsis* genes under drought and cold stresses by using a full-length cDNA microarray. Plant Cell 13:61-72

Shi J, Kim KN, Ritz O, Albrecht V,Gupta R, Harter K, Luan S and Kudla J (1999) Novel protein kinases associated with calcineurin B-like calcium sensors in *Arabidopsis*. Plant Cell 11:2393-2406

Shi H, Ishitani M, Kim C and Zhu J-K (2000) The *Arabidopsis thaliana* salt tolerance gene *SOS1* encodes a putative Na^+/H^+ antiporter. Proc Natl Acad Sci USA 97:6896-6901
Shinozaki K and Yamaguchi-Shinozaki K (2000) Molecular responses to dehydration and low temperature: differences and cross-talk between two stress signaling pathways. Curr Opin Plant Biol 3:217-223
Shinwari ZK, Nakashima K, Miura S, Kasuga M, Seki M, Yamaguchi-Shinozaki K and Shinozaki K (1998) An *Arabidopsis* gene family encoding DRE/CRT binding proteins involved in low-temperature-responsive gene expression. Biochem Biophys Res Commun 250:161-170
Steponkus PL (1984) Role of the plasma membrane in freezing injury and cold acclimation. Annu Rev Plant Physiol 35:543-584
Steponkus PL, Uemura M and Webb MS (1993) Membrane destabilization during freeze-induced dehydration. Curr Topics Plant Physiol 10:37-47
Stockinger EJ, Gilmour SJ and Thomashow MF (1997) *Arabidopsis thaliana CBF1* encodes an AP2 domain-containing transcriptional activator that binds to the C-repeat/DRE, a cis-acting DNA regulatory element that stimulates transcription in response to low temperature and water deficit. Proc Natl Acad Sci USA 94:1035-1040
Stockinger EJ, Mao Y, Regier MK, Triezenberg SJ and Thomashow MF (2001) Transcriptional adaptor and histone acetyltransferase proteins in *Arabidopsis* and their interactions with CBF1, a transcriptional activator involved in cold-regulated gene expression. Nucl Acids Res 29:1524-1533
Suzuki I, Los DA, Kanesaki Y, Mikami K and Murata N (2000) The pathway for perception and transduction of low-temperature signals in *Synechocystis*. EMBO J 19:1327-1334
Suzuki I, Kanesaki Y, Mikami K, Kanehisa M and Murata N (2001) Cold-regulated genes under control of the cold sensor Hik33 in *Synecocystis*. Mol Microbiol 40:235-244
Tähtiharju S and Palva E T (2001) Antisense inhibition of protein phosphatase *2C* accelerates cold acclimation in *Arabidopsis thaliana*. Plant J 26: 461-470
Tähtiharju S, Sangwan V, Monroy AF, Dhindsa RS and Borg M (1997) The induction of *kin* genes in cold-acclimating *Arabidopsis thaliana*. Evidence for a role of calcium. Planta 203: 442-447
Tepperman JM, Zhu T, Chang H-S, Wang X and Quail PH (2001) Multiple transcription-factor genes are early targets for phytochrome A signaling. Proc Natl Acad Sci USA 98:9437-9442
Thomasow MF (1998) Role of cold-responsive genes in plant freezing tolerance. Plant Physiol 118:1-7
Thomashow MF (1999) Plant cold acclimation: freezing tolerance genes and regulatory mechanisms. Annu Rev Plant Physiol 50:571-599
Thomashow MF (2001) So what's new in the field of cold acclimation? Lots! Plant Physiol 125:89-93
Townley HH and Knight MR (2002) Calmodulin as potential a negative regulator of *Arabidopsis COR* gene expression. Plant Physiol 128:1169-1172
Trewavas AJ and Malhó R (1997) Signal perception and transduction: the origin of the phenotype. Plant Cell 9:1181-1195
Uno Y, Furihata T, Abe H, Yoshida R, Shinozaki K and Yamaguchi-Shinozaki K (2000) *Arabidopsis* basic leucine zipper transcription factors involved in an abscisic acid-dependent signal transduction pathway under drought and high-salinity conditions. Proc Natl. Acad Sci USA 97:11632-11637

Urao T, Yamaguchi-Shinozaki, K, Urao S and Shinozaki K (1993) An *Arabidopsis myb* homolog is induced by dehydration stress and its gene product binds to the conserved MYB recognition sequence. Plant Cell 5:1529-1539

Urao T, Yakubov B, Satoh R, Yamaguchi-Shinozaki K, Seki M, Hirayama T and Shinozaki K (1999) A transmembrane hybrid-type histidine kinase in *Arabidopsis thaliana* functions as an osmosensor. Plant Cell 11:1743-1754

Urao T, Yamaguchi-Shinozaki K and Shinozaki K (2000) Two-component systems in plant signal transduction. Trends Plant Sci 5:67-74

Vigh L, Los D, Horváth I and Murata N (1993) The primary signal in the biological perception of temperature: Pd-catalyzed hydrogenation of membrane lipids stimulated the expression of the *desA* gene in *Synecocystis* PCC6803. Proc Natl Acad Sci USA 90:9090-9094

Viswanathan C and Zhu J-K (2002) Molecular genetic analysis of cold-regulated gene transcription. Phil Trans R Soc Lond B 357:877-886

Wanner LA and Junttila O (1999) Cold-induced freezing tolerance in *Arabidopsis*. Plant Physiol 120:391-399

Warren G, McKown R, Martin AL and Teutonico V (1996) Isolation of mutations affecting the development of freezing tolerance in *Arabidopsis* thaliana (L.) Heynh. Plant Physiol 111:1011-1019

Weber MHW, Fricke I, Doll N and Marahiel MA (2002) CSDBase: an interactive database for cold shock domain-containing proteins and bacterial cold shock response. Nucl Acids Res 30:375-378

Weiser CJ (1970) Cold resistance and injury in woody plants. Science 169:1269-1278

Welling A, Kaikuranta P and Rinne P (1997) Photoperiodic induction of dormancy and freezing tolerance in Betula pubescens. Involvement of ABA and dehydrins. Physiol Plant 100:119-125

Welling A, Moriz T, Palva ET and Junttila O (2002) Independent activation of cold acclimation by low temperature and short photoperiod in hybrid aspen. Plant Physiol 129:1633-1641

Wu Y, Kuzma J, Maréchal E, Graeff R, Lee HC, Foster R and Chua N-H (1997) Abscisic acid signaling in plants through cyclic ADP-ribose in plants. Science 278:2126-2130

Xin Z and Browse J (1998) *eskimo1* mutants of *Arabidopsis* are constitutively freezing-tolerant. Proc Natl Acad Sci USA 95:7799-7804

Xin Z and Browse J (2000) Cold comfort farm: the acclimation of plants to freezing temperatures. Plant Cell Environ 23:893-902

Xiong L, Ishitani M, Lee H and Zhu J-K (2001a) The *Arabidopsis* LOS5/ABA3 locus encodes a molybdenum cofactor sulfurase and modulates cold stress- and osmotic stress-responsive gene expression. Plant Cell 13:2063-2083

Xiong L, Lee BH, Ishitani M, Lee H, Zhang C and Zhu J-K (2001b) *FIERY1* encoding an inositol polyphosphate 1-phosphatase is a negative regulator of abscisic acid and stress signaling in *Arabidopsis*. Genes Dev 15:1971-1984

Xiong L, Lee H, Ishitani M, Tanaka Y, Stevenson Becky, Koiwa H, Bressan RA, Hasegawa PM and Zhu J-K (2002) Repression of stress-responsive genes by FIERY2, a novel transcriptional regulator in *Arabidopsis*. Proc Natl Acad Sci USA 99:10899-10904

Yamaguchi-Shinozaki K and Shinozaki K (1993) *Arabidopsis* DNA encoding two desiccation-responsive *rd29* genes. Plant Physiol 101:1119-1120

Yamaguchi-Shinozaki K and Shinozaki K (1994) A novel *cis*-acting element in an *Arabidopsis* gene is involved in responsiveness to drought, low-temperature, or high-salt stress. Plant Cell 6:251-264

Yang Y, Fang S, Jensen JP, Weissman AM and Ashwell JD (2000) Ubiquitin protein ligase activity of IAPs and their degradation in proteasomes in response to apoptotic stimuli. Science 288:874-877

Zhu J-K (2001) Cell signaling under salt, water and cold stresses. Curr Op in Plant Biol 4: 401-406

Zhu J-K (2002) Salt and drought stress signal transduction in plants. Annu Rev Plant Biol 53:247-273

Örvar BL, Sangwan V, Omann F and Dhindsa RS (2000) Early steps in cold sensing by plant cells: the role of actin cytoskeleton and membrane fluidity. Plant J 23:785-794

7 Heavy metal signalling in plants: linking cellular and organismic responses

Andrea Polle and Andres Schützendübel

Abstract

Heavy metals are required in plants as essential micronutrients or act as toxic compounds. How do plants perceive heavy metals and which signalling cascades are triggered leading to plant adaptation or injury? Copper (Cu) and cadmium (Cd) are reviewed as examples for heavy metals with contrasting physicochemical properties and functions in plants. Cu is an essential ligand for the catalytic activity of many enzymes. Its uptake and trafficking are tightly regulated and mediated by specific transporters and chaperones. Cu serves as a signalling intermediate for ethylene reception. Excess Cu is sensed by binding to transcription factors, thereby, activating an arsenal of abiotic stress defences including increased expression of metallothioneins, phytochelatins, and antioxidants which contribute to remove "free" Cu and to re-establish cellular ion and redox homeostasis. In contrast to Cu, no specific uptake systems are known for Cd. Cd enters cells by metal transporters with broad substrate specificities and probably also via Ca channels. It is toxic because of its high reactivity with sulphhydryl groups and causes oxidative stress by depletion of antioxidative systems and stimulation of H_2O_2-producing enzymes. As a result, Cd triggers stress signalling pathways similar to those activated by Cu including cascades leading to programmed cell death. Important cross-talk exists between heavy metal and other abiotic stress signalling pathways (drought, oxidative stress). Excess heavy metals affect root functions at multiple levels and cause accumulation of abscisic acid (ABA). We propose a model how ABA and Cd signalling may interact at the organismic level to influence plant water status. Cytokinins act as antagonists of Cd indicating that the plant internal hormonal status may critically affect heavy metal tolerance.

7.1 Introduction

Heavy metals are defined as metals with a density higher than 5 g cm^{-3} (Weast 1984). From a biological perspective, this definition is not very useful because it comprises the majority of naturally occurring elements. However, only a limited number of these elements is soluble under physiological conditions and, thus, may become available for living cells. Among them are elements which serve plant

Topics in Current Genetics, Vol. 4
H. Hirt, K. Shinozaki (Eds.) Plant Responses To Abiotic Stress

Table 1. Regulatory limits for heavy metal concentrations in soil[1] (EU-Richtlinie 91/692/EWG, ABI EG, 31. Dec. 1991 Nr. L377, p. 48) and KSVO[2], 1992

Element	Regulatory limit1 (mg kg^{-1} DM)	mean soil concentration[2] (mg kg^{-1} DM)	limit in sludge for farm land[2] (mg kg^{-1} DM)
Cadmium	1-3	0.1-0.5	1.5
Copper	50-140	2-40	60
Nickel	30-75	5-50	50
Lead	50-130	2-60	100
Zinc	150-300	10-80	200
Mercury	1-1.5	0.02-0.5	1
Chromium	not determined	5-100	100

metabolism as micronutrients or trace elements (Fe, Mo, Mn, Zn, Ni, Cu, V, Co, W, Cr) and which become toxic when present in excess, as well as others with no known biological functions and high phytotoxicity such as As, Hg, Ag, Sb, Cd, Pb, and U (Godbold and Hüttermann 1985, Breckle 1991, Nies 1999).

The regulatory limits of heavy metals in the environment are defined by national legislation. Current regulations in the European communities have been summarised in table 1. Heavy metal concentrations in soils show regional differences and may locally exceed regulatory limits 10- to 50-fold (Lantzy and Mackensie 1979, Galloway et al. 1982, Jackson and Alloway 1991, Wagner 1993, Angelone and Bini 1992, Haag-Kerwer et al. 1999). Soils covering ore-bearing rocks or near slag heaps naturally contain heavy metals in amounts, which are toxic to most plant species. On such sites, specialised plant communities of "chemoecotypes" have evolved providing opportunities to investigate traits of heavy metal resistance. Well-known examples are *Silene vulgaris, Caradaminopsis halleri* (now *Arabidopsis halleri), Agrostis tenuis*, and *Minuartia verna* (Ernst 1990). However, apart from such confined natural habitats, there is growing concern about an increasing release of heavy metals into the environment.

Sources of heavy metals include traffic, refuse dumps, and sewage sludge. Emissions of dust, aerosols, and fly ashes from metal processing industries, e.g. in electroplating and galvanising, or metal-mining and smelting lead to spreading of heavy metals into rural areas. In agricultural soils heavy metal pollution is an increasing problem because of soil-amendment with municipal sewage sludge (Table 1) and intense use of phosphate fertilisers, which contain Cd as a contaminant (Hüttermann et al. 1999). The long biological lifetime and retention in soils favours heavy metal accumulation in the food web with potentially negative effects for human health (Wagner 1993). The bioavailability for heavy metals is plant specific and depends on the demand of specific metals as micronutrients and on the plant's ability to regulate actively metal mobilisation by exudation of organic acids or protons into the rhizosphere (Marschner 1995, Hinsinger 1998, McLaughlin et al. 1998, Hinsinger 2001). In addition, soil properties influence the chemical mobility of metals, thereby, regulating their release into the soil solution (Juste et al. 1985, Juste and Solida 1988).

The ability of plants to extract metals from soil, plant internal metal allocation, and cellular detoxification mechanisms are research areas currently attracting increasing attention. These topics have recently been covered by excellent reviews (Salt et al. 1998, Sanita di Toppi and Gabrielli 1999a, Clemens 2001, Cobbett and Goldsbrough 2002, Hall 2002; Schützendübel and Polle 2002) and will only be treated briefly here. In the present paper, we have chosen to focus on metals with contrasting action in plants cells. We will discuss chemical properties of these metals with respect to their toxicity and summarise current knowledge how heavy metals may interfere with cellular signalling and which signalling cascades they trigger leading to plant adaptation or injury. An attempt will be made to integrate cellular and long distance signalling in the context of organismic reactions.

7.2 Chemical properties, toxicity, and stress signalling of heavy metals with contrasting functions in plants

7.2.1 Copper

Cu is an important micronutrient with a critical deficiency level in the range of 1 - 5 mg kg^{-1}, an adequate range from 6 - 12 mg kg^{-1}, and toxicity above 20 – 30 mg kg^{-1} dry mass (Marschner 1995). However, there are huge species-specific differences in the ability to tolerate Cu. Certain metallophytes may even accumulate up to 1000 mg Cu kg^{-1} in leaves (Morrison et al. 1981).

Cu is a transition metal with an electrochemical potential of -260 mV, which is well within the cellular redox range of aerobic cells from -420 mV to +800 mV. It participates as an important redox component in cellular electron transport chains, for example as cofactor of enzymes, especially of oxidases. Autoxidation of "free" Cu^+ results in $O_2^{\cdot-}$ formation and subsequently in H_2O_2 and $OH^\bullet$ production via Fenton-type reactions. Oxidative injury by transitions metals such as Cu^+ and Fe^{2+} is well documented in microbes, plants, and animals (reviewed by Stohs and Baghi 1995, Nies 1999, Schützendübel and Polle 2002). Excess Cu seems to induce programmed cell death because markers (HIN1, HSR203J), whose expression is typically activated in cells committed to hypersensitive cell death, are also found in response to excess Cu (Pontier et al. 1999).

Since free metals are potentially dangerous, their uptake and cellular concentrations must be strictly regulated. Cu has a high affinity for peptide and sulphhydryl, carboxylic, and phenolic groups. Therefore, Cu is usually present in living cells in bound forms. Recent calculations led to the conclusion that the concentration of free copper in yeast is less than one ion per cell in the face of an overall concentration of 70 μM; in other words, unbound Cu does essentially not exist (Rae et al. 1999). This implies a tight co-regulation between Cu uptake and the provision of Cu-binding sites. Inside the cell, Cu is transported via chaperones (see section 7.3.2). It is possible that superoxide dismutases (SOD), which contain Cu/Zn, Fe, or Mn in their reaction centre, play a dual role in preventing metal toxicity: on the one hand they scavenge $O_2^{\cdot-}$ radicals, thus, maintaining reactive oxygen species at low concentrations and on the other hand they seem to be involved in preventing accumulation of free metals. Yeast lacking Fe-SODs and/or Mn SODs showed elevated levels of "free" iron (Srinivasan et al. 2000) emphasising the importance of SODs for Fe-binding. In a similar line of evidence it was shown that, exposure to excess Cu caused increases in Cu/Zn SOD activities (Chongpraditnum et al. 1992, Kurepa et al. 1997). Kurepa et al. (1997) suggested that Cu/Zn-SOD was under Cu-mediated transcriptional control. Whether these increases in SOD, which have so far solely been interpreted as a means to prevent oxidative injury, are also important to control the level of "free" copper needs to be investigated.

Cu and Cd (see below) activate the formation of phytochelatins (PC) and metallothioneins (MT), both cysteinyl-rich compounds with functions in heavy metal sequestration (Cobbett and Goldsbrough 2002). MTs are a family of ubiquitous small proteins. The promoter regions of MT carry MRE elements (metal responsive elements: GCGCGCA) leading to increased MT accumulation because of heavy metal exposure (Cobbett and Goldsbrough 2002). PCs are produced enzymatically from the tri-peptide precursor GSH (γ-glutamyl cysteinyl glycine). In response to heavy metals γECS (γ-glutamyl cysteinyl synthase), the first and limiting enzyme in GSH biosynthesis (Noctor et al. 1998) is activated transcriptionally (Lee and Korban 2002) and leads to PC accumulation (Rauser 1999). Metals bound to GSH or PCs are transported into the vacuole via ABC transporters (Rea 1999). Analysis of mutants and transgenic plants provided compelling evidence that the ability to synthesise glutathione is crucial for protection from heavy metals and failure to do so leads to increased sensitivity (Howden et al. 1995, Zhu et al. 1999a,b). Still, the significance of metal sequestration by MTs and PCs as a universal defence mechanism to protect the cell against these metals is controversial. It has frequently been shown that cellular concentrations of GSH and/or PC are not correlated with heavy metal tolerance. It is likely that additional independent traits also contribute to mediate heavy metal tolerance. For example, when PC and GSH synthesis was blocked in hypertolerant species of *Silene vulgaris, Thlaspi caerulescens, Holcus lanatus*, and *Agrostis castellana*, Cu sensitivity was not increased (Schat et al. 2002). This suggests that Cu-sequestration by PCs is not essential for constitutive tolerance or hypertolerance. With respect to Cd, a differential behaviour was observed. Cd sensitivity was increased in non-hypertolerant but not in hypertolerant plants indicating that adaptive hypertolerance is not based on PC-sequestration of Cd (Schat et al. 2002).

7.2.2 Cadmium

In contrast to copper, Cd has no known biological function in higher plants. The critical tissue concentration, at which the metal causes decreases in biomass, is in the range of 3 – 10 mg kg^{-1} dry mass (Bahlsberg-Pahlson 1989).

Cd has higher affinity to thiol groups than other metallic micronutrients, e.g. a three-fold higher affinity for sulphhydryls than Cu (Schützendübel and Polle 2002). This feature is probably also the major basis for its toxicity. Cadmium directly affects the sulphhydryl homeostasis and inhibits SH-bearing, redox regulated enzymes in living organisms (Canesi et al. 1998, Chrestensen et al. 2000, Hall 2002, Schützendübel and Polle 2002). Cd can also bind to other functional groups containing nitrogen or oxygen (Nieboer and Richardson 1980). When Cd binding was analysed by x-ray absorption spectroscopy interaction with O and N-ligands was found in the xylem and with S in roots (Salt et al. 1995).

Although Cd does not participate directly in cellular redox reactions, it is well established that Cd exposure leads to oxidative injury such as lipid peroxidation and protein carbonylation (Gallego et al. 1996, Chaoui et al. 1997, Romero-Pueras et al. 2002, Schützendübel et al. 2002). Cd-exposed *Arabidopsis* show a fast upregulation of HSP70, a chaperone involved in re-folding of denatured protein, probably in an attempt to rescue cell metabolism (Suzuki et al. 2001). Protein denaturation and also displacement of other divalent cations such as Zn and Fe from proteins cause the release of "free" ions (Stohs et al. 2000). It is, therefore, conceivable that Cd may increase the level of free transition metals and cause oxidative injury via free Fe/Cu-catalysed Fenton reactions (Fig. 1).

Cd disturbs the cellular redox balance. One of the most prominent responses to Cd, and also to other heavy metals, is an initial transient depletion in GSH, which is probably because of an increased demand of this precursor for PC synthesis (Grill et al. 1987, De Vos et al. 1992, Vögeli-Lange and Wagner 1996, Gallego et al. 1996, Sanita di Toppi et al. 1998, Xiang and Oliver 1998, Madhava Rao and Sresty 2000, Schützendübel et al. 2001). During prolonged Cd exposure, the GSH pools recover. The reason is that under such conditions the demand for sulphur increases, which in turn leads to increased expression of a high affinity sulphate transporter and, thus, in increased sulphate uptake (Nocito et al. 2002).

The initial depletion of the GSH level was temporally correlated with an inhibition of antioxidant systems (Schützendübel and Polle 2002). Furthermore, it was shown that Cd triggers H_2O_2 accumulation (Piqueras et al. 1999, Romero-Pueras et al. 1999, Schützendübel et al. 2001, Schützendübel et al. 2002). The sources of this H_2O_2 are not known yet. Model calculations suggest that the inhibition of antioxidant systems by Cd would be sufficient to cause significant H_2O_2 accumulation (Schützendübel and Polle 2002). However, stimulation of H_2O_2-producing enzymes analogous to pathogen responses may also be involved. As a defence against pathogens, plasma membrane localised NADPH-oxidases are activated to trigger an oxidative burst resulting in transient H_2O_2 accumulation (Levine et al. 1994, Tenhaken and Rübel 1999). To date, evidence is still lacking whether this system is also involved in Cd-responses. However, it is tempting to speculate about a participation of NADPH-oxidases in Cd-responses, because the plant $gp91^{phox}$ NADPH oxidase homologue is regulated by Ca^{2+} (Sagi and Fluhr 2001). Ca^{2+} is readily displaced by Cd^{2+} (Das et al. 1997) suggesting that effects of Cd on Ca-dependent enzymes systems are likely (Fig. 1). This is also supported by the observation that the Cd-induced oxidative burst was abolished by Ca in BY2 tobacco cell cultures (Piqueras et al. 1999).

In contrast to NADPH-oxidases, whose role in Cd-responses is speculative, the participation of oxalate oxidases in Cd-mediated H_2O_2 formation has been shown

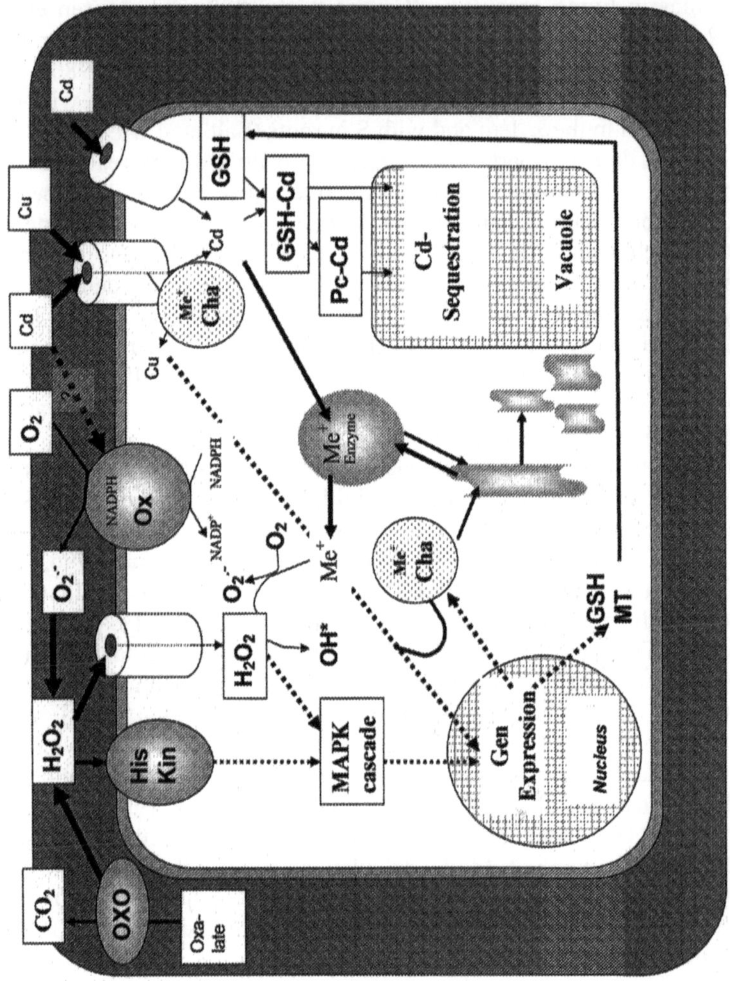

Fig. 1. Possible routes of cadmium and copper stress signalling. Cd causes H_2O_2 accumulation either by stimulation of H_2O_2-forming enzymes (OXO = oxalate oxidase, NADPH OX = NADPH oxidase) or indirectly by displacement of transition metals from chaperones (Me^+ Cha) or enzymes (Me^+ enzymes). This leads to unfolding of the enzymes and, if not rescued by metallochaperones, to protein degradation. The released transition metals will result in oxidative stress. H_2O_2 triggers the MAPK cascade probably involving histidine kinases (His Kin) and activates transcription of defence genes. Free transition metals also activate genes required for protection such as chaperones, metallothioneins (MT), and enzymes for GSH biosynthesis. GSH binds Cd directly or will be used for phytochelatin synthesis (PC) and transport of sequestered Cd into the vacuole. The MTs and metallochaperones combat the effects of metal displacement and oxidative stress. Excess Cu causes oxidative stress by stimulation of oxalate oxidase or by electron transfer to molecular oxygen. Subsequent cellular responses are similar to those caused by Cd.

in transgenic tobacco expressing a wheat germin gene (Berna and Bernier 1999). Oxalate oxidase, gf-2.8, a member of the germin gene family was stimulated upon exposure to heavy metals including copper and cadmium. The gf-2.8 expression wasalso upregulated by pathogens, developmental and hormonal signals (Berna and Bernier 1999). Thus, it is evident that cross-talk between developmental, other stress signalling pathways and Cd, respective heavy metal signalling, must exist. gf-2.8 homologues have also been identified in *Arabidopsis* (Membré et al. 1997). Nevertheless, it is still an open question whether the Cd-induced H_2O_2 formation *in planta* is generally mediated by members of this enzyme family or by other not yet identified systems.

For a long time H_2O_2 has been considered mainly a harmful oxidant, whose accumulation in response to stresses leads to unspecific oxidation and necrosis. However, now it has been recognised that H_2O_2 also acts as a secondary signalling compound inducing defence pathway including e.g. the MAPK cascade (Kovtum et al. 2000). A comparison of global transcriptome analysis of *Arabidopsis* genes-responding to H_2O_2 (microarray data, Desikan et al. 2001) and Cd-responsive genes (differential display data, Suzuki et al. 2001) shows that a suite of common genes responded to both stimuli. Among the genes identified were transcription factors, e.g. DREB2A, rd29A, and OBF5 (Suzuki et al. 2001), which have roles in abiotic stress responses. OBF5 is a DNA binding protein, which can recognise the upstream region of a glutathione-S-transferase (GST, Chen et al. 1996), an enzyme also activated via the MAPK signalling cascade by H_2O_2 (Kovtun et al. 2000). GSTs are necessary for xenobiotic detoxification and may be involved in the transport of GSH-conjugates to vacuole (Marrs and Walbot 1997). It is not known whether GSTs are important for Cd-detoxification. It is surprising that microarray analysis of *Arabidopsis* challenged by H_2O_2 has shown little evidence for significant upregulation of oxidative stress related genes, e.g. those encoding enzymes of the glutathione-ascorbate pathway (Desikan et al. 2001). Xiang and Oliver (1998) suggested an independent transcriptional regulation of genes encoding GSH-synthesising enzymes by H_2O_2 and Cd. The GSH biosynthetic pathway, however, was activated by jasmonate pointing to cross-talk between general stress signalling and heavy metal signalling, respectively.

7.3 Uptake and sensing of heavy metals: regulation of metal homeostasis

7.3.1 Extracellular cellular processes and biotrophic interactions

In whole plants, roots are the primary site to which heavy metals gain access. The dissolved ions move apoplastically with the inflowing water. In general, a large fraction of Cd or Cu is retained by the roots and only comparatively small amounts (about 10%) are transported to the shoots (Hogan and Rauser 1981, Cataldo et al. 1983, Lolkema and Vooijs 1986, Arduini et al. 1996, 1998, Simon 1998, Liao et al. 2000, Vassilev et. 1999). Although Cd is not a nutrient, an active transport of this element across the Casparian strip must be postulated since cell walls in the vascular bundle of roots contained higher Cd concentrations than those of cortex cells (Polle and Fritz, unpublished data). Analysis of the subcellular localisation of Cd by analytical electron microscopy showed that Cd and Cu were enriched in cell walls compared with the cytosol (Arduini et al. 1996). Because of their negative charges, cell walls have significant capacities for retention of Cd and Cu (Weigel and Jäger 1980, Lolkema and Vooijs 1986, Rauser 1987, Hart et al. 1998). The binding properties of cell walls and their role for metal tolerance are still a controversial issue. Taking cell walls as a "dead" compartment, it is clear that their chemical binding capacity would be limited and, thus, protection against excess heavy metals restricted. In contrast to this view, cell walls have been recognised in recent years as a compartment of active metabolism, e.g. as a source of signalling molecules in pathogen interactions, and as a location, where heavy metals can be bound to protein or silicates (Bringezu et al. 1999). Blinda et al. (1997) showed that exposure to heavy metals (Cd, Ni) was followed by a significant de novo synthesis of proteins released into the apoplastic space. This suggests that walls have more than a passive role in environmental sensing. However, more work needs to be done to elucidate the function of the extracellular compartment in heavy metal signal transmission and detoxification.

In natural environments, most plants have symbiotic associations with mycobionts, which modify host nutrient relationships (Jenschke and Godbold 2000). These symbiotic interactions are important since they generally increase heavy metal tolerance (Jenschke and Godbold 2000, Schützendübel and Polle 2002). Several mechanisms have been suggested to explain the ameliorative influence of mycorrhiza on plants exposed to heavy metals and all involve exclusion or restriction of metal movement by the fungus to host roots. Since mycorrhizal fungi form a hyphal mantle around the root tip, they prevent the physical contact of the root tip with the surrounding medium. With their large surface area, mycorrhizal fungi can immobilise significant concentrations of Cd in cell walls, thus, decreasing the portion available to plants. This was found for ectomycorrhizal as well as for arbuscular mycorrhizal symbioses (Jenschke et al. 1999, Frey et al. 2000, Rivera-Becerril et al. 2002). Furthermore, mycorrhizal fungi such as *Paxillus involutus* sequester huge concentrations of Cd in the vacuole (Blaudez et al. 2000), which correlate with the vacuolar sulphur concentrations (Ott et al. 2002). Mycorrhizal fungi also activate MTs upon heavy metal exposure. Heterologuous complementation assays with yeast confirmed that GmarMT1, a MT-like polypetide, conferred tolerance against Cu and Cd (Lanfranco et al. 2002).

Mycorrhizal symbiots probably also affect plant-inherent tolerance. For example, Schützendübel and Polle (2002) reported that mycorrhiza showed a significant increase in host-derived phenolics. Phenolics can act as a pre-formed defences because heavy metals, e.g. Cu, have high affinities to such secondary metabolites. With respect to aluminium, a protective function of phenolics has already been demonstrated (Yamamoto et al. 1998). Thus, in addition to providing a barrier against excess heavy metals mycorrhiza can also stimulate the host defence and may contribute to increase physiological metal tolerance. It will be a challenging future task to unravel signalling networks between symbiotic organisms, which lead to synergistic increases in heavy metal tolerance of the host plant.

7.3.2 Cellular signalling of copper – means to maintain homeostasis

Uptake of essential micronutrients like Cu, Zn, or Fe is normally regulated at transcriptional and posttranscriptional levels involving selective and non selective cation channels. These include the heavy metal CPx ATPases (members of P-type ATPase, which generate a proton gradient across the membrane required to drive efflux carrier), Nramps (natural resistance macrophage protein with the function of multispecific metal transporters with roles in manganese and iron homeostasis), the CDF (cation diffusion facilitator, which mediate cation efflux and contain His-rich regions which affect metal specificity and have major roles in Zn and Co resistance), and the ZIP family (Zn and iron transporters, with metal binding His-rich sequences) (reviewed by Guerinot 2000, Williams et al. 2002). Analysis of bacteria, fungi, plants, and animals suggests that the systems involved in metal uptake and sensing are evolutionary conserved (Mäser et al. 2001). In plants, the picture of metal uptake and sensing is complicated and incomplete due to the high number of putative transporters identified by genomic approaches (Mäser et al. 2001). At an organismic level, the situation will be more complex because of biotrophic interactions as outlined above. Virtually nothing is known about the molecular biology of metal transporters in mycorrhizal fungi and how they interact with the plant regulatory network for metal uptake. However, yeast systems are well characterised. Because of the similarities of transport systems across the microbial and plant kingdom, yeast Cu-transport and intracellular trafficking systems will be compared with those of plants.

In *S. cerevisae*, copper transporters (CTR) have been detected, which are regulated at the level of gene transcription, by posttranscriptional events and protein trafficking (Puig et al. 2002). A homologue COPT1 has been found in *Arabidopsis*, which reconstituted a yeast Cu-uptake mutant (Kampfenkel et al. 1995). This indicates that COPT1 encodes a copper transporter (Kampfenkel et al. 1995). Since its expression has only been found in stems and leaves but not in roots, specific systems for Cu uptake from the soil remain to be identified. It is also possible that root uptake occurs mainly via non selective cation channels, and that intracellular metal concentrations are regulated by activation of efflux carriers like in bacteria (Silver 1996).

In cells, "free" copper is apparently absent (see above). Under normal, i.e. Cu-limiting conditions, three Cu-chaperones have been identified in yeast, which are essential for intracellular binding and transport of Cu: ATX1 and Cox17 deliver Cu into the Golgi vesicle and mitochondria, respectively, while Lys7 is important for the Cu metallation of cytoplasmatic superoxide dismutase (SOD) (Lin et al. 1997, Glerum et al. 1997, Culotta et al. 1997, Portnoy et al. 2001). ATX1 is a ubiquitous metallochaperone with homologues found in plants, microbes, and animals that functions in copper transfer to an integral membrane cation transporting P-type ATPase (Huffman and O'Halloran 2000). In yeast, this ATPase,

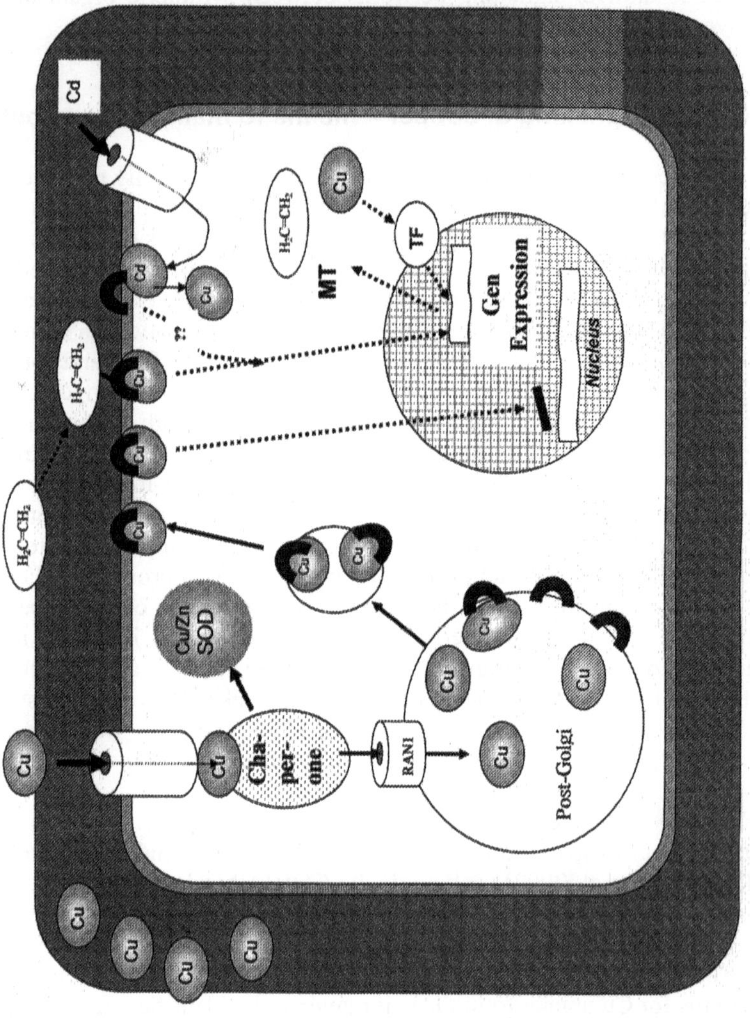

Fig. 2. Model for the role of Cu in ethylene signalling and hypothetical interference of Cd with this pathway. Cu is taken up by a Cu-specific transporter, transported by chaperones to various destinations e.g. to superoxide dismutases (Cu/Zn-SOD), mitochondria, and is pumped into post-Golgi vesicles. In these vesicles, complementation of ethylene receptors takes place and the functional units are delivered to the plasma membrane. In the absence of ethylene, they function as negative regulators of ethylene-activated genes. When ethylene binding inactivates the receptor, downstream signalling pathways are derepressed and activate hormone responses (after Hirayama et al. 1999) such as metallothionein (MT) transcription. Excess Cu mediates MT expression by activation of transcription factors. The role of Cd is speculative. By its ability to displace cations like Cu, Cd might inactivate the functional receptor and, thereby, activate the constitutive ethylene phenotype like in RAN1 mutants.

denominated as Ccc2, pumps copper into the lumen of the Golgi vesicles (Culotta et al. 1997). The *Arabidopsis* homologue of ATX1 was involved in the intracellular trafficking of copper facilitating the connection to an acceptor molecule (Himelblau et al. 1998). RAN1, the *Arabidopsis* homologue to Ccc2, is involved in the transfer of copper into Golgi vesicles (Hirayama et al. 1999). It was surprising that RAN1-cosuppressed plants showed a constitutive ethylene response phenotype indicating a link between Cu and ethylene signalling. Hirayama et al. (1999) proposed a model according to which RAN1 was necessary to deliver Cu to form functional ethylene receptors. The functional receptors are targeted to the plasma membrane (Fig. 2). Receptors containing Cu are active in the absence of the hormone and negatively regulate down-stream signalling pathways. In the presence of ethylene, the receptors are inactivated, presumably by suppressing histidine kinase/phosphatase activity. This in turn results in a suppression of the downstream pathway controlling components activating the ethylene phenotype. This means that Cu is not only triggering signalling pathways but is also a signal transducing molecule. It is notable that the promotor of a MT gene *in Lycopersicon esculentum* contains ethylene- and MRE motifs (Whitelaw et al. 1997) indicating that this protein can be regulated independently by excess Cu and other stress signals.

In yeast, signalling of excess Cu involves activation of transcription factors (TFs). The TF Ace1 was activated directly by free copper binding to cysteinyl residues within a Cu-regulatory binding domain (Beaudouin and Labbe 2001). The transcriptional activation of downstream genes involves cis-acting metal-responsive elements found in multiple copies in gene promoters, and metal-responsive transcription factors (Thiele 1992). Downstream located genes activated by these TFs have several functions in protecting cells against copper toxicity among them activation of genes encoding metallothioneins and superoxide dismutase. The co-regulation of metallothioneins and SOD is suspicious because both are important to prevent oxidative injury catalysed by free transitions metals (see above). These results indicate that the Cu level in yeast cells will be maintained stable by sensing Cu directly via metal responsive TFs. Under normal conditions copper will be always be bound and transported by proteins to its place of action. Excess Cu will switch on defences. Comparable signalling networks have been elucidated for iron and zinc sensing in yeast, emphasising the strict regulation of intercellular levels of "free" metals.

It is not known yet which TFs are involved in sensing of Cu and regulation of its cellular concentrations in plants. However, it can be expected that genomic analyses of Cu-responsive gene expression will give answers soon. The isolation of several MT genes from different plant species induced by heavy metals like Cu, Zn and also Cd suggests that regulatory pathways similar those operating in yeast may exist in plant cells (Tommey et al 1990, Zhou and Goldsbrough 1994, Robinson et al. 1996). It will be interesting to see whether plant Cu/Zn-SODs are activated by Cu-metallochaperones analogous to those of yeast and mammalians (Schmidt et al. 1999). Post-transcriptional regulation of plant and fungal SODs would explain why discrepancies between measured SOD activities and transcript levels have repeatedly been observed after heavy metal challenge (Kampfenkel et al. 1995, Jacob et al. 2001, Schützendübel et al. 2001).

7.3.3 Cellular signalling of cadmium

In contrast to essential metals, specific transporters for Cd have not been unequivocally demonstrated in plants although biochemical evidence suggests that such systems may exist in some specialised ecotypes (Zhao et al. 2002). Cd uptake occurs via Zn and Fe-transporters, which also have low affinities for Cd. ZRC1, a member of the CDF family of yeast involved in Zn transport, is localised in vacuolar membrane suggesting that this protein may also be involved in effluxing Cd from the cytosol into the vacuole (Li and Kaplan 1998). In plants, a zinc transporter (ZNT1) also mediated low affinity Cd-uptake (Pence et al. 2000). Yeast ZRC1 deletion mutants showed increased sensitivity to Zn and Cd (Conklin et al 1994). Complementation of these mutants with homologues from the Zn-hyperaccumulator plant, *Thaspi goesingense*, increased the resistance to Cd, Co, Ni, and Zn (Mäser et al. 2001). In *Schizosaccharomyces pombe*, deletion of Zhf, a CDF involved in Zn transfer to the endoplasmatic reticulum, rendered the mutants significantly more Cd tolerant but Zn sensitive (Clemens et al. 2002). The protective effect against Cd was independent of the phytochelatin pathway since PC synthase-deficient cells also showed significant increases in Cd tolerance when Zhf was inactivated.

Ectopic expression of *Arabidopsis* AtNramps (contribute to iron homeostasis) in yeast increased Cd sensitivity and accumulation (Thomine et al. 2000). In *Arabidopsis*, disruption of AtNramp3 leads to increased Cd resistance, whereas overexpression confers slightly higher Cd sensitivity (Thomine et al. 2000). IRT1, an *Arabidopsis* transporter of the ZIP family, which is expressed in roots of Fe-deficient plants (Korshunova et al 1999), is inhibited by Cd. Expression of IRT1 in yeast results in increased Cd sensitivity suggesting that IRT1 also mediates Cd uptake (Rogers et al. 2000). At the organismic level, it has been shown that sufficient Fe supply had beneficial effect on the Cd tolerance of plants, whereas Fe deficiency increased Cd susceptibility (Siedlecka and Krupa 1999).

Ca^{2+} channels have also been suggested to be involved in Cd uptake (White, 2000). Clemens et al. (1998) reported that a wheat Ca-transporter (LCT1) expressed in yeast also mediated Cd uptake. However, this uptake system may be species-specific, since homologues have not been found in *Arabidopsis*. Because specific transport systems for Cd seem to be lacking, one can assume that no Cd-specific signalling mechanisms exist to control its uptake. Nevertheless, Cd is immediately sensed because it affects the cellular redox status; it interferes with Ca signalling pathways, and disturbs uptake of other divalent cations such as Zn or Fe.

Identification of Cd-responsive genes in *Arabidopsis* by differential display revealed 31 clones among them 8 with no homologies to known functions of proteins (Suzuki et al. 2001). The others were assigned the following functions: signal transduction (protein kinases, transcription factors, calcium binding), protein folding, sulphur metabolism, metal binding, and abiotic stress responding. The temporal profiles of transcript accumulation showed early responses for kinases and transcription factors and with some delay also upregulation of genes encoding stress responsive proteins (chaperones, metal transporters) (Suzuki et al. 2001). This suggests that Cd rapidly activates signal transduction pathways including the protein phosphorylation cascade. Among the kinases, identified MEKK1 is of particular interest because it conferred increased Cd resistance to transfected yeast (Suzuki et al 2001). Cross-talk exists between Cd signalling and pathogen defence signalling because increased transcript levels of homologues to the transcription factors bZIP and WRKY were also found (Suzuki et al. 2001). Although pathways of Cd signalling in plants are not complete, the emerging picture suggests that plants employ a net of existing signalling cascades to "report" imbalances in cellular homeostasis to the nucleus, where a diverse array of responses will be activated. It is likely that the necessity of plants to cope with ever changing environmental conditions and co-evolving micro-organisms makes it more advantageous to transmit specific stress signals into a net of pleiotropic responses than to channel these signals to specific defence responses with a greater likelihood of failure.

7.4 Stress signals triggering plant growth and development at the organismic level

7.4.1 Links between cellular heavy metal signalling and inhibition of root growth

A common response to heavy metal exposure is a significant reduction in plant growth (Balsberg Pahlsson 1989, Kahle 1993, Sanita di Toppi and Gabrielli 1999a). Normal growth is the result of cell division, elongation, and differentiation including also programmed cell death in certain tissues like the xylem. Numerous reports show that heavy metals almost instantaneously affect root elongation (Hunter and Welkie 1976, Hogan and Rauser 1981, Godbold and Hüttermann 1985, Liao et al. 2000, Schützendübel et al. 2001) accompanied by significant decreases in mitotic activity (Jiang et al. 2001) and damage to nucleoli in the tip meristem (Liu et al. 1995). It has recently been shown that the cellular redox state, especially the concentration of GSH, regulates cell division (May et al 1998). The cell cycle consists of alternating phases of DNA replication (S phase) and mitotic stadiums (M phase) separated by gaps (G phase). An important checkpoint, the transition between the gap phase 1 (G1) and the S phase, is regulated by the intracellular GSH level (Vernoux et al. 2000). Evidence was obtained by showing that the ROOT MERISTEMLESS *Arabidopsis* mutant lacked a functional gene for γ-ECS, whose activity is decisive for cellular GSH concentrations (Noctor et al. 1998). Blockers of GSH synthesis also abolished cell division (Vernoux et al. 2000). In situ, analysis confirmed that proliferating root cells contain high GSH concentrations, whereas cellular GSH levels declined towards the quiescent centre in root tips (Sanchez-Fernandez et al. 1997). Cd and excess Cu caused an immediate decline in the overall GSH concentration of roots tips (Rauser et al. 1991, Meuwly and Rauser 1992, De Vos et al. 1992, Heiss et al. 1999, Schützendübel et al. 2001, Schützendübel et al. 2002). Addition of GSH reduces the inhibition of root growth (Chen and Kao 1995). Therefore, one likely mechanism of heavy metals is to block the cell cycle via effects on the GSH status. However, this effect will not persist because GSH concentrations recover during prolonged Cd exposure, whereas growth does not (Schützendübel et al. 2001).

Cd also suppresses cell expansion. In shoots, Cd inhibited a proton pump responsible to build up turgor (Aidid and Okamoto 1992). It is likely that this occurs in other plant organs as well. Furthermore, roots exposed to Cd show increased ethylene production, a hormone, which inhibits cell expansion (reviewed by Johnson and Ecker 1998). Cd also leads to significant accumulation of H_2O_2 (see section 7.2.1), which causes cell wall stiffening (Ros Barcelo 1997), thus preventing further extensibility of the walls. Thus, growth inhibition of roots is probably a pleiotropic effect caused by direct inhibition of important enzymes as well as by interference of Cd with cellular signalling.

H_2O_2, which accumulates in response to heavy metals, is involved as secondary messenger in abiotic and biotic stress signalling pathways leading to cellular suicide (reviewed by Beers and McDowell 2001). Suspension cultures of tobacco cells exposed to Cd show apoptotic-like symptoms (Fojtova and Kovarik 2000). Anatomical analysis of Cd-exposed roots indicates that the response may be cell specific and that only "competent" cells may undergo PCD because the tips showed no evidence for a general increase in cell death but formation of protoxylem elements in the zone, which normally constitutes the elongation zone (Schützendübel et al. 2001). First, this observation indicates that only localised cell death takes place. Second, it may afford an explanation for the finding that the inhibition of elongation persists, even when the plants are transferred to Cd-free medium. If cells in the elongation zone were already committed to differentiate according to their future functions (e.g. as cortex cells, endodermis, xylem, etc), the loss of turgor necessary for elongation would stop growth but apparently not the ability to develop further according to their destination. Consequently, xylematic structures differentiate in the root tip and the vital functions of the root tip are lost. Apparently, the morphogenetic gradient of hormones (auxin, gibberillins) is also destroyed because further symptoms developed at sub-lethal Cd-concentrations resemble those of root tip decapitation, i.e., significant formation of side roots (Greger and Lindberg 1986, Schützendübel, unpublished data). The advantage for the plant is obvious and the strategy resembles that against pathogens. An attacked plant sacrifices a small part of an infested organ by switching to the cellular suicide programme. These cells then form a barrier preventing spreading of the invading organism and protect the remaining parts. At the same time, immunisation is found (Alvarez et al. 1998). For Cd and Cu, increased tolerance has also been observed after pre-treatment with low concentrations of these metals (Talanova et al. 2000). It will be a challenging future task to analyse the molecular basis of acquired resistance to heavy metals.

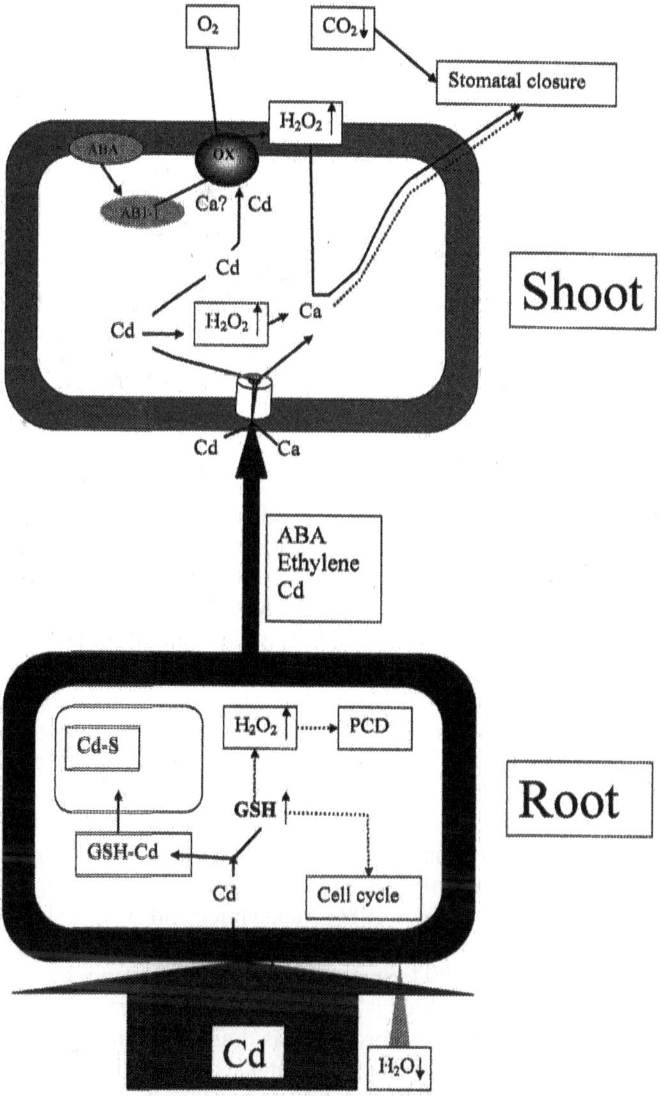

Fig. 3. A tentative model for the integration of cellular and long distance signalling of Cd. Cd is taken up and strongly retained by roots. Its detoxification and sequestration in the vacuole consumes GSH. The depletion in GSH leads to a halt of the cell cycle and to H_2O_2 accumulation, which triggers programmed cell death (PCD). Water uptake becomes limiting causing abscisic acid (ABA) formation. ABA, Cd, and perhaps others signalling molecules are transported to the leaves. ABA mediates stomatal closure via a signalling pathway involving H_2O_2 formation by NADPH-oxidase (OX) and activation of Ca channels. Cd is transported into the cells by Ca channels, where it will cause additional H_2O_2 formation, thus, aggravating the ABA response.

7.4.2 Long distance signalling and shoot responses to heavy metals

Long distance signals mediate the communication between roots and shoots (Fig. 3). Plant hormones (auxins, ethylene, gibberillins, abscisic acid (ABA)) as well as nutrient supply (carbohydrates, nitrogen) play decisive roles in this respect. The complex network of interactions of hormone and nutrient factors is not fully understood but there is ample evidence that both Cd and excess Cu have significant effects on most of these compounds. For example, Cd induces the biosynthesis of ABA and ethylene in roots (Fuhrer 1982, Poschenrieder et al. 1989, Chen and Kao 1995, Hollenbach et al. 1997, Schlagnhaufer et al. 1997, Sanita di Toppi et al. 1998, Munoz et al. 1998, Sanita di Toppi et al. 1999b, Chen et al. 2001). These are transmittable signals, which evoke stress responses in the shoot. Ethylene inhibits cell expansion and plays a role in positional signalling of cells (reviewed by Johnson and Ecker 1998). ABA plays a major role in plant adaptation to drought stress promoting stomatal closure by altering ion fluxes in guard cells (reviewed by Leung and Giraudat 1998). Plants exposed to Cd or excess Cu show responses, which can typically also be evoked by plant "stress" hormones such as significant reduction in expansion growth of leaves and diminished cell size (ethylene response) as well as symptoms of water deficit such as decreased stomatal conductance, and diminution of transpiration (ABA response) (Lolkema and Vooijs 1986, Barcelo and Poschenrieder 1990, Costa and Morel 1994, Moustakas et al. 1997, Haag-Kerwer et al. 1999, Perfus-Barbeoch et al. 2002).

It is still a matter of debate to what extent direct toxic effects of heavy metals or transmitted signals and cross-talk with other stress reactions evoke these symptoms. In the case of Cd, water uptake in roots is disturbed, the hydraulic conductivity decreased and, thereby, water supply to the shoots diminished (Marchiol et al. 1996). The transport of Cd to the shoot is driven by transpiration and can be reduced by application of ABA (Rubio et al 1994, Salt et al. 1995). The influence of excess Cu on water relations is less clear but symptoms such as loss in water use efficiency and accumulation of proline, a general marker of drought stress, have been reported (Lolkema and Voijs 1986, Maksymiec and Baszynski 1996, Chen et al. 2001, Vinit-Dunant et al. 2002). Proline biosynthesis was also found in Cd-stressed plants (Schat et al. 1997, Sha and Dubey 1998, Talanova et al. 2000). The accumulation of these metabolites is important for Cd-tolerance, because the survival rate of algae overexpressing proline was drastically enhanced (Siripornadulsil et al. 2002). Glutathione rescued photosynthesis (El Shintinavy 1999).

Cross-talk exists between drought-induced and Cd-induced signalling pathways and involves ABA signalling because independently osmotic stress, ABA, and Cd induced the formation of MTs in chicken pea (Munoz et al. 1998). At the first glance, induction of MTs by drought stress might appear surprising. However, Moran et al. (1994) observed in drought-stressed pea seedlings a release of transition metals, which would on the one hand induce oxidative stress and on the other hand result in activation of MT-encoding genes as outlined before (see section 7.3.2). MTs contribute to control the concentration of "free" metals and reactive oxygen species would activate defences, e.g. via the MAPK cascade (Fig. 1). These responses would help to regain cellular oxidant and metal homeostasis.

The ABA signalling pathway in guard cells, which leads to stomatal closure, has been shown to occur via induction of H_2O_2 synthesis, which in turn activates Ca-channels and blocks K^+-inward current (Pei et al. 2000, Murata et al. 2001). In the ABA-insensitive ABI1-1 mutant, the stimulation of H_2O_2 was interrupted and, thus, signal transduction resulting in stomatal closure was blocked (Pei et al. 2000). Cd-induced stomatal closure is independent from ABA-signalling because it occurs in the ABI1-1 mutant (Perfus-Barbeoch et al. 2002). Perfus-Barbeoch et al. (2002) provided evidence that Cd enters the guard cells via Ca-channels and that this leads to stomatal closure. Ca-channel blockers abolished Cd-induced stomatal closure, whereas the ABI1-1 mutant displayed stomatal closure upon Cd-exposure similar to that found in controls (Perfus-Barbeoch et al. 2002). Since Cd causes H_2O_2 accumulation (see section 7.2.2), and H_2O_2 is a necessary signal transducer for stomatal closure, we can infer that Cd must be taken up by the cell and acts inside to stimulate H_2O_2-producing systems. Whether H_2O_2 itself is accumulated outside, inside, or at multiple sites is not known. Chloroplasts have been discussed as potential H_2O_2-sources for stomatal closure (Neill et al. 2002). However, this is highly unlikely because the chloroplasts are equipped with powerful antioxidant systems (Polle 2001). In addition, there is no evidence for injury to the light-driven reactions of photosynthesis (Haag-Kerver et al. 1999, Baryla et al. 2001, Vinit-Dunand et al. 2002). This means that NADPH production in chloroplasts is unlikely to be limited. Taken together, the data strongly suggest that Cd acts downstream of the ABA-signal and prior to the H_2O_2 signal. The observation that application of ABA and Cd together aggravate the effects on plant performance compared with Cd alone (Moya et al. 1995) supports the idea that both ABA and Cd act synergistically. In addition, this finding shows that not all plant responses to heavy metals are "strategically" directed to counteract negative consequences of toxic compounds.

There is probably no general answer to the question whether stomatal closure and the associated losses in water use efficiency and net photosynthesis are primarily a result of direct negative effects of toxic ions or an indirect effect mediated via ABA or other long-distance signals. Both mechanisms are likely to occur. Which of them is the first to evoke responses will depend on the capacity of roots to retain heavy metals, the sensitivity of the systems to produce ABA (and other hormones?), the transport kinetics of these compounds, and the sensitivity of the target organs. Detailed ecophysiological studies have shown that the effects of Cd and Cu in shoots depend on the growth stage and physiological age of leaves (Skorzynska-Polit and Baszynski 1997, Krupa and Moniak 1998, Vinit-Dunand et al. 2002). For example, stomatal conductance, net photosynthetic activity, and also the maximal photochemical yield remained unaffected in young leaves of *Cucumis sativa*, even though expansion growth was inhibited by Cu (Vinit-Dunand et al. 2002). Mature leaves accumulated less copper, maintained maximum photochemical yield but nevertheless showed strong diminution of stomatal conductance and a corresponding decline in net photosynthesis as well as stronger accumulation of starch than the expanding leaves (Vinit-Dunand et al. 2002). Similar observations have been reported for cadmium: smaller cell size, less leaf area, starch accumulation in chloroplasts and diminished stomatal conductance but no effects on the properties of the photosynthetic electron transport (Moya et al. 1995, Haag-Kerver et al. 1999, Baryla et al. 2001). Starch and sucrose accumulation lead to inhibition of photosynthesis via feedback mechanisms (Koch 1996, Morcuende et al. 1997, Paul and Pellny 2003). Application of gibberillins reversed the inhibitory effect of Cd on growth and resulted in remobilization of carbohydrates (Moya et al. 1995, Ghorbanli et al. 1999), whereas the auxin indolacetic acid had no protective effect (Hunter and Welkie 1977, Moya et al. 1995). Molecular analyses of these interactions are yet completely missing. However, the observation that the juvenility of an organ affects its sensitivity towards heavy metals is intriguing and deserves further attention.

7.5 Conclusions and implication for future research

Copper and cadmium are heavy metals with contrasting physicochemical properties and functions in plants. At the organismic level, uptake of heavy metals into plant cells is modulated by biotrophic interactions and by plant-inherent features such as their capacity to retain heavy metals in the roots, for example by binding to cell wall components. Little is known about the physiology and molecular biology of these processes despite their importance for mediating metal tolerance in natural environments. Mycorrhizal fungi are especially intriguing in this respect. Given the similarities of fungal (yeast) and plant copper uptake and intercellular trafficking, the time seems ripe to find out how these systems are regulated in symbiotic associations affording higher protection to the host.

Plant cells take up Cu by specific transport systems. Inside the cell, chaperones serve intracellular Cu transport to vesicular storage sites and to target enzymes such as Cu/Zn-SOD, ethylene receptors, etc. "Free" Cu is extremely dangerous because it will reduce molecular oxygen leading to increased formation of superoxide, hydrogen peroxide, and hydroxyl radicals switching normal metabolism to programmed cell death.

Specific uptake systems for Cd have not been found. Cd seems to enter the cell via Fe and Zn transporters and probably also via Ca channels. It does not participate directly in cellular redox reactions but inactivates redox sensitive enzymes by binding to thiol-group. Its strong affinity to sulphhydryl-groups leads to a depletion in GSH similar to that induced by excess Cu and results in H_2O_2 accumulation. Since Cd is known to displace divalent cations such as Ca, Cu, Fe, we suspect that Cd may also cause oxidative stress by increasing "free" transition metal concentrations. This would explain that sensing systems which report redox imbalances caused by excess transition metals can also be activated by Cd. Free Cu probably binds to TFs, which in turn activate transcription of metal-binding ligands such as MTs and enzymes required for GSH and phytochelatin biosynthesis. The latter compounds serve sequestration of free metals, thereby, re-establishing the cellular ion homeostasis. The protection afforded by this reaction seems to be limited as there is increasing evidence that hypertolerance is mediated by additional independent traits with unknown molecular basis. First data obtained with mutants in metal transporters suggest that limitation of metal entry into cells may contribute to tolerance. However, the large number of putative transporters identified by genome analysis together with their suspected functions in micronutrient uptake will make it difficult to increase Cd tolerance via modulation of transport systems.

Despite different uptake routes and properties, Cd and Cu stimulate partly the same signalling cascades leading to activation of abiotic stress defences. Cross-talk exists between heavy metal and other stress signalling pathways (drought, oxidative stress), probably employing H_2O_2 and ABA as signal transducing compounds. Current data suggest that plants employ a net of existing signalling cascades to "report" imbalances in cellular homeostasis to the nucleus, where a diverse array of responses will be activated. Perhaps, it is an evolutionary advantage to cope with ever-changing environmental conditions, if specific stress signals were "translated" into common cellular response signals. These can be transduced in a net of multiple signalling pathways and evoke pleiotrophic defences. Such a defence system may be less prone to failure but implies that not all responses observed upon stress impact must be essential for adaptation and survival.

H_2O_2 seems to play a central role as signalling intermediate for heavy metal stress. Functional analysis of H_2O_2 during heavy metal signal transduction is yet missing. It will be an important goal of future research to unravel the identity of heavy metal-induced H_2O_2 sources and to analyse their functional role in mutants. The combination of molecular and physiological data led us to propose a tentative model integrating cellular and organismic responses to heavy metals. Not yet included in this model is the surprising observation that the hormonal status of a leaf critically determines its heavy metal susceptibility. To date, some physiological and pharmacological experiments suggest that cytokinins are major antagonistic players. These observations open interesting perspectives for future research.

Acknowledgements

The authors are grateful to the European Community and the German Science Foundation for continuous support.

References

Aidid SB, Okamoto H (1992) Effects of lead, cadmium and zinc on the electric membrane potential at the xylem /symplast interface and cell elongation of *Impatiens balsamina.* Environm Exp Bot 32:439-448

Alvarez ME, Pennell R, Meijer P-J, Ishikawa A, Dixon RA, Lamb C (1998) Reactive oxygen intermediates mediate a systemic signal network in the establishment of plant immunity. Cell 92:773-784

Angelone M, Bini C (1992) Trace elements concentrations in soils and plants of western europe. In:Adriano DC (ed.) Biogeochemistry of Trace Metals. Lewis Publishers, Boca Raton, FL, pp 19–60

Arduini I, Godbold DL, Onnis A (1996) Cadmium and copper uptake and distribution in Mediterranean tree seedlings. Physiol Plant 97:111-117

Arduini I, Godbold DL, Onnis A, Stefani A (1998) Heavy metals influence mineral nutrition of tree seedlings. Chemosphere 36:739-744

Balsberg Pahlsson A-M (1989) Toxicity of heavy metals (Zn, Cu, Cd, Pb) to vascular plants. Water Air Soil Poll 47:287-319

Barcelo J, Poschenrieder C (1990) Plant water relations as affected by heavy metal stress:a review. J Plant Nutr 13:1-37

Baryla A, Carrier P, Franck F, Coulomb C, Sahut C, Havaux M (2001) Leaf chlorosis in oil seed rape plants (*Brassica napus*) grown on cadmium-polluted soil:causes and consequences for photosynthesis and growth. Planta 212:696-709

Beaudoin J, Labbe S (2001) The fission yeast copper-sensing transcription factor Cufl regulates the copper transporter gene expression through an Ace1/Amt1-like recognition sequence. J Biol Chem 276:15472-15480

Beers E, McDowell JM (2001) Regulation and execution of programmed cell death in response to pathogens, stress and environmental cues. Curr Opin Plant Biol 4:561-567

Berna A, Bernier F (1999) Regulation by biotic and abiotic stress of a wheat germin gene encoding oxalate oxidase, a H_2O_2 producing enzyme. Plant Mol Biol 39:539-549

Blaudez D, Botton B, Chalot M (2000) Cadmium uptake and subcellular compartmentation in the ectomycorrhizal fungus *Paxillus involutus*. Microbiol 146:1109-1117

Blinda A, Koch B, Ramanjulu S, Dietz KJ (1997) De novo synthesis and accumulation of apoplastic proteins in leaves of heavy metal exposed barley seedlings. Plant Cell Environm 20:969-981

Breckle CW (1991) Growth under heavy metals. In:Waisel Y, Eshel A, Kafkafi U (eds.) Plant roots:the hidden half. Marcel Dekker, New York, NY, pp 351 –3 73

Bringezu K, Lichtenberger O, Leopold I, Neimann D (1999) Heavy metal tolerance of *Silene vulgaris.* J Plant Physiol 154:536-546

Canesi L, Ciacci C, Piccoli G, Stocchi V, Viarengo A, Gallo G (1998) In vitro and in vivo effects of heavy metals on mussel digestive gland hexokinase activity:The role of glutathione. Comp Biochem Physiol Pharmacol Toxicol Endocrinol 120:261-268

Cataldo DA, Garland TR, Wildung RE (1983) Cadmium uptake kinetics in intact soybean plants. Plant Physiol 73:844-848

Chaoui A, Mazhoudi S, Ghorbal MH, El Ferjani E (1997) Cadmium and zinc induction of lipid peroxidation and effects on antioxidant enzyme activities in bean (*Phaseolus vulgaris* L.). Plant Sci 127:139-147

Chen SL, Kao CH (1995) Glutathione reduces the inhibition of rice seedling root growth caused by cadmium. Plant Growth Reg 16:249-252

Chen SL, Kao CH (1995) Prior temperature exposure affects subsequent Cd-induced ethylene production in rice leaves. Plant Sci 104:135-138

Chen W, Chao G, Singh KB (1999) The promotor of a H_2O_2-inducible, *Arabidopsis* glutathione-S-transferase gene contains closely linked OBF- and OBP1-binding sites. Plant J 6:955-966

Chen CT, Chen LM, Lin CC, Kao CH (2001) Regulation of proline accumulation in detached rice leaves exposed to excess copper. Plant Sci 160:283-290

Chongpraditnum P, Mori S, Chino M (1992) Excess copper induces a cytosolic Cu, Zn-superoxide dismutase in soybean root. Plant Cell Physiol 33:239 – 244

Chrestensen CA, Starke DW, Mieyal JJ (2000) Acute cadmium exposure inactivates thioltransferase (glutaredoxin), inhibits intracellular reduction of protein-glutathionyl-mixed disulfides, and initiates apoptosis. J Biol Chem 275:26556 – 26565

Clemens S (2001) Molecular mechanisms of plant metal tolerance and homeostasis. Planta 212:475-486

Clemens S, Antosiewicz DM, Ward JM, Schachtman DP, Schroeder JI (1998) The plant cDNA LCT1 mediates the uptake of calcium and cadmium in yeast. Proc Natl Acad Sci, USA 95:12043-12048

Clemens S, Bloss T, Vess C, Neumann D, Nies DH, zu Nieden U (2002) A transporter in the endoplasmatic reticulum of *Schizosaccharomyces pombe* cells mediates zinc storage and differentially affects transition metal tolerance. J Biol Chem 277:18215-18221

Cobbett C, Goldsbrough PB (2002) Phytochelatins and metallothioneins:roles in Heavy Metal Detoxification and Homeostasis. Annu. Rev. Plant Biol. 53:159–182

Conklin DS, Culbertson MR, Kung C (1994) Interactions between gene products involved in divalent cation transport in *Saccharomyces cervisae*. Mol Gen Genet 244:303-311

Costa G and Morel JL (1994) Water relations, gas exchange and amino acid content in Cd-treated lettuce. Plant Physiol Biochem 32:561-570

Culotta VC, Klomp LWJ, Strain J, Casareno RLB, Krems B, Gitlin, JD (1997) The copper chaperone for superoxide dismutase. J Biol Chem 272:23469-23472

Das P, Samantaray S, Rout GR (1997) Studies on cadmium toxicity in plants:a review. Environm Poll 98:29-36

De Vos RCH, Vonk MJ, Vooijs R, Schat H (1992) Glutathione depletion due to copper-induced phytochelatin synthesis causes oxidative stress in *Silene cucubalus*. Plant Physiol 98:853-858

Desikan R, A-H-Mackerness S, Hancock J, Neill S (2001) Regulation of the *Arabidopsis* transcriptome by oxidative stress. Plant Physiol 127:159-172

El Shintinavy (1999) Glutathione counteracts the inhibitory effect induced by cadmium on photosynthetic process in soy bean. Photosynth 36:171-179

Ernst WHO (1990) Mine vegetation in Europe. Heavy metal tolerance in plants:evolutionary aspects. In:Shaw AJ (ed.) CRC Press, Boca Raton, p.21

Fojtova M, Kovarik A (2000) Genotoxic effect of cadmium is associated with apoptotic changes in tobacco cells. Plant Cell Environm 23:531-537

Frey B, Zierold K, Brunner I (2000) Extracellular complexation of Cd in the Hartig net and cytosolic Zn sequstration in the fungal mantle *of Picea abies – Hebeloma crustuliniforme* ectomycorrhizas. Plant Cell Environm 23:1257-1265

Fuhrer J (1982) Ethylene biosynthesis and cadmium toxicity in leaf tissue of bean (*Phaseolus vulgaris* L.). Plant Physiol 70:162-167

Gallego SM, Benavides MP, Tomaro ML (1996) Effect of heavy metal ion excess on sunflower leaves:evidence for involvement of oxidative stress. Plant Sci 121:151-159

Galloway JN, Thornton JD, Norton SA, Volcho HL, McLean RA (1982) Trace metals in atmospheric deposition:a review and assessment. Atm Environm 16:1677

Ghorbanli M, Kaveh SH, Sepehr MF (1999) Effects of cadmium and gibberellin on growth and photosynthesis of *Glycine max*. Photosynth 37:627-631

Glerum DM, Shtanko A, Tzagoloff A (1997) Characterization of *C0X17*, a yeast gene involved in copper metabolism and assembly of cytochrome oxidase. J Biol Chem 271:14504-14509

Godbold DL, Hüttermann A (1985) Effect of zinc, cadmium and mercury on root elongation of *Picea abies* (Karst.) seedlings, and the significance of these metals to forest dieback. Environm Poll 38:375-381

Greger M, Lindberg S (1986) Effects of Cd and EDTA on young suger beets (*Beta vulgaris*) I. Cd uptake and sugar accumulation. Physiol Plant 66:69-74

Grill E, Winnacker E-L, Zenk MH (1987) Phytochelatins, a class heavy-metal-binding peptides from plants, are functional analogous to metallothioneins. Proc Natl Acad Sci USA 8:439-443

Guerinot ML (2000) The ZIP family of metal transporters. Biochim Biophy Ac 1465:190-198

Haag-Kerwer A, Schäfer HJ, Heiss S, Walter C, Rausch T (1999) Cadmium exposure in *Brassica juncea* causes a decline in transpiration rate and leaf expansion without effect on photosynthesis. J Exp Bot 50:1827-1835

Hall JL (2002) Cellular mechanisms for heavy metal detoxification and tolerance. J Exp Bot 366:1-11

Hart JJ, Welch RM, Norvell WA, Sullivan LA, Kochian LV (1998) Characterization of cadmium binding, uptake, and translocation in intact seedlings of bread and durum wheat cultivars. *Plant Physiol* 116:1413-1420

Heiss S, Schaefer HJ, Kerwer AH, Rausch T (1999) Cloning sulfur assimilation genes of *Brassica juncea* L.:cadmium differentially affects the expression of a putative low affinity sulfate transporter and isoforms of APS sulfurylase and APS reductase. Plant Mol Biol 39:847-857

Himelblau E, Mira H, Lin SJ, Culotta V, Penarrubia L, Amasino RM (1998) Identification of a functional homolog of the yeast copper homeostasis gene *ATX1* from *Arabidopsis.* Plant Physiol 117:1227–1234

Hinsinger P (1998) How do plant roots acquire mineral nutrients? Chemical processes involved in the rhizosphere. Adv Agron 64:225-265

Hinsinger P (2001) Bioavailability of soil inorganic P in the rhizosphere as affected by root induced chemical changes:a review. Plant and Soil 237:173-195

Hirayama T, Kieber JJ, Hirayama N, Kogan M, Guzman P, Nourizadeh S, Alonso JM, Dailey WP, Dancis A, Ecker JR (1999) Responsive-to-antagonist1, a Menkes/Wilson disease-related copper transporter, is required for ethylene signaling in *Arabidopsis.* Cell 97:383-393

Hogan GD, Rauser WF (1981) Role of copper binding, absorption, and translocation in copper tolerance of *Agrostis-gigantea* Roth. J Exp Bot 32:27-36

Hollenbach B, Schreiber L, Hartung W, Dietz KJ (1997) Cadmium leads to stimulated expression of the lipid transfer protein genes in barley:implication for the involvement of lipid tansfer proteins in wax assembly. Planta 203:9-19

Howden R, Andersen C, Goldsbrough PB, Cobett CS (1995) A cadmium-sensitive, glutathione deficient mutant of *Arabidopsis thaliana.* Plant Physiol 107:1067-1073

Huffman DL, O'Halloran TV (2000) Energetics of copper trafficking between the Atx1 metallochaperone and the intracellular copper transporter, Ccc2. J Biol Chem 275:18611-18614

Hunter R, Welkie GW (1977) Growth of copper treated corn roots as affected by EDTA, IAA, succinic acid-2,2dimethyl hydrazide, vitamins and potassium. Environm Exp Bot 17:19-26

Hüttermann A, Arduini I, Godbold DL (1999) Metall pollution and Forest decline. In:Prasad MNV, Hagemeyer J (eds.) Heavy metall stress in plants. Springer Verlag, Berlin. pp. 253-272

Jackson AP, Alloway BJ (1991) The transfer of cadmium from sewage sludge amended soils into the edible component of food crops. Water Air Soil Poll 57:873-881

Jacob C, Courbot M, Brun A, Steinman HM, Jacquot JP, Botton B, Chalot M (2001) Molecular cloning characterization and regulation by cadmium of a superoxide dismutase from the ectomycorrhizal fungus *Paxillus involutus*. Eur J Biochem 268:3223-3233

Jenschke G, Godbold DL (2000) Metal toxicity and ectomycorrhizas. Physiol Plant 109:107-116

Jenschke G, Winter S, Godbold DL (1999) Ectomycorrhizas and cadmium toxicity in Norway spruce seedlings. Tree Physiol 19:23-30

Jiang W, Liu D, Liu X (2001) Effects of copper on root growth, cell division, and nucleolus of *Zea mays*. Biologia Plant 44:105-109

Johnson P, Ecker J (1998) The ethylene gas signal transduction pathway:a molecular perspective. Annu Rev Genet 32:227-254

Juste C, Soldan P, L'Hermite P (1985) Factors influencing heavy metal availability in field experiments with sewage sludges. In:Leschber R, Davis RD (eds.); Chemical methods for assessing bioavailable metals in sludge and soils. Elsevier Applied Science Publishers; Barking, Essex; pp. 82-89

Juste C, Solida P (1988) Influence de l'addition de differentes matieres fertilisantes sur la biodisponibilite du cadmium, du manganese, du nickel et du zinc contenus dans un sol sableux amende par des boues de station d'epuration. Agron 8:897-904

Kahle H (1993) Response of roots to heavy metals. Environm Exp Bot 33:99-119

Kampfenkel K, Kushnir S, Babiychuk E, Inze D, Van Montagu M (1995) Molecular characterization of a putative *Arabidopsis thaliana* copper transporter and its yeast homologue. J Biol Chem 270:28479-28486

Koch KE (1996) Carbohydrate modulated gene expression in plants. Ann Rev Plant Physiol Mol Biol 47:509-540

Korshunova YO, Eide D, Clark G, Guerinot ML, Pakrasi HB (1999) The IRT1 protein from *Arabidopsis thaliana* is a metal transporter with a broad substrate range. Plant Mol Biol 40:37-44

Kovtun Y, Chiu WL, Tena G, Sheen J (2000) Functional analysis of oxidative stress-activated mitogen-activated protein kinase cascade in plants. Proc Natl Acad Sci, USA 97:2940-2945

Krupa Z, Moniak M (1998) The stage of leaf maturity implicates the response of the photosynthetic apparatus to cadmium toxicity. Plant Sci 138:149-156

Kurepa J, van Montagu M, Inze D (1997) Expression of of sodCp and sodB genes in *Nicotiana tabacum*:effects of light and copper excess. J Exp Bot 48:2007-2014

Lanfranco L, Bolchi A, Ros EC, Ottonello S, Bonfante P (2002) Differential expression of a metallothionein gene during the presymbiotic versus the symbiotic phase of an arbuscular mycorrhizal fungus. Plant Physiol 130:58-67

Lantzy RJ, Mackenzie FT (1979) Atmospheric trace metals:global cycles and assessment of man's impact. Geochim Cosmochim Ac 43:511

Lee S, Korban SS (2002) The trancriptional regulation of *Arabidopsis thaliana (*L.) Heynh. Planta 215:689-693

Leita L, DeNobili M, Cesco S, Mondini C (1996) Analysis of intercellular cadmium forms in roots and leaves of bush bean. J Plant Nutr 19:527-533

Leung J, Giraudaut J (1998) Abscisic acid signal transduction. Annu Rev Plant Physiol Plant Mol Biol 49:199-222

Levine A, Tenhaken R, Dixon R, Lamb C (1994) H_2O_2 from the oxidative burst orchestrates the plant hypersensitive disease resistance response. Cell 79:583-593

Li L, Kaplan J (1998) Defects in the yeast high affinity iron transport system result in increased metal sensitivity because of the increased expression of transporters with a broad transition metal specificity. J Biol Chem 273:22181-22187

Liao MT, Hedley MJ, Woolley DJ, Brooks RR, Nichols MA (2000) Copper uptake and translocation in chicory (*Cichorium intybus* L. cv. Grasslands Puna) and tomato (*Lycopersicon esculentum Mill. cv.* Rondy) plants grown in NFT system. I. Copper uptake and distribution in plants. Plant Soil 221:135-142

Lin S-J, Pufahl RA, Dancis A, O'Halloran TVO, Culotta VC (1997) A role for the *Saccharomyces cerevisiae ATX1* gene is copper trafficking and iron transport. J. Biol. Chem 272:9215-9220

Liu D, Jiang W, Wang W, Zhai L (1995) Evaluation of metal toxicity on root tip cells by the *Allium* test. Is J Plant Sci 43:125-133

Lolkema PC, Vooijs R (1986) Copper tolerance in *Silene-Cucubalus* – subcellular distribution of copper and its effects on chloroplasts and plastocyanin synthesis. Planta 167:30-36

Madhava Rao KV, Sresty TVS (2000) Antioxidative parameters in the seedlings of pigeonpea (*Cajanus cajan* (L.) Millspaugh) in response to Zn an Ni stresses. Plant Sci 157:113-128

Maksymiec W, Baszynski T (1996) Chlorophyll fluorescence in primary leaves of excess Cu treated runner bean plants depends on their growth stages and the duration of Cu-action. J Plant Physiol 149:196-200

Marchiol L, Leita L, Martin M, Peresotti A, Zerbi G (1996) Physiological responses of two soybean cultivars to cadmium. J Environm Qual 25:562-566

Marrs KA, Walbot V (1997) Expression and RNA splicing of the maize glutathione-S-transferase Bronze2 gene is regulted by cadmium and other stresses. Plant Physiol 113:93-102

Marschner H (1995) Mineral nutrition of higher plants. Academic Press, New York, pp. 889

Mäser P, Thomine S, Schroeder J, Ward JM, Hirschi K, Sze H, Talke IN, Amtmann A, Maathius FJM, Sanders D, Harper JF, Tchieu J, Gribskov M, Persans MW, Salt DE, Kim SA, Guerinot ML (2001) Phylogenetic relationships within cation tranporter families of *Arabidopsis*. Plant Physiol 126:1646-1667

May M, Vernoux T, Leaver C, van Montagu M, Inze D (1998) Glutathione homeostasis in plants:implications for environmental sensing and plant development. J Exp Bot 49:649-667

McLaughlin MJ, Smolders E, Merckx R, Checkai RT (1998) Soil-root interface:physicochemical processes. In:Huang PM, Adriano DC, Logan TJ (eds.) Soil chemistry and ecosystem health. SSSA Special Publication Number 52; Soil Science Society of America Inc. Madison, pp. 233-277

Membré AN, Berna A, Neutelings A, David H, Staiger D, Vasquez JS, Raynal M, Delseny M, Bernier F 1997. cDNA sequence, genome organization and differential expression of three *Arabidopsis* genes for germin/oxalate oxidase-like proteins. Plant Mol Biol 35:459–469

Meuwly P, Rauser W (1992) Alteration of thiol pools in roots and shoots of maize seedlings exposed to cadmium. Plant Physiol 99:8-15

Moran JF, Becana Iturbe-Ormaetxe I, Frechilla S, Klucas R, Aparicio-Tejo P (1994) Drought induces oxidative stress in pea plants. Planta 194:346-352

Morcuende R, Perez P, Martinez-Carrasco R (1998) Short-term feedback inhibition of photosynthesis in wheat leaves supplied with sucrose and glycerol at two temperatures. Photosynth 33:179-188

Morrison RS, Brooks RD, Reeves RD, Malaise F, Horowitz P, Aronson M, Merriam GR (1981) The diverse chemical forms of heavy metals in tissue extracts of some metallophytes from Shaba province, Zaire. Phytochem 20:455-458

Moustakas M, Ouzounidou G, Symeeonidis L, Karatagalis S (1997) Field studies of the effect of excess copper on wheat photsynthesis and productivity. Soil Sci Plant Nutr 43:531-539

Moya JL, Ros R, Picazo I (1995) Heavy metal-hormone interactions in rice plants – effects on growth, net photosynthesis, and carbohydrate distribution. J Plant Growth Reg 14:61-67.

Munoz FJ, Ullan RV, Labrador E, Dopico B (1998) Increased expression of two cDNAs encoding metallothionein-like proteins during growth of *Cicer arietinum* epicotyls Physiol Plant 104:273-279

Murata Y, Pei ZM, Mori IC, Schroeder J (2001) Abscisic acid activation of plasma membrane Ca^{2+} channels in guard cells require cytosolic NAD(P)H and is differentially disrupted upstream and down stream of reactive oxygen species production in abi1-1 and abi2-1 phosphatase 2C mutants. Plant Cell 13:2513-2523

Neill S, Desikan R, Hancock J (2002) Hydrogen peroxide signaling. Curr Op Plant Biol 5:388-395

Nieboer E, Richardson DHS (1980) The replacement of the nondescript term " heavy metal" by a biologically significant and chemically significant classification of metal ions. Environm Poll B1:3-26

Nies DH (1999) Microbial heavy-metal resistance. Appl Microbiol Biotechnol 51:730-750

Nocito FF, Pirovano L, Coccuci M, Sacchi GA (2002) Cadmium induced sulfate uptake in maize roots. Plant Physiol 129:1872-1879

Noctor G, Arisi ACM, Jouanin L, Kunert KJ, Rennenberg H (1998) Glutathione:biosynthesis, metabolism and relationship to stress tolerance explored in transformed plants. J Exp Bot 49:623-647

Ott T, Fritz E, Polle A, Schützendübel A (2002) Characterisation of antioxidative systems in the ectomycorrhiza-building basidiomycete *Paxillus involutus* (Bartsch.) FR. and its reaction to cadmium. FEMS Microbiol Ecol 42:359-366

Paul M, Pellny TK (2003) Carbon metabolite feedback regulation of leaf photosynthesis and development. J Exp Bot 54:539-547

Pei ZM, Murata Y, Benning G, Thomine S, Klusener B, Allen GJ, Grill E, Schroeder JM (2000) Calcium channels activated by hydrogen peroxide mediate abscisic acid signaling in guard cells. Nature 406:731-734

Pence NS, Larson PB, Ebbs SD, Letham DLD Lasat MM Garvin DF Eide D, Kochian LV (2000) The molecular physiology of heavy metal transport in the Zn/Cd hyperaccumulator *Thlaspi caerulescens*. Proc Natl Acad Sci, USA 97:4956-4960

Perfus-Barbeoch L, Leonhardt N, Vavasseur A, Forestier C (2002) Heavy metal toxicity:cadmium permeates through calcium channels and disturbs the plant water status. Plant J 32:539-548

Piqueras A, Olmos, E, Martinez-Solano JR, Hellin E (1999) Cd-induced oxidative burst in Tobacco BY2 Cells:Time course, subcellular location and antioxidant response. Free Rad Res 31:33-38

Polle A (2001) Dissecting the superoxide dismutase-ascorbate-glutathione pathway by metabolic modeling:computer analysis as a step towards flux analysis. Plant Physiol 126:445-462

Pontier D, Gan S, Amasino R, Roby D, Lam E (1999) Markers for hypersensitive response and senescence show distinct pattern of expression. Plant Mol Biol 39:1243-1255

Portnoy ME, Schmidt PJ, Rogers RS, Culotta VC (2001) Metal transporters that contribute copper to metallochaperones in *Saccharomyces cerevisiae*. MGG Mol Gen Gen 265:873-882

Poschenrieder C, Gunse B, Barcelo J (1989) Influence of cadmium on water relations, stomatal resistance and abscisic acid content in expanding bean leaves. Plant Physiol 90:1365-1371

Puig S Lee J Lau M Thiele DJ (2002) Biochemical and genetic analyses of yeast and human high affinity copper transporters suggest a conserved mechanism for copper uptake. J Biol Chem 277:26021-26030

Rae TD, Schmidt PJ, Pufahl RA, Culotta VC, O'Halloran TV (1999) Undetectable intracellular free copper: the requirement of a copper chaperone for superoxide dismutase. Science 284:805-808

Rauser W (1999) Structure and function of metal chelators produced by plants. Cell Biochem Biophys 31:19-48

Rauser W, Schupp R, Rennenberg H (1991) Cysteine, γ-glutamylcysteine, and glutathione levels in maize seedlings. Distribution and translocation in normal and cadmium-exposed plants. Plant Physiol 97:128-138

Rauser WE (1987) Compartmental efflux analysis and removal of extracellular cadmium from roots. Plant Physiol 85:62-65

Rea P (1999) MRP subfamily ABC transporters from plants and yeast. J Exp Bot 50:895-913

Rivera-Becerril F, Calantzi C, Turnau K, Caussanel JP, Belimov AA, Gianazzi S, Strasser R, Gianinazzi-Pearson V (2002) Cadmium accumulation and buffering of cadmium-induced stress by arbuscular mycorrhiza in three *Pisum sativum* L. genotypes. J Exp Bot 53:1177-1185

Robinson NJ, Wilson JR, Turner JS (1996) Expression of the type 2 metallonthionein-like gene MT2 from *Arabidopsis thaliana* in Zn2C-metallothionein-deficient *Synechococcus* PCC 7942:putative role for MT2 in Zn2C metabolism. Plant Mol. Biol. 30:1169–1179

Rogers EE, Eide DJ, Guerinot ML (2000) Altered selectivity in an *Arabidopsis* metal transporter. Proc Natl Acad Sci USA 97:12356-12360

Romero-Pueras MC, Palma JM, Gomez LA, del Rio LA, Sandalio LM (2002) Cadmium causes oxidative modification of proteins in plants. Plant Cell Environm 25:677-686

Romero-Puertas MC, McCarthy I, Sandalio LM, Palma JM, Corpas FJ, Gomez M, Del Rio LA. 1999. Cadmium toxicity and oxidative metabolism of pea leaf peroxisomes. Free Rad Res 31:S25-S31

Ros Barcelo A (1997) Lignification in plant cell walls. Int Rev Cytol 176:87-132

Rubio MI, Escrig I, Martinezcortina C, Lopezbenet FJ, Sanz A (1994) Cadmium and nickel accumulation in rice plants - effects on mineral-nutrition and possible interactions of abscisic and gibberellic acids. Plant Growth Reg 14:151-157

Sagi and Fluhr R (2001) Superoxide production by plant homologues of the gp91phox NADPH oxidase. Modulation of activity by calcium and by tobacco mosaic virus infection. Plant Physiol 126:1281-1290

Salt DE, Prince RC, Pickering IJ, Raskin I (1995) Mechanism of cadmium mobility and accumulation in Indian Mustard. *Plant Physiol* 109:1472-1433

Salt DE, Smith RD, Raskin I (1998) Phytoremediation. Annu Rev Plant Physiol Plant Mol Biol. 49:643-668

Sanchez-Fernandez R, Fricker M, Corben LB, White NS, Sheard N, Leaver CJ, Van Montagu M, Inze D, May MJ (1997) Cell proliferation and hair tip growth in the *Arabidopsis* roots are under mechanistically different forms of redox control. Proc Natl Acad Sci USA 94:2745-2750

Sanita di Toppi L and Gabrielli R (1999a) Response to cadmium in higher plants. Environm Exp Bot 41:105-130

Sanita di Toppi L, Lambardi M, Pecchioni N, Pazzagli L, Durante M, Gabrielli M, di Toppi LS (1999b) Effects of cadmium stress on hairy roots of *Daucus carota.* J Plant Physiol 1999:385-391

Sanita di Toppi LS, Lambardi M, Pazzagli L, Cappugi G, Durante M, Gabbrielli R (1998) Response to cadmium in carrot in vitro plants and cell suspension cultures. Plant Sci 137:119-129

Schat H, Llugany M, Vooijs R, Hartley-Whitaker J, Bleeker PM (2002) The role of phytochelatins in constituive and adaptive heavy metal tolerrance in in hyperaccumulator and non-hyperaccumulator metallophytes. J Exp Botany 53:2381-2392

Schat H, Sharma SS, Vooijs R (1997) Heavy metal-induced accumulation of free proline in a metal tolerant and a non-tolerant ecotype of *Silene vulgaris*. Physiol Plant 101:477-482

Schlagnhaufer CD, Arteca R, Pell EJ (1997) Sequential expression of two 1-aminocyclopropane-1-carboxylate synthase genes in response to biotic and abiotic stresses in potato (*Solanum tuberosum* L.) leaves. Plant Mol Biol 35:683-688

Schmidt PJ, Ramos GM, Culotta VC (1999) A gain of superoxide dismutase (SOD) activity obtained with CCS, the copper metallochaperone for SOD1. J Biol Chem 274:36952-36956

Schützendübel A, Polle A (2002) Plant responses to abiotic stresses:heavy metal-induced oxidative stress and protection by mycorrhization. J Exp Bot 53:1351-1365

Schützendübel A, Nikolova P, Rudolf C, Polle A (2002) Cadmium and H_2O_2-induced oxidative stress in *Populus x canescens* roots. Plant Physiol Biochem 40:577-584

Schützendübel A, Schwanz P, Teichmann T, Gross K, Langenfeld-Heyser R, Godbold D, Polle A (2001) Cadmium–induced changes in antioxidative systems, H_2O_2 content and differentiation in pine (*Pinus sylvestris)* roots. Plant Physiol 127:887-898

Sha K, Dubey RS (1998) Effect of cadmium on proline accumulation and ribonuclease activity in rice seedlings:role of proline as possible enzyme protectant. Biol Plant 40:121-130

Siedlecka A, Krupa Z (1999) Cd/Fe interaction in higher plants – its consequences for the photosynthetic apparatus. Photosynth 36:321-331

Silver S (1996) Bacterial resistance to toxic metal ions – a review. Gene 179:9-19

Simon L (1998) Cadmium accumulation and distribution in the sunflower plant. J Plant Nutrition 21:341-352

Siripornadulsil S, Traina S, Verma DP, Sayre RT (2002) Molecular mechanisms of proline-mediated tolerance to toxic heavy metals in transgenic microalgae. Plant Cell 14:2837-2847

Skórzynska-Polit E, Baszynski T (1997) Differences in sensitivity of the photosynthetic apparatus in Cd-stressed runner bean plants in relation to their age. Plant Sci 128:11-21

Srinivasan C, Liba A, Imlay J, Valentine JS, Gralla EB (2000) Yeast lacking superoxide dismutase(s) show elevated levels of "free iron" as measured by whole cell electron paramagnetic resonance. J Biol Chem 275:29187-29192

Stohs SJ, Bagchi D (1995) Oxidative mechanisms in the toxicity of metal ions. Free Rad Biol Med 18:321-336

Stohs SJ, Bagchi D, Hassoun E, Bagchi M (2000) Oxidative mechanisms in the toxicity of chromium and cadmium ions. J Environm Pathol Toxicol Oncol 19:201-213

Suzuki N, Koizumi N, Sano H (2001) Screening of cadmium-responsive genes in *Arabidopsis thaliana*. Plant Cell Environm 24:1177-1188

Talanova VV, Titov AF, Boeva NP (2000) Effect of increasing concentrations of lead and cadmium on cucumber seedlings. Biologia Plant 43:441-444

Tenhaken R and Rübel C (1999) Cloning of putative subunits of the soy bean plasma membrane NADPH Oxidase involved in the oxidative burst by antibody expression screening. Protoplasma 205:21-28

Thiele DJ (1992) Metal-regulated transcription in eukaryotes. Nucleic Acids Res. 20:1183–1191

Thomine S, Wang R, Ward JM, Crawford NM. Schroeder JI (2000) Cadmium and iron transport by members of a plant metal transporter family in *Arabidopsis* with homology to Nramp genes. Proc Natl Acad Sci USA 97:4991-4996

Tommey AM, Shi J, Lindsay WP, Urwin PE, Robinson NJ (1991) Expression of the pea gene PsMTa in *E. coli*—metal binding properties of the expressed protein. FEBS Lett. 292:48-52

Vassilev A, Tsonev T, Yordanov I (1999) Physiological response of barley *(Hordeum vulgare)* to cadmium containing soil during ontogensis. Environm Poll 103:287-293

Vernoux T, Wilson RC;, Seeley KA, Reichheld JP, Muroy S, Brown S, Maughan SC, Cobbett CS, Van Montagu M, Inze D, May MJ, Sung ZR. 2000. The ROOT MERISTEMLESS1/CADMIUM SENSITIVE2 gene defines a glutathione-dependent pathway involved in initiation and maintenance of cell division during postembryonic root development. Plant Cell 12:97-109

Vinit-Dunand F, Epron D, Alaoui-Sosse B, Badot PM (2002) Effects of copper on growth and on photosynthesis of mature and expanding leaves in cucumber plant. Plant Sci 163:53-58

Vögeli-Lange R, Wagner GW (1996) Relationship between cadmium, glutathione and cadmium-binding peptides (phytochelatins) in leaves of intact tobacco seedlings. Plant Sci 114:11-18

Wagner GJ (1993) Accumulation of cadmium in crop plants and its consequences to human health. Adv Agron 51:173-212

Weast RC (1984) CRC Handbook of Chemistry and Physics, 64 edn., Boca Raton, CRC

Weigel HJ, Jäger HJ (1980) Subcellular distribution and chemical form of cadmium in bean plants. Plant Physiol 65:480-482

White P (2000) Calcium channels in higher plants. Biophys Biochim Ac 1465:171-189

Whitelaw CA, Le Huquet JA, Thurman DA, Tomsett AB (1997) The isolation and characterisation of type II metallothionein-like genes from tomato (*Lycopersicon esculentum* L.) Plant Mol Biol 33:503-511

Williams LE, Pittman JK, Hall JL (2002) Emerging mechanisms for heavy metal transport in plants. Biochim Biophys Ac Biomembranes 1465:104-126

Xiang C, Oliver DJ (1998) Glutathione metabolic genes coordinately respond to heavy metals and jasmonic acid in *Arabidopsis*. Plant Cell 10:1539-1550

Yamamoto Y, Hachia A, Hamada H, Matsumoto H (1998) Phenylpropanoids as a protectant of aluminium toxicity in cultured tobacco cells. Plant Cell Physiol 39:950-957

Zhao FJ, Hamon RE, Lombi E, McLaughlin MJ, McGrath SP (2002) Characteristics of Cadmimum uptake in two contrasting ecotypes of the hyperaccumulator *Thlaspi caerulescens*. J Exp Bot 53:535-543

Zhou J, Goldsbrough PB (1994) Functional homologs of animal and fungal metallothionein genes from *Arabidopsis*. Plant Cell 6:875-884

Zhu YL, Pilon-Smits E, Jouanin L, Terry N (1999a) Overexpression of glutathione synthetase in Indian mustard enhances cadmium accumulation and tolerance. Plant Physiol 119:73-79

Zhu YL, Pilon-Smits E, Tarun AS, Weber S, Jouanin L, Terry N (1999b) Cadmium tolerance and accumulation in Indian mustard is enhanced by overexpressing gamma-glutamylcysteine synthetase. Plant Physiol 121:1169-1177

8 Molecular genetics of genotoxic stress signalling in plants

Roman Ulm

Abstract

Cells are under constant threat by endogenous and exogenous factors affecting DNA integrity. In response, complex signalling networks are activated and appropriate countermeasures are taken. Although plants are inevitably exposed to diverse DNA damaging agents (genotoxins) due to their sessile life-style and dependence on sunlight for photosynthesis, plant signalling components activated by confronting genotoxic stress are largely unknown. However, recent genetic and biochemical analyses have advanced our understanding of genotoxic stress signalling. In particular, as deduced from mammalian model systems, players of both the postulated "nuclear"- and "non-nuclear"-target-mediated signal transduction chains were identified. Importantly, components of both pathways are crucial for plant tolerance to genotoxic stress.

8.1 Introduction

All organisms have the capacity to dynamically respond to environmental challenges as a result of the activation of complex signalling networks. One of the most extreme challenges is damage to the genetic information itself. The genomes of all living organisms are under continuous assault by environmental agents (e.g. UV irradiation and reactive chemicals) as well as by-products of endogenous metabolic processes (e.g. reactive oxygen species and erroneous DNA replication). As a result of the perception of the genotoxic stress, the cell cycle is halted to gain the time necessary for DNA repair, and genes required for repair and protection of other cellular components endangered by the genotoxic treatment are activated. Alternatively, particularly in multicellular eucaryotes, cells may respond by undergoing apoptosis, thereby eliminating damaged cells.

Research on genotoxic stress perception and signalling in mammalian cells is of particular importance due to its implications in human health and disease, including carcinogenesis. In plants, however, owing to the static nature of their cells anchored by cell walls, tumourous tissue cannot metastasise and plants do not die of cancer. On the other hand, their reproductive tissues are derived from cells that went through many rounds of DNA replication producing the entire organism, before forming gametes. This feature makes plants particularly sensitive to the potential accumulation of mutations in the germline, which finally opens the way for

Topics in Current Genetics, Vol. 4
H. Hirt, K. Shinozaki (Eds.) Plant Responses To Abiotic Stress

the passage of somatic mutations to the next generation (Walbot 1996). As the somatic phenotype might be influenced by inherent or environmentally induced genomic change, beneficial mutations may in some cases be directly made use of, thus, a kind of selection might occur before the gametes are formed. Moreover, plants possess a characteristic life cycle that includes a diploid sporophytic and a haploid gametophytic phase, the latter providing a mechanism to eliminate deleterious alleles, even when recessive. In further contrast to animals, plants are sessile organisms that depend on solar radiation as the vital source of biological energy and thus are continuously exposed to environmental mutagens, including ultraviolet-B (UV-B) radiation, and tolerance to this abiotic stress factor is critical for plant fitness (Rozema et al. 1997; Jansen et al. 1998).

Repair of DNA damage is essential for the maintenance of genomic integrity and substantial information is available on DNA repair processes in plants (e.g. Britt 1996; Gorbunova and Levy 1999; Tuteja et al. 2001; Britt 2002), including genetically defined roles in *Arabidopsis* of components involved in the major repair pathways: photoreactivation (PHR), base excision repair (BER), nucleotide excision repair (NER), non-homologous end-joining (NHEJ) and homologous recombination (HR) (see Table 1). In contrast, knowledge on perception and signalling of DNA-damaging threats in plants is rather limited and genetic support for proteins involved in genotoxic signalling in *Arabidopsis* is only emerging. Importantly, as deduced from the mammalian system, they might include signalling components engaged by both "nuclear" and "non-nuclear" targets of genotoxic agents. This review will be focused on recent advances in the identification of genetically defined components in genotoxic stress signalling in plants and will not address the topic of DNA damage repair, for which the reader is referred to several recent reviews (e.g. Britt 1996; Gorbunova and Levy 1999; Tuteja et al. 2001; Britt 2002).

8.2 What is genotoxic stress?

Diverse modifications of the molecular structure of the genetic material can arise as a result of errors introduced during replication, recombination, and repair itself. Other base alterations can result from the intrinsic instability of the specific chemical bonds and from the ability of DNA to readily react with a wide range of chemical and physical agents. Genotoxic stress results from agents (so-called genotoxins or mutagens) that are capable of damaging the nuclear and extranuclear genetic material of cells, i.e. they are "toxic to the genome". Thus, the unifying characteristic of genotoxins is the ability to damage DNA. The agents used in the laboratory to analyse the response of organisms to this type of stress are of different physical and chemical nature with varying DNA-damaging capabilities, making cross-comparisons particularly difficult. They include, for example, ultraviolet (UV) radiation (particularly UV-B and UV-C), the alkylating agent methyl methanesulfonate (MMS), reactive oxygen species (ROS), and ionizing radiation

Table 1. A non-exhaustive list of components required for genotoxic stress responses in *Arabidopsis*

Mutant	Affected gene	Sensitivity	Reference
"Sunscreen"			
tt4	*CHS* (chalcone synthase)	UV-B	Landry et a. 1995
tt5	*CHI* (chalcone isomerase)	UV-B	Landry et a. 1995
fah1	*FAH1* (ferulic acid hydroxylase 1)	UV-B	Landry et a. 1995
DNA repair			
uvr2	*PHR1* (CPD photolyase, PHR)	UV-B	Landry et al. 1997
uvr3	(6-4 photolyase, PHR)	UV-B	Nakajima et al. 1998
ros1	*ROS1* (DNA glycosylase/lyase, BER)	MMS, ROS	Gong et al. 2002
uvh1	*RAD1/XPF* (NER)	UV-B, UV-C, IR	Fidantsef et al. 2000; Gallego et al. 2000; Liu et al. 2000
uvh3/uvr1	*RAD2/XPG* (NER)	UV-B, UV-C, IR, ROS	Liu et al. 2001
xpb1	*XPB1/RAD25* (NER)	MMS	Costa et al. 2001
ku70	*KU70* (NHEJ)	MMS, IR	Bundock et al. 2002; Riha et al. 2002
ku80	*KU80* (NHEJ)	ROS, bleomycin	West et al. 2002
rad50	*RAD50* (HR/NHEJ)	MMS	Gallego et al. 2001
mre11	*MRE11* (HR/NHEJ)	IR, MMS	Bundock and Hooykaas 2002
mim	*MIM* (SMC-like, HR)	MMS, MMC, UV-C, IR	Mengiste et al. 1999
Signalling			
mkp1	*MKP1* (MAP kinase phosphatase)	UV-C, MMS	Ulm et al. 2001
atm	*ATM* (PI3K-like)	MMS, IR	Garcia et al. 2003
myb4	*MYB4* (Myb transcription factor)	UV-B	Jin et al. 2000
uvr8	*UVR8* (RCC1-like)	UV-B	Kliebenstein et al. 2002

(IR). All these agents cause a wide array of different DNA lesions, the most prevalent of which are briefly introduced below.

MMS is a monofunctional alkylating agent that induces mostly N-methylpurines, the removal of which results in apurinic sites preventing DNA replication (Friedberg et al. 1995). Furthermore, they can indirectly lead to double-strand breaks, for example as a result of repair processes, hence the radiation mimicking effect of MMS (e.g. Menke et al. 2001).

IR damages DNA as a consequence of both direct and indirect effects, that is, either as a result of direct interaction of the radiation energy with DNA or as a result of the interaction of DNA with radiation-generated ROS. IR can evoke damage to all cellular components and causes a variety of DNA lesions, such as various types of base damage and, particularly, DNA strand breaks (Friedberg et al. 1995).

DNA is considered a major cellular target for UV radiation, with peak absorption at around 260 nm determined by its component nucleotides. UV radiation induces oxidative damage (pyrimidine hydrates), DNA-protein and DNA-DNA crosslinks and most prevalently various pyrimidine dimers, in particular cyclobutane pyrimidine dimers (CPD) that constitute about 75% of UV-induced DNA lesions and pyrimidine [6-4] pyrimidinone dimers (6-4 photoproduct) that make up the majority of the remainder (Britt 1996). The UV radiation spectrum has been subdivided into three wavelength bands designated as UV-C (<280 nm), UV-B (280-320 nm), and UV-A (320-400 nm). Solar UV radiation reaching the earth consists only of UV-A and UV-B, since penetration of the atmospheric ozone layer drops dramatically for wavelengths below 320 nm and declines to zero below 295 nm. However, UV-C induces at high rate lesions equivalent to those generated by UV-B and is extensively used to explore biological responses to this class of DNA damaging radiation (Friedberg et al. 1995).

Oxidative damage to DNA due to attack by ROS must be considered as an important source of spontaneous DNA damage (Marnett and Plastaras 2001). There are various intra- and extracellular sources for ROS. Radiation in particular has, in addition to the direct interaction of radiation energy with DNA, an indirect effect on genetic information through the formation of ROS and their potential to damage DNA and other cellular constituents. Free radicals may cripple DNA in a variety of ways, resulting, for example, in fragmentation, base loss, base changes and strand breaks (Friedberg et al. 1995).

However, these genotoxic agents by no means damage exclusively DNA ("mutagenic effect"). Rather, they have a complex impact on cellular metabolism ("cytotoxic effect") as a consequence of damage conferred to other cellular constituents, including proteins and lipids. It should also be noted that living organisms are rarely exposed to the DNA-damaging agents that are most conveniently studied in the laboratory. Nonetheless, these agents have proved to be instrumental in deciphering genotoxic stress responses including perception, signalling, and repair in all organisms, among them *Arabidopsis* (Table 1).

The multitude of modifications evoked by genotoxic agents constitutes the substrate for a manifold of cellular responses particularly well-known in yeast and animal systems that will be outlined briefly here, in order to provide a frame of reference for recent advances in plant systems. For more detailed information on non-plant systems, the reader is referred to the literature cited and references therein. Concerning responses in the plant system, the effects of UV-B responses will be discussed in part separately from the other genotoxic stresses.

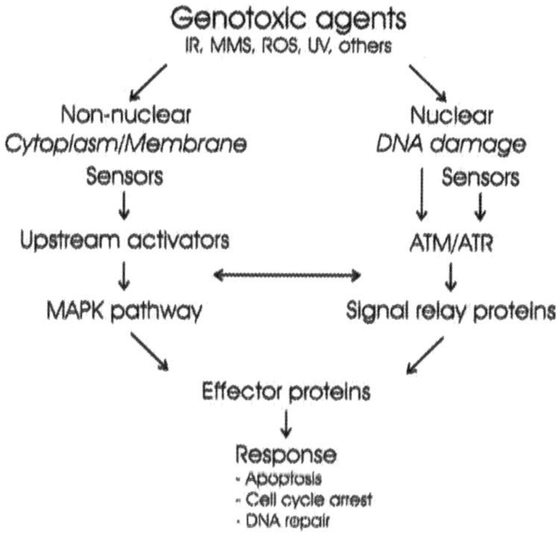

Fig. 1. Signalling in response to genotoxic agents.

8.3 Genotoxic stress signalling

In mammals, two major signalling pathways link genotoxic stress perception to adequate responses (Fig.1, Liu et al. 1998). The cellular responses result either directly from DNA damage ("nuclear" target-mediated) or are initiated outside the nucleus and do not involve DNA damage directly ("non-nuclear"). In the former case, key players include transcription factor p53 and the ATM/ATR sensor kinases. In the latter case, a major signalling pathway involves the activation of members of the mitogen-activated protein (MAP) kinase family. The separation of genotoxic signalling into DNA-damage-mediated and non-DNA-damage-mediated pathways will be followed in this review as a guideline; however, numerous exceptions to the generalities can be found. Moreover, it should be noted that the two pathways are not isolated from each other but rather interact at several levels (Fig. 1, Rotman and Shiloh 1999).

8.3.1 From inside the nucleus

8.3.1.1 Nuclear-target-mediated signalling in non-plant systems

The sensor proteins that directly recognize DNA damage are not yet precisely known. However, prime suspects are members of a group of checkpoint proteins and a pair of large protein kinases, the phosphatidylinositol 3-kinase (PI3K)-like ATM (Ataxia telangiectasia mutated) and ATR (ATM and Rad3-related; Rotman and Shiloh 1999; Kastan and Lim 2000; Abraham 2001; Melo and Toczyski 2002, and references therein). In spite of their PI3K-like domains, ATM, and ATM-related proteins (see Table 2) are not lipid kinases but serine/threonine protein kinases (Kim et al. 1999). Defects in ATM give rise to ataxia telangiectasia (A-T), a rare human neurodegenerative, and cancer predisposition disease with a complex clinical phenotype. ATM has both nuclear and cytoplasmic functions that may contribute to the pleiotropic nature of A-T (Abraham 2001). However, its nuclear role is central to very early stages of DNA damage signalling (Rotman and Shiloh 1999; Kastan and Lim 2000; Abraham 2001). Consistently, cells from A-T patients show increased sensitivity to IR and radiomimetic chemicals, but are proficient in their response to UV radiation. Thus, ATM is crucial in signalling DNA double-strand breaks that may occur as a consequence of cellular metabolism during replication and repair, or after exposure to specific DNA-damaging agents (Rotman and Shiloh 1999; Abraham 2001). ATR on the other hand, seems to be involved in the response to other types of DNA damage as well, such as those evoked by exposure to UV radiation. Mutations in ATR are associated with embryonic lethality and chromosomal fragmentation (Brown and Baltimore 2000). In mammalian cells, activation of these "sensor kinases" is central to DNA damage-induced checkpoint responses (Abraham 2001). It is presently not clear why these kinases share homology to PI3Ks; their activation in response to DNA damage, however, initiates a protein phosphorylation cascade resulting in the activation of the effector checkpoint kinases (CHK1, CHK2) and other key players, including the Nijmegen breakage syndrome protein NBS1 (a member of the MRE11-nuclease complex), the breast cancer associated protein BRCA1, the non-receptor tyrosine kinase c-ABL, the tumour suppressor gene product p53, and its regulator MDM2 (reviewed by Rotman and Shiloh 1999; Colman et al. 2000; Kastan and Lim 2000; Abraham 2001; Appella and Anderson 2001; Melo and Toczyski 2002). This complex DNA damage-responsive signal transduction pathway finally regulates cell cycle transitions, apoptosis and DNA repair, altogether increasing the faithful transmission of genetic information and, consequently, survival of the organism.

The DNA damage response pathways mediated by kinase cascades have been conserved through eukaryotic evolution, as shown by the existence and function of the yeast ATM and ATR orthologs Tel1p (*S. cerevisiae* and *S. pombe*) and Mec1p/Rad3p (*S. cerevisiae/ S. pombe*), respectively (Melo and Toczyski 2002, see also Table 2). *mec1* single mutant strains, but not *tel1*, are sensitive to DNA-damaging agents and fail to arrest the cell cycle in response to DNA damage. Global gene expression analysis revealed the requirement of Mec1p function in the regulation of several genes in response to DNA damaging agents (Gasch et al. 2001). Tel1p function appears to be redundant with Mec1p as *tel1mec1* double mutant strains are more sensitive to genotoxic agents than *mec1* single mutants and *TEL1* overexpression partially suppresses the *mec1* hypersensitive phenotype in response to DNA damage (Morrow et al. 1995; Craven et al. 2002). Furthermore, similar to the situation in mammals, these two large members of the PI3K family phosphorylate multiple replication, repair, and checkpoint proteins. Amongst these are the two checkpoint kinases Rad53p/Cds1p (*S. cerevisiae/ S. pombe*) and Chk1p (*S. cerevisiae* and *S. pombe*) (Melo and Toczyski 2002). Yeast cells mutated in components of this pathway are impaired in cell-cycle check-

Table 2. ATM and related proteins in eukaryotes

S. cerevisiae	*S. pombe*	*C. elegans*	*Drosophila*	**Vertebrate**	*Arabidopsis*	**Ref**
Tel1p	Tel1p	ATM	TEL1	ATM	ATM	(1)
Mec1p	Rad3p	ATL1	MEI41	ATR	ATR	(2)
				DNA-PK_{CS}		
Tor1p Tor2p	Tor1p Tor2p	TOR	TOR	TOR/FRAP	TOR	(3)

References for *Arabidopsis* homologs : (1) (Garcia et al. 2000); (2) (Perry and Kleckner 2003); (3) (Menand et al. 2002)

points and gene expression changes, and exhibit hypersensitivity to genotoxic stress treatments (Gasch et al. 2001; Melo and Toczyski 2002).

8.3.1.2 Nuclear-target-mediated signalling in plant systems

In plants, ATM and ATR homologs are encoded in the fully sequenced *Arabidopsis* genome (Table 2); however, genetic data on genotoxic signalling presumably initiated in the nucleus through the perception of damaged DNA is presently limited to the *Arabidopsis* ATM homolog (Garçia et al. 2003). The 440 kDa protein encoded by *AtATM* contains a carboxy-terminal PI3K-like domain flanked by two loosely conserved domains termed FAT (FRAP, ATM, TRRAP) and FATC (the "C" indicates carboxy-terminal) (Garcia et al. 2000; Garcia et al. 2003), as found in other members of the ATM family as well (Abraham 2001). Moreover, recent protein sequence analysis described the non-kinase domain of AtATM as composed of HEAT (huntingtin, elongation factor 3, A subunit of PP2A and TOR1) repeats, similarly identified in ATMs, ATRs and related proteins from diverse organisms, suggesting a conserved biochemical mechanism of action amongst the different family members (Perry and Kleckner 2003).

A reverse genetic approach identified two T-DNA tagged *atm* mutants whose phenotype suggests that ATM function is conserved in *Arabidopsis* and that AtATM may play a critical role in DNA damage-responsive cell cycle checkpoints (Garcia et al. 2003). Consistently, the *atm* mutants are hypersensitive to both IR and the radiomimicking MMS, but not to UV-B, suggesting a critical function of AtATM in the response to DNA strand breaks. In addition, the IR-mediated transcriptional induction of genes involved in the cellular response to DNA strand breaks (*AtRAD51*, *AtPARP1*, *AtGR1* and *AtLIG4*) is defective in *atm* mutant plants. This indicates an important role of AtATM in the signal transduction cascade resulting in the transcriptional gene activation of a group of genes following exposure to IR (Garcia et al. 2003). AtATM itself is constitutively expressed and not induced by IR (Garcia et al. 2000). In contrast to hypersensitivity to genotoxic stress, *atm* mutant plants apparently possess a normal vegetative development, a feature that differs from the complex pleiotropic phenotype of human A-T patients. However, *Arabidopsis atm* mutants are partially sterile due to a defect in female gametophyte development resulting from post-meiotic arrest. In addition, *atm* mutants have a reduced number of viable pollen grains. The meiotic defect in *atm* meiocytes was shown to include a number of anomalies, such as bridges between paired chromosomes and chromosome fragmentation (Garcia et al. 2003). This defect corroborates a common meiotic function of ATM conserved among eukaryotes. However, apparent meiotic recombination frequencies in *atm* are similar to wild type and the developmentally regulated meiotic recombination genes (*AtRAD51*, *AtSPO11* and *AtDMC1*) are expressed at normal levels (Garcia et al. 2003). Thus, at present, the exact function of AtATM in meiosis and meiotic recombination remains to be determined. Moreover, as noted by the authors, the absence of meiotic arrest in various other meiotic mutants in *Arabidopsis* puts to doubt the existence of a functional meiotic checkpoint (Garcia et al. 2003). Thus, it remains to be established if an AtATM-dependent meiotic checkpoint exists in *Arabidopsis*.

It will be particularly interesting to determine the further conservation and plant specificity of the nuclear genotoxic stress pathways. Analysis of the *Arabidopsis* genome indicates the existence of several conserved components (http://www.tigr.org/~jeisen/Arabidopsis_Repair/Repair.table.html and http://ag.arizona.edu/dnametab/tables/ESTTable.xls), including an ATR homolog (Table 2). The possible involvement of AtATR in genotoxic stress signalling (particularly UV-B) remains to be established. However, a prominent absentee in *Arabidopsis* is a p53 homolog, and the existence of p53-related pathways is a matter of debate (Whittle et al. 2001). The tumour suppressor protein p53 is a transcription factor that integrates DNA damage checkpoint signals with normal and aberrant (oncogenic) mitogenic signalling in animals, deciding if the cells will grow, arrest or die (Appella and Anderson 2001; Wahl and Carr 2001). It is noteworthy, that the presence and absence of p53 in animals and yeast, respectively, is suggested to reflect the different priorities of multi- versus unicellular organisms in response to DNA damage. In other words, members of the animal kingdom - in contrast to unicellular organisms - can tolerate a dead cell, but one proliferating uncontrollably may be lethal (Wahl and Carr 2001). Plants, on the other hand, are not killed by uncontrolled tumourous cell growth. Thus, it will be interesting to determine whether or not plants have a functional equivalent to p53, and if they do, how that pathway differs from its animal counterpart. In this regard, however, it should be noted that apoptosis-like responses after UV irradiation have been detected in plants (Danon and Gallois 1998; Mitsuhara et al. 1999), indicating the conservation of programmed removal of cells in response to unbearable levels of genotoxic stress.

Substrates of AtATM and their function in plants remain to be identified. Mutations of known components of the predicted MRE11/RAD50/NBS1-complex implicated in DNA repair processes lead to genotoxic stress hypersensitivity in *Arabidopsis* (Gallego et al. 2001; Bundock and Hooykaas 2002). NBS1 and its homolog Xrs2p are substrates of ATM and Mec1p in mammals and yeast, respectively. However, no *Arabidopsis* homolog of NBS1/Xrs2p is currently known, most likely due to low sequence conservation as already indicated by the low homology between mammalian NBS1 and yeast Xrs2p (D'Amours and Jackson 2002). Furthermore, the *Arabidopsis mre11* and *rad50* mutants exhibit aberrant telomere length regulation (Gallego et al. 2001; Gallego and White 2001; Bundock and Hooykaas 2002), as do their counterparts in other eukaryotes (D'Amours and Jackson 2002). Hence, AtATM might be involved in telomere maintenance, and general conservation of ATM signalling to the MRE11 complex during meiosis and DNA damage response is suggested.

8.3.2 From the cell periphery

8.3.2.1 Non-nuclear-target-mediated signalling in non-plant systems

A second pathway triggered by genotoxic agents originates outside the nucleus and exploits signal transduction cascades normally used for other cellular responses, including growth factor signalling (reviewed by Herrlich et al. 1997; Liu et al. 1998; Shaulian and Karin 2002). In the case of UV radiation, this pathway involves clustering and activation of several growth factor and cytokine receptor tyrosine kinases at the cell membrane (Devary et al. 1992; Sachsenmaier et al. 1994; Rosette and Karin 1996) or, as in the case of MMS, an unknown downstream component (Liu et al. 1996). UV irradiation indeed induces tyrosine phosphorylation of receptor tyrosine kinases (e.g. Sachsenmaier et al. 1994; Knebel et al. 1996; Rosette and Karin 1996; Kitagawa et al. 2002). However, it is not exactly known how UV activates the cell surface receptors in mammalian cells. It was suggested that UV irradiation might perturb the cell surface or alter receptor conformation leading to receptor multimerization, clustering, and activation (Rosette and Karin 1996). Alternatively, the inhibition of a membrane-bound protein tyrosine phosphatase was proposed as a genotoxic stress target, thereby preventing dephosphorylation of the receptor tyrosine kinases (Knebel et al. 1996). Furthermore, a central element in the response to diverse genotoxins appears to be the generation of ROS in the affected cells (Friedberg et al. 1995) and considerable evidence suggests that many genotoxic agents trigger their effects on signalling pathways through a mechanism involving oxidative stress (Devary et al. 1992; Liu et al. 1998). Whatever the exact mechanism of signal initiation, this signal transduction pathway involves activation of MAP kinase cascades. The core of MAP kinase pathways that is conserved in all eukaryotes consists of a three-tiered kinase module, composed of a MAP kinase kinase kinase (MAPKKK), a MAP kinase kinase (MAPKK) and a terminal MAP kinase (MAPK) that relays signals by sequential phosphorylation and activation (Fig. 2). Upon their activation, a pool of the MAP kinases is translocated from the cytoplasm to the nucleus. Thus, downstream targets comprise cytoplasmic and nuclear proteins, including other protein kinases, phosphatases, phospholipases, cytoskeleton-associated proteins, and a number of transcription factors (reviewed by Widmann et al. 1999; Kyriakis and Avruch 2001). Phosphorylation of transcription factors leads to appropriate reprogramming of gene expression in response to activating stimuli, contributing to a particular readout of the MAPK cascade.

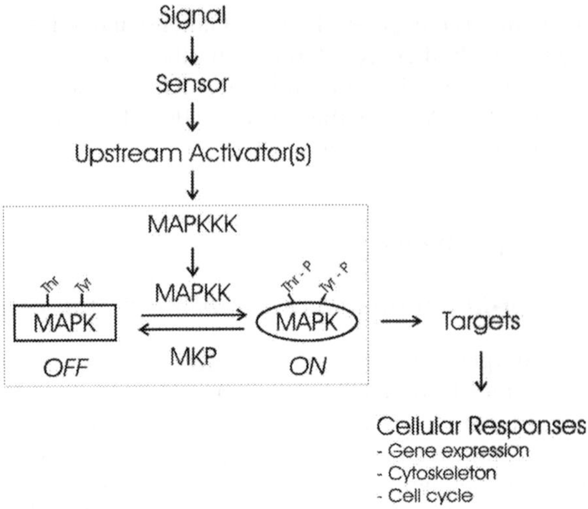

Fig. 2. MAP kinase signalling.

Members of three distinct MAP kinase subfamilies, namely the extracellular signal-regulated kinases (ERK), c-jun N-terminal kinases (JNKs), and p38 kinases have been implicated in response to genotoxic insults, resulting in the induction of many target genes through the activation of transcription factors such as activator protein-1 (AP-1; Jun/Fos) (reviewed, for example, by Liu et al. 1998; Shaulian and Karin 2002). The activation of JNK and the induction of two UV-responsive transcription factors, AP-1 and nuclear factor (NF)-κB, were fully functional in enucleated cells, providing evidence that the UV response does not require a signal generated in the nucleus, but is likely to be initiated by membrane proximal events (Devary et al. 1993). It is, however, not excluded that there is additional retrograde DNA damage-dependent activation of the growth factor signalling components, including MAP kinases (discussed in Liu et al. 1998). Intriguingly, the activation of NF-κB by UV irradiation is mediated sequentially by DNA damage-independent and -dependent mechanisms (Bender et al. 1998).

An important feature that determines the outcome of the cellular reaction is the magnitude and duration of MAPK activation (Marshall 1995), which is regulated by the balance between phosphorylating and activating MAPKKs and dephosphorylating and deactivating phosphatases (Fig. 2). These negative regulators include serine-threonine phosphatases, tyrosine phosphatases, and the dual-specificity phosphatases known as MAP kinase phosphatases (MKPs). Members of the MKP family are potent and specific inactivators of MAPKs through dephosphorylation of both the threonine and the tyrosine in the Thr-x-Tyr motif in their activation loop (Camps et al. 2000; Theodosiou and Ashworth 2002).

A recognized function of JNK in response to UV radiation includes regulation of the stress-induced apoptotic signalling mechanism. Consistently, simultaneous disruption of all functional *Jnk* genes or their upstream activators (*Mkk4* and *Mkk7*) resulted in protection against UV- and MMS-induced apoptosis in fibroblasts, due to a defect in an early mitochondrial response to JNK activation (Tournier et al. 2000; Tournier et al. 2001). Given the central role of MAP kinases in genotoxic stress responses, their inactivation might be crucial for signal attenuation (Liu et al. 1998). It has been hypothesized that UV-induced transient activation of JNK leads to stress relief, while sustained activity results in apoptotic cell death (Chen et al. 1996; Franklin et al. 1998). In agreement with this assumption, conditional expression of human *MKP-1* influenced genotoxic stress relief in a stably-transformed leukemic cell line through the inhibition of UV-induced JNK activity resulting in cytoprotection against UV-induced apoptosis (Franklin et al. 1998). Consistently with this function, *MKP-1* gene expression is induced by UV-C radiation as well as by MMS treatment, temporarily coinciding with a decline in JNK activity. In addition, the constitutive expression of *MKP-1* inhibits JNK activity and reduces the UV-C- and MMS-induced activation of transcription factor AP-1-dependent reporter genes (Liu et al. 1995; Liu et al. 1998). However, genetic ablation of MKP-1 function in mouse did not have any reported phenotypic consequences, which is most likely attributable to functional redundancy (Dorfman et al. 1996).

What is the role of MAP kinases in the non-apoptotic UV response? While details of similarities of the genes and signalling pathways that are activated by UV radiation and mitogens are rapidly emerging, the understanding of the exact physiological function of the UV response remained only fragmentary. Recent evidence, however, points to an important role of p38 in the initiation of a G_2/M checkpoint after UV irradiation (Bulavin et al. 2001), indicating that the DNA integrity checkpoint pathways initiated as a result of DNA damage perception and the stress kinase (SAPK/JNK and p38) pathways activated through membrane proximal events interact to modulate cell cycle control (Pearce and Humphrey 2001; Bulavin et al. 2002). It is speculated that this provides a prophylactic response under conditions in which DNA damage is inevitable and thereby serves to minimize genetic damage resulting from cell cycle progression under these circumstances. As a result, cells are allowed to enter mitosis only after appropriate stress recovery (Bulavin et al. 2001). In addition, a recent report on c-Jun, the eponymous target protein of JNK kinases, shed some light on the paradoxical activation of a mitogenic signalling pathway by agents that inhibit rather than stimulate cell proliferation. As for p38 described above, a direct impact on the cell cycle mechanism and DNA damage checkpoints was postulated. In particular, it was found that the function of c-Jun lies in its stimulation of cell cycle re-entry after p53-imposed growth arrest in response to UV irradiation (Shaulian et al. 2000; Shaulian and Karin 2002).

Interestingly, a homologous UV response pathway independent of DNA damage response exists in yeast (Engelberg et al. 1994; Herrlich et al. 1997). In *S. cerevisiae*, the transcriptional activation of the histidine biosynthesis genes *HIS3* and *HIS4* by the Gcn4 transcription factor (AP-1 family member, functional homolog of c-Jun) is triggered by UV irradiation in a Ras-dependent fashion (Ras is small GTP-binding protein, an upstream activator of MAP kinase pathways). Importantly, the UV-resistance of yeast correlates with the level of Ras activity and Gcn4 function (Engelberg et al. 1994). Furthermore, in fission yeast, UV irradiation activates the Spc1 MAP kinase, and *spc1* mutants are hypersensitive to killing by UV and MMS. In contrast to different checkpoint mutants known in yeast that fail to arrest cell division in response to DNA damage, *spc1* mutants are defective in resuming division after the genotoxic insult (Degols and Russell 1997), demonstrating a key function of this MAP kinase in survival of UV-irradiated cells.

8.3.2.2 Non-nuclear-target-mediated signalling in plant systems

Similar to the case of yeast and mammals, it was recently shown that the induction of plant signal transduction pathways by genotoxic stress is most probably initiated at the cell periphery and includes activation of MAP kinases (Stratmann et al. 2000; Ulm et al. 2001; Miles et al. 2002), lending strong support to the existence of a UV response equivalent in plants, as debated before (Herrlich et al. 1997). An initial piece of evidence was the finding that UV irradiation induces the expression of several plant defence genes (including proteinase inhibitors I and II) in a octadecanoid pathway-dependent manner in tomato, as indicated by using the JL-5 mutant that is deficient in jasmonate synthesis (Conconi et al. 1996). Similarly to the UV response in mammals and yeast, a MAP kinase-like activity was detected after UV irradiation in tomato, suggesting the activation of MAP kinase pathways in response to this environmental cue in plants as well (Stratmann et al. 2000). In a direct approach to test whether a UV response equivalent exists in plants, the expression of the *Arabidopsis HIS4* homolog *HDH* (encoding a histidinol dehydrogenase) was monitored and found to be UV-B-inducible in *Arabidopsis* seedlings (Zimmermann et al. 1999). Furthermore, an *Arabidopsis* cDNA encoding a nucleotide diphosphate kinase (NDPKIa) was identified that complements the *gcn4* yeast mutant. NDPKIa specifically binds the *HIS4* promoter in vitro and induces *HIS4* transcription in yeast. In addition, the expression of *NDPKIa* is induced by UV-B in *Arabidopsis* (Zimmermann et al. 1999). Interestingly, a recent report linked this NDPK to oxidative stress and MAP kinase signalling in *Arabidopsis* (Moon et al. 2003)

Genetic data on genotoxic stress signalling in *Arabidopsis* presumably initiated at the cell periphery is limited to the MAP kinase phosphatase homolog AtMKP1 (Ulm et al. 2001; Ulm et al. 2002). The T-DNA knock-out mutant *mkp1*, deficient in AtMKP1, was recently identified in a forward genetic screen for genotoxic stress hypersensitivity. *mkp1* is affected in response to both MMS and UV-C, indicating a role of AtMKP1 in the signal transduction pathway activated by these rather distinct genotoxic agents (Ulm et al. 2001).

It was postulated that AtMKP1 regulates a 49 kD MAP kinase in planta, a notion that was further supported by the identification of the AtMKP1-interacting MAP kinase AtMPK6 and its AtMKP1-dependent genotoxic stress-responsive activity in vivo. Activation of AtMPK6, as determined by immunokinase assays, was highest in the AtMKP1-deficient *mkp1*, intermediate in the wild type control and suppressed in an *AtMKP1*-overexpressing line (Ulm et al. 2002). Thus, the level of AtMKP1 determines the activation level of AtMPK6, establishing AtMKP1 as a regulator of AtMPK6 in response to genotoxic stress in vivo. Extrapolating from mammals (see Sect. 3.1.2), transient and low-level MAPK activation may contribute to genotoxic stress relief, whereas prolonged and high-level activation may trigger cell death. It is thus tempting to speculate that the loss of phosphatase in *mkp1* and the resulting elevated activation of AtMPK6 might tip the balance towards cell death at a stress level permissive for the wild type. The genotoxic stress hypersensitivity of *mkp1* and the elevated activation of AtMPK6 are in agreement with this notion (Ulm et al. 2002). However, next to AtMPK6, yeast two-hybrid analysis also identified AtMPK3 and AtMPK4 as possible AtMKP1-interacting MAPKs, among nine AtMPKs 1-9 tested (Ulm et al. 2002). As there is a total of 20 MPKs encoded in the *Arabidopsis* genome (Ichimura et al. 2002), the panel of AtMKP1-interactors and genotoxic stress-activated MAP kinases might be even larger and the basis for the cellular response might be more complex.

In plants, several intra- and extracellular cues activate MAP kinase pathways and a number of pathway components have been identified on the basis of sequence conservation (Tena et al. 2001; Ichimura et al. 2002; Jonak et al. 2002). Surprisingly, all plant MAP kinases are classified as PERKs (plant ERKs) belonging to the extracellular signal-regulated (ERK) subfamily (Tena et al. 2001), the mammalian members of which are mainly responsible for the transduction of mitogenic signals (Widmann et al. 1999). Thus, no other classes comprising stress-activated protein kinases (SAPK; JNK or p38 kinases) present in other organisms are recognizable in the genomic sequence of *Arabidopsis*.

At present, putative MAP kinase functions in *Arabidopsis* are assigned only to the stress-activated AtMPKs 3, 4, and 6 (Tena et al. 2001; Jonak et al. 2002). For example, immunokinase assays revealed that AtMPK4 and 6 are enzymatically activated by similar environmental stresses such as cold, low humidity, touch, wounding, high salinity, and osmolarity, but also by microbial elicitors (Tena et al. 2001; Jonak et al. 2002, and references therein). In addition, AtMPK6 but not AtMPK4 activity is increased in response to ROS (Yuasa et al. 2001). Furthermore, transient expression experiments in protoplasts indicated the involvement of AtMPK3 and AtMPK6 in responses to oxidative stress (Kovtun et al. 2000). Thus, compelling evidence points to diverse stress-related functions of plant MAPKs, suggesting that certain plant ERKs have evolved the capacity to signal adverse environmental conditions to compensate for the absence of plant SAPKs. Interestingly, AtMKP1 interacts specifically with the three stress-related MAP kinases in *Arabidopsis*, suggesting a function as a regulator of diverse environmental stresses. Supporting this notion is the increased resistance of the *mkp1* mutant to elevated salinity, making AtMKP1 a negative regulator of plant salt tolerance next to its positive regulatory role in genotoxic stress responses (Ulm et al. 2002). However, the exact mechanism of the apparent salt resistance in *mkp1* remains to be identified. It is of note that a connection between genotoxic stress responses and salt signalling pathways had already been postulated based on the phenotypes of the *uvs66* mutant (Albinsky et al. 1999). However, as the responsible mutation is not identified yet, the insight on this phenotypic link at the molecular level is limited.

In addition to the post-translational activation of MAP kinases, it is known that a subgroup is also regulated at the gene expression level in plants (Ichimura et al.

2002). *AtMPK3* is transcriptionally induced in response to various abiotic stresses (Mizoguchi et al. 1996), including ROS and UV-B (Desikan et al. 2001). Furthermore, it was shown that *AtMPK3* and, to a lesser extent, *AtMPK4* are transcriptionally activated in response to UV-C radiation; this is, however, independent of AtMKP1 (Ulm et al. 2002). At present, it is not known how AtMKP1 itself is regulated, but unlike the majority of mammalian MKPs that are immediate early genes, no transcriptional regulation of AtMKP1 has been found.

It is quite interesting that, in a way distinct from the case of other environmental stress treatments, AtMPK6 is activated by genotoxic stress treatments in *Arabidopsis* without the concomitant triggering of AtMPK4 (Ulm et al. 2002). In tobacco, it was found that UV-C activates the AtMPK6 homolog NtSIPK, but not the AtMPK3 homolog NtWIPK (Miles et al. 2002). It is remarkable that ROS activate AtMPK6, but not AtMPK4 (Yuasa et al. 2001), and furthermore, NtSIPK activation by UV-C was found to be reliant on ROS (Miles et al. 2002), indicative of a prominent role of ROS generation in the genotoxic stress signalling in plants, similarly to the situation in mammals (Devary et al. 1992; Liu et al. 1998). It is noteworthy, that both the overexpression and the suppression of *NtSIPK* in tobacco render plants sensitive to ROS-stress (Samuel and Ellis 2002); however, altered resistance to UV-C or other genotoxic agents was not reported. On the other hand, the *mkp1* mutant is hypersensitive to UV-C and MMS, but displays wild type phenotype in response to exogenous ROS (Ulm et al. 2001; Ulm et al. 2002).

In addition to ROS-dependence, it is suggested that NtSIPK activation in response to UV-C in tobacco cell cultures relies on Ca^{2+} ions as well, as indicated by a pharmacological approach (Miles et al. 2002). In addition, suramin, an inhibitor of growth factor/cytokine interactions with their corresponding cell surface receptor tyrosine kinases in animal cells, blocks the activation of MAPK by UV-C, ozone and hydrogen peroxide to a similar extent, suggesting a common signalling pathway (Miles et al. 2002). Similarly, it was shown that suramin blocks UV-B-mediated alkalisation and MAP kinase activation in tomato suspension-cultured cells (Yalamanchili and Stratmann 2002). This is comparable to the situation in mammalian model systems (Sachsenmaier et al. 1994, see also Sect. 3.2.1.) and indicates signal initiation at the cell membrane; however, the possible receptor targets of suramin in plants are elusive.

The exact cellular function of genotoxic stress-responsive MAP kinase pathways in plants is not known at present; however, like in other eukaryotes, it might have a prophylactic function under conditions when DNA damage is to be anticipated. Importantly, mutant analysis indicates that this pathway is required for genotoxic stress tolerance in *Arabidopsis* (Ulm et al. 2001).

8.3.3 Transcriptional response to genotoxic stress in plants

An important output of signal transduction pathways includes changes in gene expression. Not surprisingly, exposure of plant cells to treatments that damage DNA equally evokes gene induction and repression (e.g. Conconi et al. 1996; Ulm et al. 2002; Garcia et al. 2003). However, it remains to be determined which of the aforementioned signalling pathways induces which particular transcriptional responses to genotoxic stress treatments. For example, a crucial role for AtATM was suggested in the signalling pathway that leads to the transcriptional response to IR in the case of a particular set of genes (Garcia et al. 2003, see also Sect. 3.1.2.). Moreover, gene expression profiling of wild type, *mkp1* mutant, and an *AtMKP1*-overexpressing line identified a subset of UV-C-induced genes that probably represents targets of AtMKP1 regulation. It was suggested that AtMKP1 is a negative regulator of this specific group of genes under standard growth conditions, whereas it is a positive regulator of gene induction in response to UV-C (Ulm et al. 2002). It remains to be determined whether this transcriptional readout is due to deregulated activity of AtMPK6 or to another, not yet identified mechanism.

Furthermore, it was found that UV-C irradiation transcriptionally induces *EDS5* (a member of the multidrug and toxin extrusion [MATE] transporter family) and induction is dependent on the pathogen response lipase-like proteins EDS1, PAD4 and the small membrane-associated NDR1 (Nawrath et al. 2002). Thus, it is indicated that the signal transduction pathways of the responses to UV-C irradiation and pathogen infection share common elements. However, even though UV-C irradiation results in the accumulation of salicylic acid and this plant hormone induces *EDS5* transcription, the UV-C responsive *EDS5* induction was not altered in a salicylic acid induction-deficient mutant or in transgenic plants engineered to degrade salicylic acid (Nawrath et al. 2002).

8.3.4 UV-B signalling

In a natural setting, plants are exposed to chronically high levels of sunlight, including its UV-B portion. Depletion of the stratospheric ozone layer leads to elevated terrestrial UV-B levels. Increases in solar UV-B radiation may have substantial effects on the growth and development of many plant species (Rozema et al. 1997; Jansen et al. 1998). On the molecular level, several transcriptionally induced plant genes are known (e.g. Zimmermann et al. 1999; Desikan et al. 2001; Jenkins et al. 2001; Brosche and Strid 2003). Components involved in transducing UV-B perception to gene expression include ROS, Ca^{2+}/calmodulin, reversible protein phosphorylation, and various plant hormones (reviewed, for example, in Brosche and Strid 2003). In addition, in tomato suspension-cultured cells, MAP kinases and other, molecularly unidentified signalling elements of the polypeptide wound signal systemin are suggested to be employed in the response to UV-B (Yalamanchili and Stratmann 2002).

Different genetic screens were carried out using genotoxic levels of UV-B, as indicated by a number of mutants impaired in DNA repair components (Table 1, and references therein). Genetic support for specific signal transduction pathways by which UV-B regulates gene expression is rather limited at present. However, a number of *Arabidopsis* mutants are known that fail to synthesize UV-protective phenylpropanoid pigments due to defects in their biosynthesis, resulting in UV hypersensitivity (Table 1). The importance of signalling UV-B stress perception to increase "sunscreen" components is supported by the hypersensitive *uvr8* and the hyposensitive *myb4* mutants, identifying UVR8 functions as positive regulator and the transcription factor AtMYB4 as negative regulator of phenylpropanoid metabolism in response to UV-B (Jin et al. 2000; Kliebenstein et al. 2002). The *myb4* mutant is deficient in the R2R3 MYB transcription factor AtMYB4 resulting in enhanced UV-B tolerance due to elevated production of sinapate esters, an important UV-B protecting sunscreen (Jin et al. 2000). AtMYB4 functions as a repressor of its target gene *C4H* encoding cinnamate 4-hydroxylase, whereas UV-B-mediated downregulation of *AtMYB4* expression leads to derepression of *C4H* and consequently synthesis of protecting sinapate esters. Consistently, *AtMYB4*-overexpression results in UV-B sensitivity (Jin et al. 2000). In contrast, a mutation in the *UVR8* gene reduces the induction of flavonoids and *chalcone synthase* expression in response to UV-B. The *uvr8* mutant is hypersensitive to UV-B stress and shows enhanced induction of PR1 and PR5 proteins. Interestingly, the *UVR8* gene encodes a protein similar to the human regulator of chromatin condensation 1 (RCC1) (Kliebenstein et al. 2002), a nuclear guanine nucleotide exchange factor for the small GTPase Ran, which is involved in several essential cellular processes, including nucleocytoplasmic transport, mitotic spindle formation, the regulation of cell cycle progression and nuclear envelope assembly (Dasso 2002). Ran GTPases and interacting proteins are present in *Arabidopsis*, indicating conservation of the Ran regulatory mechanism (Haizel et al. 1997). However, it will be interesting to elucidate the mechanism of action of UVR8 in response to UV-B stress and the functional significance of the similarity between UVR8 and RCC1 also remains to be established.

It is noteworthy that a specific UV-B photoreceptor is postulated in plants but not yet identified at the molecular level (e.g. Jenkins et al. 2001; Brosche and Strid 2003), which conceptually provides another parallel to the mammalian and yeast "non-DNA" pathways. Related support is provided by the photomorphogenic responses to UV-B that are independent of the damaging effects (Kim et al. 1998; Boccalandro et al. 2001). It remains to be determined how these non-damaging responses are initiated and signalled and how these events relate to UV-B stress responses.

8.4 Rapid genomic change in plants?

In addition to the obvious mutagenic effect of genotoxic agents, a variety of these factors seem to have a rather indirect impact on the plant's genome as well, owing to the induction of somatic recombination (e.g. Lebel et al. 1993; Puchta et al. 1995; Ries et al. 2000; Kovalchuk et al. 2001). Surprisingly, general abiotic stress factors like heat and salinity, as well as biotic stress seem to increase somatic recombination in plants (Lebel et al. 1993; Puchta et al. 1995; Lucht et al. 2002). In addition, several instances have been documented in which the genome does alter in response to the environment (McClintock 1984; Walbot and Cullis 1985; Walbot 1999). These observations pose the question if these environmental stresses also have the capacity to damage DNA. Alternatively, the plant's genomic response to environmental threats might be programmed, involving particular signal transduction chains culminating in genomic change, possibly enabling a fast and flexible adaptation of plant populations to the changing environment. It is tempting to speculate that the components involved in signalling genotoxic stress in plants may provide a clue to signalling diverse environmental stresses to heritable DNA alterations.

8.5 Conclusions

The response of cells to genotoxic agents is multifaceted and closely linked with regulatory pathways controlling cell growth and division. It is well conceivable that the specific combination of pathways that is induced by a particular genotoxin dictates the fate of the cell after genotoxic stress. In addition, the integration with diverse internal and environmental signals will influence the outcome. Not surprisingly, the first glances into genotoxic stress signalling venues in plants imply similar complexity and impact on basic cellular functions. The *Arabidopsis* mutants impaired in genotoxic stress responses provide the tools to dissect the underlying signalling pathways and their interactions in this genetically tractable model organism. Furthermore, they provide an entry point into understanding plant responses to DNA damaging threats and their links to other stress responses. Given the fundamentally different life strategies of plants versus higher animals, the study of genotoxic stress perception and signalling will certainly profit from a comparison of the two systems.

Acknowledgements

I apologize to researchers in the field for not being able to include all relevant papers. I would like to thank Alexander Baumann, Erzsebet Fejes, Nikki Holbrook, Ferenc Nagy, Jurek Paszkowski, and Alain Tissier for critical reading and helpful comments on the manuscript. This work was supported by the Wolfgang Paul Award to Ferenc Nagy.

References

Abraham RT (2001) Cell cycle checkpoint signaling through the ATM and ATR kinases. Genes Dev 15:2177-2196

Albinsky D, Masson JE, Bogucki A, Afsar K, Vass I, Nagy F, Paszkowski J (1999) Plant responses to genotoxic stress are linked to an ABA/salinity signaling pathway. Plant J 17:73-82

Appella E, Anderson CW (2001) Post-translational modifications and activation of p53 by genotoxic stresses. Eur J Biochem 268:2764-2772

Bender K, Gottlicher M, Whiteside S, Rahmsdorf HJ, Herrlich P (1998) Sequential DNA damage-independent and -dependent activation of NF-kappaB by UV. EMBO J 17:5170-5181

Boccalandro HE, Mazza CA, Mazzella MA, Casal JJ, Ballare CL (2001) Ultraviolet B radiation enhances a phytochrome-B-mediated photomorphogenic response in Arabidopsis. Plant Physiol 126:780-788

Britt A (2002) Repair of damaged bases. In: Somerville CR, Meyerowitz EM (eds) The Arabidopsis Book. American Society of Plant Biologists, Rockville, MD, doi/10.1199/tab.0005, http://www.aspb.org/publications/arabidopsis

Britt AB (1996) DNA damage and repair in plants. Annu Rev Plant Physiol Plant Mol Biol 47:75-100

Brosche M, Strid A (2003) Molecular events following perception of ultraviolet-B radiation by plants. Physiol Plant 117:1-10

Brown EJ, Baltimore D (2000) ATR disruption leads to chromosomal fragmentation and early embryonic lethality. Genes Dev 14:397-402

Bulavin DV, Amundson SA, Fornace AJ (2002) p38 and Chk1 kinases: different conductors for the G(2)/M checkpoint symphony. Curr Opin Genet Dev 12:92-97

Bulavin DV, Higashimoto Y, Popoff IJ, Gaarde WA, Basrur V, Potapova O, Appella E, Fornace AJ, Jr. (2001) Initiation of a G2/M checkpoint after ultraviolet radiation requires p38 kinase. Nature 411:102-107

Bundock P, Hooykaas P (2002) Severe developmental defects, hypersensitivity to DNA-damaging agents, and lengthened telomeres in *Arabidopsis* MRE11 mutants. Plant Cell 14:2451-2462

Bundock P, van Attikum H, Hooykaas P (2002) Increased telomere length and hypersensitivity to DNA damaging agents in an *Arabidopsis* KU70 mutant. Nucleic Acids Res 30:3395-3400

Camps M, Nichols A, Arkinstall S (2000) Dual specificity phosphatases: a gene family for control of MAP kinase function. FASEB J 14:6-16

Chen YR, Wang X, Templeton D, Davis RJ, Tan TH (1996) The role of c-Jun N-terminal kinase (JNK) in apoptosis induced by ultraviolet C and gamma radiation. Duration of JNK activation may determine cell death and proliferation. J Biol Chem 271:31929-31936

Colman MS, Afshari CA, Barrett JC (2000) Regulation of p53 stability and activity in response to genotoxic stress. Mutat Res 462:179-188

Conconi A, Smerdon MJ, Howe GA, Ryan CA (1996) The octadecanoid signalling pathway in plants mediates a response to ultraviolet radiation. Nature 383:826-829

Costa RM, Morgante PG, Berra CM, Nakabashi M, Bruneau D, Bouchez D, Sweder KS, Van Sluys MA, Menck CF (2001) The participation of AtXPB1, the XPB/RAD25 homologue gene from *Arabidopsis thaliana*, in DNA repair and plant development. Plant J 28:385-395

Craven RJ, Greenwell PW, Dominska M, Petes TD (2002) Regulation of genome stability by TEL1 and MEC1, yeast homologs of the mammalian ATM and ATR genes. Genetics 161:493-507

D'Amours D, Jackson SP (2002) The Mre11 complex: at the crossroads of DNA repair and checkpoint signalling. Nat Rev Mol Cell Biol 3:317-327

Danon A, Gallois P (1998) UV-C radiation induces apoptotic-like changes in Arabidopsis thaliana. FEBS Lett 437:131-136

Dasso M (2002) The Ran GTPase: theme and variations. Curr Biol 12:R502-R508

Degols G, Russell P (1997) Discrete roles of the Spc1 kinase and the Atf1 transcription factor in the UV response of *Schizosaccharomyces pombe*. Mol Cell Biol 17:3356-3363

Desikan R, A-H-Mackerness S, Hancock JT, Neill SJ (2001) Regulation of the *Arabidopsis* transcriptome by oxidative stress. Plant Physiol 127:159-172

Devary Y, Gottlieb RA, Smeal T, Karin M (1992) The mammalian ultraviolet response is triggered by activation of Src tyrosine kinases. Cell 71:1081-1091

Devary Y, Rosette C, DiDonato JA, Karin M (1993) NF-kappa B activation by ultraviolet light not dependent on a nuclear signal. Science 261:1442-1445

Dorfman K, Carrasco D, Gruda M, Ryan C, Lira SA, Bravo R (1996) Disruption of the erk/mkp-1 gene does not affect mouse development: normal MAP kinase activity in ERP/MKP-1-deficient fibroblasts. Oncogene 13:925-931

Engelberg D, Klein C, Martinetto H, Struhl K, Karin M (1994) The UV response involving the Ras signaling pathway and AP-1 transcription factors is conserved between yeast and mammals. Cell 77:381-390

Fidantsef AL, Mitchell DL, Britt AB (2000) The *Arabidopsis* UVH1 gene is a homolog of the yeast repair endonuclease RAD1. Plant Physiol 124:579-586

Franklin CC, Srikanth S, Kraft AS (1998) Conditional expression of mitogen-activated protein kinase phosphatase-1, MKP-1, is cytoprotective against UV-induced apoptosis. Proc Natl Acad Sci USA 95:3014-3019

Friedberg EC, Walker GC, Siede W (1995) DNA repair and mutagenesis. ASM Press, Washington, D.C.

Gallego F, Fleck O, Li A, Wyrzykowska J, Tinland B (2000) AtRAD1, a plant homologue of human and yeast nucleotide excision repair endonucleases, is involved in dark repair of UV damages and recombination. Plant J 21:507-518

Gallego ME, Jeanneau M, Granier F, Bouchez D, Bechtold N, White CI (2001) Disruption of the *Arabidopsis* RAD50 gene leads to plant sterility and MMS sensitivity. Plant J 25:31-41

Gallego ME, White CI (2001) RAD50 function is essential for telomere maintenance in *Arabidopsis*. Proc Natl Acad Sci USA 98:1711-1716

Garcia V, Bruchet H, Camescasse D, Granier F, Bouchez D, Tissier A (2003) AtATM is essential for meiosis and the somatic response to DNA damage in plants. Plant Cell 15:119-132

Garcia V, Salanoubat M, Choisne N, Tissier A (2000) An ATM homologue from *Arabidopsis thaliana*: complete genomic organisation and expression analysis. Nucleic Acids Res 28:1692-1699

Gasch AP, Huang M, Metzner S, Botstein D, Elledge SJ, Brown PO (2001) Genomic expression responses to DNA-damaging agents and the regulatory role of the yeast ATR homolog Mec1p. Mol Biol Cell 12:2987-3003

Gong Z, Morales-Ruiz T, Ariza RR, Roldan-Arjona T, David L, Zhu J-K (2002) ROS1, a repressor of transcriptional gene silencing in *Arabidopsis*, encodes a DNA glycosylase/lyase. Cell 111:803-814

Gorbunova VV, Levy AA (1999) How plants make ends meet: DNA double-strand break repair. Trends Plant Sci 4:263-269

Haizel T, Merkle T, Pay A, Fejes E, Nagy F (1997) Characterization of proteins that interact with the GTP-bound form of the regulatory GTPase Ran in *Arabidopsis*. Plant J 11:93-103

Herrlich P, Blattner C, Knebel A, Bender K, Rahmsdorf HJ (1997) Nuclear and non-nuclear targets of genotoxic agents in the induction of gene expression. Shared principles in yeast, rodents, man and plants. Biol Chem 378:1217-1229

Ichimura K, Shinozaki K, Tena G, Sheen J, Henry Y, Champion A, Kreis M, Zhang S, Hirt H, Wilson C, Heberle-Bors E, Ellis BE, Morris P, Innes RW, Ecker JR, Scheel D, Klessig DF, Machida Y, Mundy J, Ohashi Y, Walker JC (2002) Mitogen-activated protein kinase cascades in plants: a new nomenclature. Trends Plant Sci 7:301-308

Jansen MAK, Gaba V, Greenberg BM (1998) Higher plants and UV-B radiation: balancing damage, repair and acclimation. Trends Plant Sci 3:131-135

Jenkins GI, Long JC, Wade HK, Shenton MR, Bibikova TN (2001) UV and blue light signalling: pathways regulating chalcone synthase gene expression in *Arabidopsis*. New Phytol 151:121-131

Jin H, Cominelli E, Bailey P, Parr A, Mehrtens F, Jones J, Tonelli C, Weisshaar B, Martin C (2000) Transcriptional repression by AtMYB4 controls production of UV-protecting sunscreens in *Arabidopsis*. EMBO J 19:6150-6161

Jonak C, Okresz L, Bogre L, Hirt H (2002) Complexity, cross talk and integration of plant MAP kinase signalling. Curr Opin Plant Biol 5:415

Kastan MB, Lim D-S (2000) The many substrates and functions of ATM. Nat Rev Mol Cell Biol 1:179-186

Kim BC, Tennessen DJ, Last RL (1998) UV-B-induced photomorphogenesis in *Arabidopsis thaliana*. Plant J 15:667-674

Kim ST, Lim D-S, Canman CE, Kastan MB (1999) Substrate specificities and identification of putative substrates of ATM kinase family members. J Biol Chem 274:37538-37543

Kitagawa D, Tanemura S, Ohata S, Shimizu N, Seo J, Nishitai G, Watanabe T, Nakagawa K, Kishimoto H, Wada T, Tezuka T, Yamamoto T, Nishina H, Katada T (2002) Activation of extracellular signal-regulated kinase by ultraviolet is mediated through Src-dependent epidermal growth factor receptor phosphorylation. Its implication in an anti-apoptotic function. J Biol Chem 277:366-371

Kliebenstein DJ, Lim JE, Landry LG, Last RL (2002) *Arabidopsis* UVR8 regulates ultraviolet-B signal transduction and tolerance and contains sequence similarity to human regulator of chromatin condensation 1. Plant Physiol 130:234-243

Knebel A, Rahmsdorf HJ, Ullrich A, Herrlich P (1996) Dephosphorylation of receptor tyrosine kinases as target of regulation by radiation, oxidants or alkylating agents. EMBO J 15:5314-5325

Kovalchuk O, Titov V, Hohn B, Kovalchuk I (2001) A sensitive transgenic plant system to detect toxic inorganic compounds in the environment. Nat Biotechnol 19:568-572

Kovtun Y, Chiu WL, Tena G, Sheen J (2000) Functional analysis of oxidative stress-activated mitogen-activated protein kinase cascade in plants. Proc Natl Acad Sci USA 97:2940-2945

Kyriakis JM, Avruch J (2001) Mammalian mitogen-activated protein kinase signal transduction pathways activated by stress and inflammation. Physiol Rev 81:807-869

Landry LG, Chapple CC, Last RL (1995) *Arabidopsis* mutants lacking phenolic sunscreens exhibit enhanced ultraviolet-B injury and oxidative damage. Plant Physiol 109:1159-1166

Landry LG, Stapleton AE, Lim J, Hoffman P, Hays JB, Walbot V, Last RL (1997) An *Arabidopsis* photolyase mutant is hypersensitive to ultraviolet-B radiation. Proc Natl Acad Sci USA 94:328-332

Lebel EG, Masson J, Bogucki A, Paszkowski J (1993) Stress-induced intrachromosomal recombination in plant somatic cells. Proc Natl Acad Sci USA 90:422-426

Liu Y, Gorospe M, Holbrook NJ, Anderson CW (1998) Posttranslational mechanisms leading to mammalian gene activation in response to genotoxic stress. In: Nickoloff JA, Hoekstra MF (eds) DNA damage and repair, Vol. 2: DNA repair in higher eukaryotes. Humana Press Inc., Totowa, NJ, pp 263-298

Liu Y, Gorospe M, Yang C, Holbrook NJ (1995) Role of mitogen-activated protein kinase phosphatase during the cellular response to genotoxic stress. Inhibition of c-Jun N-terminal kinase activity and AP-1-dependent gene activation. J Biol Chem 270:8377-8380

Liu Z, Hall JD, Mount DW (2001) *Arabidopsis* UVH3 gene is a homolog of the *Saccharomyces cerevisiae* RAD2 and human XPG DNA repair genes. Plant J 26:329-338

Liu Z, Hossain GS, Islas-Osuna MA, Mitchell DL, Mount DW (2000) Repair of UV damage in plants by nucleotide excision repair: Arabidopsis UVH1 DNA repair gene is a homolog of *Saccharomyces cerevisiae* Rad1. Plant J 21:519-528

Liu ZG, Baskaran R, Lea-Chou ET, Wood LD, Chen Y, Karin M, Wang JY (1996) Three distinct signalling responses by murine fibroblasts to genotoxic stress. Nature 384:273-276

Lucht JM, Mauch-Mani B, Steiner HY, Metraux JP, Ryals J, Hohn B (2002) Pathogen stress increases somatic recombination frequency in Arabidopsis. Nat Genet 30:311-314

Marnett LJ, Plastaras JP (2001) Endogenous DNA damage and mutation. Trends Genet 17:214-221

Marshall CJ (1995) Specificity of receptor tyrosine kinase signaling: transient versus sustained extracellular signal-regulated kinase activation. Cell 80:179-185

McClintock B (1984) The significance of responses of the genome to challenge. Science 226:792-801

Melo J, Toczyski D (2002) A unified view of the DNA-damage checkpoint. Curr Opin Cell Biol 14:237-245

Menand B, Desnos T, Nussaume L, Berger F, Bouchez D, Meyer C, Robaglia C (2002) Expression and disruption of the *Arabidopsis* TOR (target of rapamycin) gene. Proc Natl Acad Sci USA 99:6422-6427

Mengiste T, Revenkova E, Bechtold N, Paszkowski J (1999) An SMC-like protein is required for efficient homologous recombination in *Arabidopsis*. EMBO J 18:4505-4512

Menke M, Chen I, Angelis KJ, Schubert I (2001) DNA damage and repair in *Arabidopsis thaliana* as measured by the comet assay after treatment with different classes of genotoxins. Mutat Res 493:87-93

Miles GP, Samuel MA, Ellis BE (2002) Suramin inhibits oxidant signalling in tobacco suspension-cultured cells. Plant Cell Environ. 25:521-527

Mitsuhara I, Malik KA, Miura M, Ohashi Y (1999) Animal cell-death suppressors Bcl-x(L) and Ced-9 inhibit cell death in tobacco plants. Curr Biol 9:775-778

Mizoguchi T, Irie K, Hirayama T, Hayashida N, Yamaguchi-Shinozaki K, Matsumoto K, Shinozaki K (1996) A gene encoding a mitogen-activated protein kinase kinase kinase is induced simultaneously with genes for a mitogen-activated protein kinase and an S6 ribosomal protein kinase by touch, cold, and water stress in *Arabidopsis thaliana*. Proc Natl Acad Sci USA 93:765-769

Moon H, Lee B, Choi G, Shin D, Prasad DT, Lee O, Kwak SS, Kim DH, Nam J, Bahk J, Hong JC, Lee SY, Cho MJ, Lim CO, Yun DJ (2003) NDP kinase 2 interacts with two oxidative stress-activated MAPKs to regulate cellular redox state and enhances multiple stress tolerance in transgenic plants. Proc Natl Acad Sci USA 100:358-363

Morrow DM, Tagle DA, Shiloh Y, Collins FS, Hieter P (1995) *TEL1*, an *S. cerevisiae* homolog of the human gene mutated in ataxia telangiectasia, is functionally related to the yeast checkpoint gene *MEC1*. Cell 82:831-840

Nakajima S, Sugiyama M, Iwai S, Hitomi K, Otoshi E, Kim ST, Jiang CZ, Todo T, Britt AB, Yamamoto K (1998) Cloning and characterization of a gene (UVR3) required for photorepair of 6-4 photoproducts in *Arabidopsis thaliana*. Nucleic Acids Res 26:638-644

Nawrath C, Heck S, Parinthawong N, Metraux JP (2002) EDS5, an essential component of salicylic acid-dependent signaling for disease resistance in *Arabidopsis*, is a member of the MATE transporter family. Plant Cell 14:275-286

Pearce AK, Humphrey TC (2001) Integrating stress-response and cell-cycle checkpoint pathways. Trends Cell Biol 11:426-433

Perry J, Kleckner N (2003) The ATRs, ATMs, and TORs are giant HEAT repeat proteins. Cell 112:151-155

Puchta H, Swoboda P, Hohn B (1995) Induction of intrachromosomal homologous recombination in whole plants. Plant J 7:203-210

Ries G, Heller W, Puchta H, Sandermann H, Seidlitz HK, Hohn B (2000) Elevated UV-B radiation reduces genome stability in plants. Nature 406:98-101

Riha K, Watson JM, Parkey J, Shippen DE (2002) Telomere length deregulation and enhanced sensitivity to genotoxic stress in *Arabidopsis* mutants deficient in Ku70. EMBO J 21:2819-2826

Rosette C, Karin M (1996) Ultraviolet light and osmotic stress: activation of the JNK cascade through multiple growth factor and cytokine receptors. Science 274:1194-1197

Rotman G, Shiloh Y (1999) ATM: a mediator of multiple responses to genotoxic stress. Oncogene 18:6135-6144

Rozema J, van de Staaij J, Björn LO, Caldwell M (1997) UV-B as an environmental factor in plant life: stress and regulation. Trends Ecol Evol 12:22-28

Sachsenmaier C, Radler-Pohl A, Zinck R, Nordheim A, Herrlich P, Rahmsdorf HJ (1994) Involvement of growth factor receptors in the mammalian UVC response. Cell 78:963-972

Samuel MA, Ellis BE (2002) Double jeopardy: both overexpression and suppression of a redox-activated plant mitogen-activated protein kinase render tobacco plants ozone sensitive. Plant Cell 14:2059-2069

Shaulian E, Karin M (2002) AP-1 as a regulator of cell life and death. Nat Cell Biol 4:E131-136

Shaulian E, Schreiber M, Piu F, Beeche M, Wagner EF, Karin M (2000) The mammalian UV response: c-Jun induction is required for exit from p53-imposed growth arrest. Cell 103:897-907

Stratmann JW, Stelmach BA, Weiler EW, Ryan CA (2000) UVB/UVA radiation activates a 48 kDa myelin basic protein kinase and potentiates wound signaling in tomato leaves. Photochem Photobiol 71:116-123

Tena G, Asai T, Chiu WL, Sheen J (2001) Plant mitogen-activated protein kinase signaling cascades. Curr Opin Plant Biol 4:392-400

Theodosiou A, Ashworth A (2002) MAP kinase phosphatases. Genome Biol 3:REVIEWS3009

Tournier C, Dong C, Turner TK, Jones SN, Flavell RA, Davis RJ (2001) MKK7 is an essential component of the JNK signal transduction pathway activated by proinflammatory cytokines. Genes Dev 15:1419-1426

Tournier C, Hess P, Yang DD, Xu J, Turner TK, Nimnual A, Bar-Sagi D, Jones SN, Flavell RA, Davis RJ (2000) Requirement of JNK for stress-induced activation of the cytochrome c-mediated death pathway. Science 288:870-874

Tuteja N, Singh MB, Misra MK, Bhalla PL, Tuteja R (2001) Molecular mechanisms of DNA damage and repair: progress in plants. Crit Rev Biochem Mol Biol 36:337-397

Ulm R, Ichimura K, Mizoguchi T, Peck SC, Zhu T, Wang X, Shinozaki K, Paszkowski J (2002) Distinct regulation of salinity and genotoxic stress responses by *Arabidopsis* MAP kinase phosphatase 1. EMBO J 21:6483-6493

Ulm R, Revenkova E, di Sansebastiano GP, Bechtold N, Paszkowski J (2001) Mitogen-activated protein kinase phosphatase is required for genotoxic stress relief in *Arabidopsis*. Genes Dev 15:699-709

Wahl GM, Carr AM (2001) The evolution of diverse biological responses to DNA damage: insights from yeast and p53. Nat Cell Biol 3:E277-286

Walbot V (1996) Sources and consequences of phenotypic and genotypic plasticity in flowering plants. Trends Plant Sci 1:27-32

Walbot V (1999) UV-B damage amplified by transposons in maize. Nature 397:398-399

Walbot V, Cullis CA (1985) Rapid genomic change in higher plants. Ann Rev Plant Physiol 36:367-396

West CE, Waterworth WM, Story GW, Sunderland PA, Jiang Q, Bray CM (2002) Disruption of the *Arabidopsis* AtKu80 gene demonstrates an essential role for AtKu80 protein in efficient repair of DNA double-strand breaks in vivo. Plant J 31:517-528

Whittle C-A, Beardmore T, Johnston MO (2001) Is G1 arrest in plant seeds induced by a p53-related pathway? Trends Plant Sci 6:248-251

Widmann C, Gibson S, Jarpe MB, Johnson GL (1999) Mitogen-activated protein kinase: conservation of a three-kinase module from yeast to human. Physiol Rev 79:143-180

Yalamanchili RD, Stratmann JW (2002) Ultraviolet-B activates components of the systemin signaling pathway in Lycopersicon peruvianum suspension-cultured cells. J Biol Chem 277:28424-28430

Yuasa T, Ichimura K, Mizoguchi T, Shinozaki K (2001) Oxidative stress activates ATMPK6, an *Arabidopsis* homologue of MAP kinase. Plant Cell Physiol 42:1012-1016

Zimmermann S, Baumann A, Jaekel K, Marbach I, Engelberg D, Frohnmeyer H (1999) UV-responsive genes of *Arabidopsis* revealed by similarity to the Gcn4-mediated UV response in yeast. J Biol Chem 274:17017-17024

Abbreviations

ATM: Ataxia telangiectasia-mutated
ATR: ATM and Rad3-related
BER: Base excision repair
HR: Homologous recombination
IR: Ionizing radiation
JNK: c-jun amino-terminal protein kinase

MAPK: Mitogen-activated protein kinase
MKP: MAP kinase phosphatase
MMS : Methyl methanesulfonate
MPK: Mitogen-activated protein kinase (gene name)
NER : Nucleotide excision repair
NHEJ: Non-homologous end-joining
PHR: Photoreactivation
PI3K: Phosphatidylinositol 3-kinase
ROS: Reactive oxygen species
SAPK : Stress-activated protein kinase
SIPK: Salicylic acid-induced protein kinase
TOR: Target of rapamycin
UV: Ultraviolet
WIPK: Wound-induced protein kinase

9 Plant salt tolerance

Viswanathan Chinnusamy and Jian-Kang Zhu

Abstract

Soil salinity adversely affects crop productivity and quality. The success of breeding programs aimed at salinity tolerant crop varieties is limited by the lack of a clear understanding of the molecular basis of salt tolerance. Recent advances in genetic analysis of *Arabidopsis* mutants defective in salt tolerance, and molecular cloning of these loci, have showed some insight into salt stress signaling and plant salt tolerance. Salt stress-induced cytosolic calcium signals are perceived by SOS3, which is a calcium sensor protein. SOS3 is constitutively myristoylated and associated with the plasma membrane. The SOS3 activates SOS2, a ser/thr protein kinase, in a calcium dependent manner. The active SOS3-SOS2 kinase complex activates SOS1, a Na^+/H^+ antiporter on the plasma membrane and also upregulates *SOS1* gene expression; this results in Na^+ efflux and ion homeostasis. Transgenic analysis showed a tonoplast-located Na^+/H^+ antiporter mediates sodium sequestration into the vacuole, and this forms an important part of the salt tolerance mechanism. Evidence also implicates a putative osmosensory histidine kinase (AtHK1)-MAPK cascade and its negative regulators (AtMKP1) in salt stress signaling that probably leads to osmotic homeostasis and ROS scavenging. ABA-mediated regulation of stress proteins and plant growth are also important for plant salt tolerance, but the signaling pathway is poorly understood.

9.1 Introduction

Soil salinity predates human civilization and is probably a cause of the breakdown of the ancient Sumerian civilization (Jacobson and Adams 1958). Today salinity remains a major abiotic stress that adversely affects crop productivity and quality (Boyer 1982). Saline soil is characterized by toxic levels of chlorides and sulfates of sodium. The electrical conductivity of saturation extracts of saline soil is more than 4.0 dS/m (≈ 40mM NaCl; Marschner 1995). The problem of soil salinity is increasing owing to 1) the use of poor quality water for irrigation, 2) improper drainage in canal-irrigated wetland agro-ecosystems, 3) entry of seawater during cyclones in coastal areas, and 4) salt accumulation in the root zone in arid and semi-arid regions due to high evaporative demand and insufficient leaching of ions as the rainfall is inadequate.

Sodium is an essential micronutrient for some of the C_4 photosynthetic plants, which import pyruvate into mesophyll chloroplasts by a Na^+/pyruvate co-

Topics in Current Genetics, Vol. 4
H. Hirt, K. Shinozaki (Eds.) Plant Responses To Abiotic Stress

transporter (Ohnishi et al. 1990). However, most crop plants are natrophobic. Salinity is detrimental to plant growth as it causes 1) nutritional constraints by decreasing uptake of phosphorus, potassium, nitrate and calcium, 2) ion cytotoxicity mainly due to Na^+, Cl^- plus SO_4^-, and 3) osmotic stress (reviewed by Zhu 2001, 2002). Na^+ competes with K^+ in biochemical reactions, which is inimical to cellular processes. Under salinity, ions like Na^+ and Cl^- penetrate the hydration shells of proteins and interfere with the non-covalent interactions between amino acids of proteins. This leads to conformational changes and loss of function of proteins. Ionic toxicity, osmotic stress, and nutritional defects under salinity may lead to metabolic imbalances, which result in oxidative stress (Zhu 2001). Plant salt tolerance mechanisms can be grouped into 1) cellular homeostasis which includes ion homeostasis and osmotic adjustment, 2) stress damage control and repair, or detoxification, and 3) growth regulation (Zhu, 2001). Considerable efforts have been made to unravel plant salt tolerance mechanisms with the ultimate goal of improving the crop productivity in saline soils. Here we discuss molecular and genetic evidence concerning the perception of salinity stress by plants, cellular signal transduction, and effectors of salt stress tolerance.

9.2 Sodium entry into plant cells

The membrane potential difference at the plasma membrane of plant cells is -140 mV, which favors passive transport of Na^+ into cells, especially with high extracellular Na^+ concentrations. Excess extracellular Na^+ enters the cell through both the transporter HKT1 and non-selective cation channels/transporters, which results in a decrease in the K^+/Na^+ ratio in the cytosol. The wheat high affinity K^+ transporter HKT1 appears to act as a low affinity Na^+ transporter (Rubio et al. 1995; Gorham et al. 1997). Expression of the *Arabidopsis* homolog of wheat *HKT1* (*ATHKT1*) in *Xenopus* oocytes mediated Na^+ influx, which suggested that ATHKT1 might be involved in Na^+ influx in plants (Uozumi et al. 2000). Eucalyptus *EcHKT1* and *EcHKT2* when expressed in oocytes showed both Na^+ and K^+ uptake, but the permeability to Na^+ was greater than that for K^+ when the extracellular concentration of Na^+ and K^+ were equal (Liu et al. 2001). These results suggest that in plants in general, HKT1 might be involved in low affinity Na^+ influx. In rice, the contribution of carrier protein-mediated Na^+ uptake is less significant than the apoplastic pathway under high salinity conditions (Yadav et al. 1996; Garcia et al. 1997). Quantitative trait loci (QTL) and inheritance analysis in rice revealed that genes that control Na^+ uptake are different from that of K^+ uptake (Garcia et al. 1997; Koyama et al. 2001). Silica deposition in the endodermis and rhizodermis, and the polymerization of silicate via colloidal silica to silica gel or polysilicic acid throughout the root apoplast, appears to block Na^+ uptake through the apoplastic pathway in the roots of rice (Yeo et al. 1999). Hence, the entry of Na^+ in rice under salinity is expected to be regulated significantly by genes that affect root development and silicon uptake. However, in wheat, sodium/potassium selectivity by carrier proteins in the root appears to be a major determinant of salt tolerance (Gorham et al. 1997). Thus, the entry of Na^+ into plant root cells can be affected by the regulation of K^+/Na^+ transporter HKT1 and non-selective cation channels, and by regulation of genes involved in root development and silicon polymerization-mediated blockage of the apoplastic route. Plant species-specific differences and the regulatory mechanism of Na^+ entry into plant roots under salinity need to be understood.

9.3 Input signals of salt stress

Salt stress affects cellular ion homeostasis as well as osmotic homeostasis. Excess Na^+ and Cl^- ions may lead to conformational changes in protein structure and/or changes in the plasma membrane electrical potential, while osmotic stress leads to turgor loss and cell volume change. Hence, excess ions (Na^+ and Cl^-) and osmotic stress-induced turgor change may act as inputs for salt stress signaling. The candidate sensors of ionic stress may include ion channels/transporters and ion binding proteins on the plasma membrane or at intracellular locations (Zhu 2002). Under High Na^+ concentrations, Na^+ may enter cells through non-specific ion channels, which might cause membrane depolarization. A change in membrane polarization could also signal salt stress, as it is known to activate Ca^{2+} channels (Sanders et al. 1999). Loss of turgor leads to a cell volume change and retraction of the plasma membrane from the cell wall. As the membrane retracts from the cell wall, membrane bound receptor kinases, ion transporters/channels, transmembrane proteins that are in contact with cell wall, and integrin-like proteins may undergo conformational changes or cluster together, and hence these proteins may act as sensors of osmotic stress. Integrins and the F-actin cytoskeleton have been implicated in the sensing of cell volume changes in mammalian cells. Regulation of microtubule organization by turgor pressure was shown in *Spirogyra* sp. (Iwata et al. 2001). Microtubules and microfilaments of the cytoskeleton have been implicated in signaling under cold stress in plants (reviewed by Viswanathan and Zhu 2002), and the pattern of microtubular organization under cold stress differs from that of ABA (Wang and Nick 2001). Since the cytoskeleton connects different organelles of the cell with the plasma membrane, it can sense cell volume change under osmotic stress and transduce it to internal Ca^{2+} channels or other signaling components. Salinity induces the biosynthesis and accumulation of the plant stress hormone abscisic acid (ABA; Jia et al. 2002) and also induces accumulation of reactive oxygen species (ROS; Smirnoff 1993; Gomez et al. 1999; Hernandez et al. 2001). Current evidence suggests that the primary salt stress signals (ionic and osmotic stress) are transduced through Ca^{2+} as well as receptor kinase pathways, while the secondary salt stress signals such as ABA and H_2O_2 also regulate plant salt tolerance.

9.3.1 Calcium signalling

Cytosolic Ca^{2+} oscillations, generated from extracellular and/or intracellular Ca^{2+} stores, act as a second messenger in cold, drought, and salt stresses (Sanders et al. 1999; Knight 2000). Calcium oscillations in plant cells vary, depending upon the type of stress (Kiegle et al. 2000), rate of stress development (Plieth et al. 1999), previous experience of stresses/cycles (Knight et al. 1997), and tissue type (Kiegle et al. 2000). Cytosolic Ca^{2+} oscillations occur within 5-10 seconds of salt stress, persist up to 1 to 10 minutes and, hence, are thought to be one of the earliest events in salt signaling (Lynch et al. 1989; Knight et al. 1997). Therefore, it is essential to analyze how such Ca^{2+} signatures are generated by a salt stress signal, and what are the components downstream that decode salt stress-specific Ca^{2+} signatures.

Cytosolic Ca^{2+} signatures can be the net result of influx and efflux of Ca^{2+}. Calcium efflux occurs through Ca^{2+} ATPases and H^+/Ca^{2+} antiporters, while influx is controlled by Ca^{2+} permeable ion channels (Sanders et al. 1999). In animal cells, ligand-gated Ca^{2+} channels are regulated by inositol (1,4,5)-triphosphate (IP_3), cyclic adenosine 5'diphosphate ribose (cADPR), and nicotinic acid adenine dinucleotide phosphate ($NAADP^+$). In plants, IP_3 and cADPR-gated Ca^{2+} channels are found in vacuolar and endoplasmic reticulum membranes (Allen et al. 1995; Wu et al. 1997), and NAADP-gated Ca^{2+} channels are found in the endoplasmic reticulum membrane (Navazio et al. 2000). In *Arabidopsis*, osmotic stress (NaCl or sorbitol) induces the synthesis of IP_3 to significantly higher levels within 1 minute of stress initiation, and it continues to increase for more than 30 minutes. Phospholipase C (PLC) hydrolyses phosphotidylinositol-4,5 bisphosphate into diacyl glycerol and IP_3, which are activators of protein kinase and calcium channels, respectively. Treatment with U-73122, an inhibitor of PLC, blocked IP_3 accumulation. The temporal pattern of IP_3 accumulation is similar to that observed for stress-induced calcium mobilization, implicating IP_3 in salt stress-induced Ca^{2+} signaling (DeWald et al. 2001; Takahashi et al. 2001). In cell cultures of *Arabidopsis,* a few seconds of osmotic stress (dehydration, mannitol or NaCl) caused a rapid and transient increase in IP_3 and expression of dehydration-inducible genes (*RD29A/LTI78/COR78* & *RD17/COR47*). This response was abolished when the cells were treated with inhibitors of PLC, such as neomycin and U73122, indicating the involvement of PLC and IP_3 in hyper-osmotic stress signaling (Takahashi et al. 2001).

Osmotic stress caused by NaCl/mannitol/sorbitol significantly increases cellular $PtdIns(4,5)P_2$ synthesis (Pical et al. 1999; DeWald et al. 2001), which is the substrate for cleavage by PLC to produce IP_3 (DeWald et al. 2001). Consistent with this, it has been shown that a *PLC* gene is also upregulated by osmotic stress (Hirayama et al. 1995). Salt stress-induced phosphatidylnositol $(4,5)P_2$ synthesis and cleavage into IP_3 may help in delayed Ca^{2+} signaling. Genetic evidence for the implication of IP_3 signaling in abiotic stresses including salinity came from the analysis of the *FRY1* locus of *Arabidopsis* (Table 1). *FRY1* encodes an inositol polyphosphate 1-phosphatase, which functions in the catabolism of IP_3. Upon ABA treatment, *fry1* mutant plants accumulated more IP_3 than did the wild type

Table 1. *Arabidopsis* mutants impaired in salt stress response

Mutant	Function	Salt tolerance	Reference
Salt overly sensitive 3 (*sos3*)	Ca^{2+} sensor	Hypersensitive	Liu and Zhu 1998
Salt overly sensitive 2 (*sos2*)	Protein kinase	Hypersensitive	Liu et al. 2000
Salt overly sensitive 1 (*sos1*)	Plasma membrane $Na^{+}H^{+}$ antiporter	Hypersensitive	Shi et al. 2000
Salt overly sensitive 4 (*sos4*)	pyridoxal kinase	Hypersensitive	Shi et al. 2002b
High affinity K transporter 1 (*hkt1*)	Plasma membrane Na^{+} transporter	Suppresses salt sensitivity of sos3 mutant	Rus et al. 2001a
Fiery1 (*fry1*)	Inositol polyphosphate 1-phosphatase	Hypersensitive	Xiong et al. 2001b
Low expression of osmotically responsive genes 5 (*los5*)/ABA deficient 3 (*aba3*)	Molybdenum cofactor sulfurase	Hypersensitive	Xiong et al. 2001a
Low expression of osmotically responsive genes 6 (*los6*) /ABA deficient 1 (*aba1*)	zeaxanthin epoxidase	Tolerant during germination	Xiong et al. 2002a
SALOBREÑO (*sañ5*)/*abi4*	ABA insensitive 4	Tolerant during germination	Quesada et al. 2000
osm1	A protein similar to SNARE type mammalian syntaxins	Hypersensitive to salt and osmotic stress	Zhu et al. 2002
mkp1	MAPK phosphatase 1	Enhanced salt tolerance	Ulm et al. 2002
photoautotrophic salt tolerance1 (*pst1*)	Not yet cloned	Enhanced salt tolerance	Tsugane et al. 1999
t365	S-adenosyl-L-methionine phosphoethanolamine N-methyltransferase	Hypersensitive	Mou et al. 2002

plants. In wild type, IP_3 accumulation was transiently induced by ABA, while in *fry1* IP_3 accumulation was sustained, which suggest that IP_3 catabolism is mediated by FRY1. The *fry1* mutant is hypersensitive to ABA and salinity stress (Xiong et al. 2001b). The *Arabidopsis SAL1* gene, a homolog of *FRY1* conferred increased salt tolerance to yeast cells (Quintero et al. 1996). These results showed that IP_3 transient induced by salt and ABA is necessary for stress tolerance. In addition to IP_3-gated Ca^{2+} channels, stretch/mechanosensitive Ca^{2+} channels may also be involved in primary Ca^{2+} oscillations (Knight et al. 1997), as these Ca^{2+} channels can be activated immediately by a change in cell volume/turgor in salt stressed cells. Hence, salt stress-induced IP_3 oscillations are an integral part of

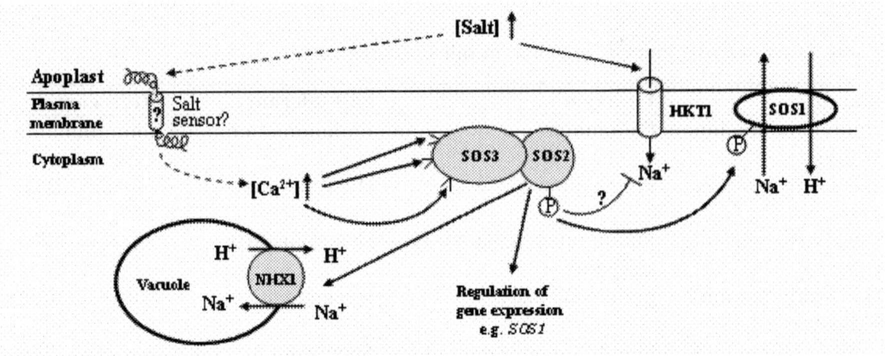

Fig. 1. The SOS pathway for ion homeostasis regulation under salt stress. Salt stress induced Ca^{2+} signals are perceived by SOS3, which activates the SOS2 kinase. Activated SOS2 kinase phosphorylates the SOS1 Na^{+}/H^{+} antiporter, which then pumps Na^{+} out of the cytosol. The SOS3-SOS2 kinase complex also regulates the transcript level of *SOS1* and other genes. The SOS3-SOS2 kinase complex may regulate Na^{+} compartmentation by activating NHX1, and also may restrict Na^{+} entry into the cytosol, e.g. by inhibiting the plasma membrane Na^{+} transporter HKT1 activity.

Ca^{2+} signaling in salt stress. Engineered alterations in intracellular Ca^{2+} levels due to overexpression of the *Arabidopsis* vacuolar Ca^{2+}/H^{+} antiporter gene (*CAX1*) in tobacco (Hirschi 1999), and the ionotrophic glutamate receptor (*GluR2*) in *Arabidopsis* (Kim et al. 2001) resulted in hypersensitivity to salt stress and other developmental abnormalities. This evidence strongly suggests that oscillations in intracellular Ca^{2+} levels form an integral part of plant salt tolerance.

9.3.2 Calcium sensors

Three major families of calcium binding proteins sense Ca^{2+} signals in plants (Liu and Zhu 1998; Harmon et al. 2000): 1) Calmodulins (CaM), which do not have enzymatic activity but transduce signals to CaM interacting proteins. Calmodulins contain four EF-hand domains responsible for Ca^{2+} binding, 2) Calcium dependent protein kinases (CDPKs), which contain CaM-like Ca^{2+}-binding domains and a kinase domain in a single protein, and 3) SOS3 and SOS3-like calcium-binding proteins (SCaBPs) (Liu and Zhu 1998; Guo et al. 2001). The specificity of Ca^{2+} signals may be achieved by the multiplicity of calcium sensors and their intracellular localization. The first genetic evidence for a calcium sensor protein mediated Ca^{2+} signaling in salt stress came from the analysis of the *salt overly sensitive 3* (*sos3*) mutant of *Arabidopsis* (Table 1, Fig. 1; Liu and Zhu 1998). The SOS3 mediated salt stress signaling in cellular ion homeostasis is discussed in the later part of this review.

The *Arabidopsis AtGSK1* gene, which encodes a protein similar to glycogen synthase kinase3, complemented yeast mutant DHT22-1a that is defective in both calcineurin (*SLN1* and *SHO1*) genes. Expression of *AtGSK1* in the yeast mutant

DHT22-1a also restored salt stress-induced expression of a Na^+-ATPase (*PMR2A*) gene. Moreover, *AtGSK1* is upregulated under ABA and salt stresses in *Arabidopsis*, suggesting the possible involvement of AtGSK1 in phosphoprotein-dependent salt stress signaling (Piao et al. 1999; Charrier et al. 2002).

The CDPK family of protein kinases contains a myristoylation and a calcium binding EF hand domain. The *Arabidopsis AtCDPK1* and *AtCDPK2* genes are induced by salinity and drought but not by low/high temperatures, suggesting that these two ATCDPKs might be involved in osmotic stress signaling (Urao et al. 1994). The involvement of CDPKs in stress-induced gene transcription was shown by Sheen (1996) in a maize leaf protoplast system transiently expressing barley *HVA1* promoter-driven synthetic green fluorescent protein. The barley *HVA1* gene, encoding a class 3 late embryogenesis-abundant protein, is induced by abiotic stresses such as drought, cold, heat, salinity, and ABA. Expression of *HVA1-SGFP* was significantly increased in the protoplasts incubated with Ca^{2+} and Ca^{2+} ionophore (ionomycin or A23187), but not by Ca^{2+} alone, indicating that Ca^{2+} entry is essential for *HVA1* expression. Further, maize protoplasts co-expressing the *HVA1-LUC* reporter and truncated *AtCDPK1* or *AtCDPK1a* (truncation of the regulatory domain of ATCDPKs results in Ca^{2+} independent, constitutive protein kinase activity) showed constitutive expression of *HVA1::LUC*. Mutated AtCDPK1, which lacks ATP binding activity, was unable to induce the *HVA1* promoter. These data indicate that protein kinase activity of AtCDPK1 is essential to activate the *HVA1* promoter. The Ca^+ requirement for induction of the abiotic stress responsive *HVA1* promoter in maize protoplasts suggests the involvement of AtCDPK1 in decoding Ca^{2+} signals under abiotic stresses in plants (Sheen 1996). Transgenic analysis showed that rice OsCDPK7 is a positive regulator of cold and salt/drought stress signaling. Rice transgenics overexpressing *OsCDPK7* under the control of the *CaMV 35S* promoter showed enhanced induction of a stress-responsive gene *RAB16A* in response to salinity/drought, and higher salt/drought stress tolerance, while transgenic lines in which OsCDPK7 was suppressed were hypersensitive to salt/drought stress (Saijo et al. 2000).

The *RAB16* (Skriver et al. 1991) and *HVA1* (Shen et al. 1996) genes have a G-box type ABRE *cis* element, which can be activated by bZIP transcription factors (Leung and Giraudat 1998). Overexpression of the catalytic domain of *ABI1 PP2C* inhibited the induction of *HVA1* transcription by ABA and ATCDPK1 (Fig. 2; Sheen 1996). These results suggest that the upregulation of *RAB16* and *HVA1* is mediated by CDPKs, probably through bZIP transcription factors, and is negatively regulated by the PP2C, ABI1. In addition to the activation of LEA-like genes, CDPKs also regulate transport proteins (aquaporins, ion channels and H^+-ATPase), which play pivotal roles in osmoregulation during osmotic and ionic stresses (Li et al. 1998; Lino et al. 1998). The Ca^{2+} requirement for activation of vacuolar chloride (VCL) and malate transporters in guard cells by CDPK was overcome by a constitutively active CDPK mutant (Pei et al. 1996).

In the common ice plant, CSP1 (a substrate protein for McCDPK1) was identified using yeast two-hybrid assays and wheat germ interaction assays. The phosphorylation of CSP1 *in vitro* by McCDPK1 required calcium. The deduced CSP1 amino acid sequence is similar to that of pseudo-response regulator-like proteins that have a highly conserved DNA binding helix-loop-helix domain and a C-terminal activation domain. Salt stress induced co-localization of McCDPK1 and CSP1 in the nucleus of ice plants, but the targets of McCDPK1-CSP1 are not known (Patharkar and Cushman 2000). This study supports the possible involvement of CDPKs in salt stress signaling, which regulates ion homeostasis and gene expression.

9.3.3 Hybrid two-component receptor kinases

Two-component systems, consisting of a sensory histidine kinase and a response regulator, function as osmosensors in bacteria and yeast. The yeast Sln1, a transmembrane osmosensory histidine kinase, transfers the phosphoryl group to a His residue in an intermediary component Ypd1, and finally to an Asp residue in the response regulator, Ssk1, to inactivate it. High osmotic stress inhibits the autophosphorylation of Sln1 and hence the active non-phosphorylated form of Ssk1 accumulates, which in turn activates the Hog1 (high-osmolarity glycerol response 1) MAPK cascade. This leads to glycerol accumulation and osmoprotection as the Hog1 MAPK cascade positively regulates genes involved in glycerol biosynthesis. In addition to Sln1, another transmembrane osmosensor, Sho1, which is not a two-component system, is also known to regulate the Hog1 MAPK cascade under osmotic stress. Sln1 and Sho1 operate at different osmotic stress levels (Wurgler-Murphy and Saito 1997; Chang and Stewart 1998).

These studies gave the impetus to clone an osmosensory hybrid histidine kinase, *ATHK1*, from *Arabidopsis* (Urao et al. 1999). ATHK1 consists of both kinase and receiver domains in the same molecule, and complements the yeast mutant *sln1-ts* defective in osmosensing. Substitution in ATHK1 of the putative phosphorylation sites, either His within the kinase domain (His-508 to Val) or Asp within the receiver domain (Asp-1074 to Glu), caused it to fail to complement the yeast *sln1-ts* mutant. *ATHK1* confers high-osmolarity tolerance to the yeast double mutant (*sln1Δ sho1Δ)* lacking both osmosensors. This demonstrates that ATHK1 is active at low osmolarity, and is changed to the inactive form in response to high osmolarity, which activates the Hog1 MAPK pathway in yeast. Thus, ATHK1 has both structural and functional similarities to the yeast Sln1, suggesting that in plants ATHK1 acts as an osmosensor to transmit the stress signal to a downstream MAPK cascade. The transcript abundance of *ATHK1* is higher in roots than in other tissues under control conditions. *ATHK1* is upregulated under salt (250 mM NaCl) and low temperature (4°C) stresses. If the mechanism of osmosensing is a high osmotic pressure induced conformational change preventing autophosphorylation of ATHK1, newly synthesized ATHK1 under stress may be in the non-phosphorylated state leading to activation of a downstream signaling pathway, which is probably a MAPK cascade (Fig. 2; Urao et al. 1999).

In *Arabidopsis*, phosphorelay intermediates with His-containing phosphotransfer domains have been cloned (*ATHP1-3*). All three *ATHPs* can complement the yeast *ypd1* mutant, which implies that all ATHPs can transfer a phosphoryl group from SLN1 to SSK1 in yeast. Further, ATHP3 (=AHP1) transfers a phosphoryl

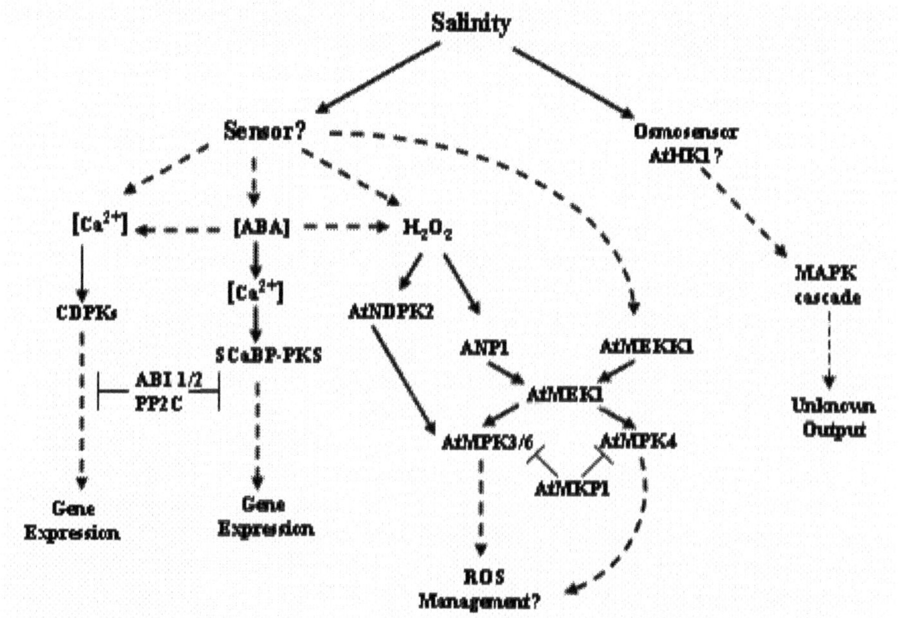

Fig. 2. Osmotic homeostasis and ROS detoxification under salt stress. Ca^{2+} signals sensed by CDPKs are transduced through unknown signaling intermediates, which induce genes encoding LEA-like proteins. ABA induced Ca^{2+} signals are perceived by SCaBPs, which activate PKS. The ABA signaling pathway upregulates osmolyte biosynthesis and genes encoding LEA-like proteins under salt stress. Ca^{2+} signaling through CDPKs and SCaBPs is under negative control of Protein Phosphatase 2C (ABI 1/2). High osmolarity may be perceived by AtHK1, which presumably transduces the signal through a MAPK pathway. Salt stress and reactive oxygen species (ROS) activated MAPK (ANP1 & AtMEKK1 = MAPKKK; AtMEK1=MAPKK; AtMPK3, 4 & 6 = MAPK) cascade may regulate oxidative stress management (Broken arrows indicate unknown signaling intermediates).

group from its His to the receiver domain of the putative two-component response regulators ARR3 and ARR4 *in vitro*. Thus, the phosphorelay intermediate, ATHP3, has the ability to accept a phosphoryl group from the osmosensor, ATHK1, and transfer its phosphoryl group to a response regulator, ARR4. Although functional complementation in yeast shows that *Arabidopsis* has ATHK1-ATHP3-ARR4 as a phosphorelay system (Miyata et al. 1998; Suzuki et al. 1998 & 2001; Urao et al. 1999), the MAPK cascade that transduces the osmotic stress signal and the target genes regulated by this putative hybrid two-component osmosensor system in higher plants are yet to be identified.

9.3.4 MAPK pathway

A canonical mitogen activated protein kinase (MAPK) module consists of a MAPK kinase kinase (MAPKKK), which activates a MAPK kinase (MAPKK) by phosphorylation of Ser or Thr residues (Ser-X-X-X-Ser/Thr) within the catalytic core. Activated MAPKK activates MAPK by phosphorylation of both Thr and Tyr within the TXY consensus sequence in MAPK. Plant MAPKs are implicated in signaling development, cell division, hormones, biotic, and abiotic stresses. Salt stress quickly (within 5-10 minutes) activates MAPKs from alfalfa (*s*alt stress *i*nduced MAPK, SIMK; Munnik et al. 1999), tobacco (*s*alicylic acid *i*nduced MAPK, SIPK; Hoyos and Zhang 2000; Mikolajczyk et al. 2000), and Arabidopsis (ATMPK3, ATMPK4 and ATMPK6; Mizoguchi et al. 1996; Ichimura et al. 2000). The SIPK is also induced by salicylic acid and osmotic stress. The activation of SIPK is calcium and ABA independent (Hoyos and Zhang 2000). Alfalfa SIMK-interacting SIMKK (=MAPKK) was identified using the yeast two hybrid screen. SIMKK activated SIMK and the activation was enhanced by salinity (Kiegerl et al. 2000). It appears that SIMK activation by SIMKK does not require an upstream MAPKKK, and SIMKK transduces both salt and pathogen elicitor stress signals (Cardinale et al. 2002). Salt stress activates *Arabidopsis* MAPKKK (ATMEKK1) and upregulates its gene expression (Ichimura et al. 1998). Co-expression of *ATMEKK1* with *MAPKKs* (*ATMKK2* and *MEK1*) complemented the growth defect of the yeast *pbs2* mutant, while co-expression of *ATMPK4* and *MEK1* complemented growth defects of the yeast *mpk1* and *bck1* mutants, suggesting that ATMEKK1, ATMKK2/MEK1, and ATMPK4 may constitute a MAP kinase cascade in *Arabidopsis* (Fig. 2; Ichimura et al. 1998).

Correlative evidence suggests that the ROS mediated signaling under salt stress occurs through MAPKs. H_2O_2 appears to act as an intermediate in ABA signaling in guard cells (Pei et al. 2000). H_2O_2 induced oxidation of Cys residues of proteins may bring conformational changes in signaling intermediates. In *Arabidopsis*, H_2O_2 induces the expression of genes involved in signaling such as calmodulin, CDPKs, His kinase, Tyr phosphatase, putative protein kinases such as ATMPK3, and genes involved in transcriptional activation such as Zn finger proteins, heat shock transcription factor, *DREB2A*, RING Zn finger protein, myb-related transcription factor, etc. (Desikan et al. 2001). ATMPK6 is activated by osmotic stresses, cold, and ROS stress imposed by H_2O_2, KO_2, paraquat and 3-amino-1,2,4-triazole (a catalase inhibitor) in *Arabidopsis* (Yuasa et al. 2001). An *Arabidopsis* MAPKKK, ANP1, is activated by H_2O_2. ANP1 initiates a phosphorylation cascade involving two MAPKs, AtMPK3 and AtMPK6. Expression of a constitutively active tobacco ANP1 orthologue, NPK1, in transgenic tobacco provided enhanced tolerance to multiple environmental stress conditions including salinity, suggesting that ANP1/NPK1 is involved in oxidative stress signaling under abiotic stresses (Fig. 2; Kovtun et al. 2000). Constitutively active ANP1 activates the MAPK cascade that activates promoters of stress-responsive genes such as *GST6* and *HSP* but not *RD29A* (Kovtun et al. 2000). This indicates that ANP1/NPK1 does not regulate DREB1, DREB2, and bZIP transcription factors, which are transcriptional activators of *RD29A* and other *COR* genes.

Nucleoside Diphosphate Kinase participates in hormone-dependent signal transduction pathways by activating guanine nucleotide-binding proteins involved in regulation of cell growth and differentiation. Transgenic analysis of NDPK2 (Nucleoside Diphosphate Kinase2) suggests that MAPK signaling regulates the oxidative stress management and growth regulation under abiotic stresses. The *NDPK2* gene in *Arabidopsis* (*AtNDPK2*) is induced by H_2O_2. Transgenic plants overexpressing *AtNDPK2* accumulated lower levels of ROS, while *AtNDPK2* mutants accumulated higher levels of ROS than wild type. AtNDPK2 interacts with ATMPK3 and ATMPK6. These two MAPKs are activated by H_2O_2 but this response was drastically reduced in an *atndpk2* mutant. Transgenic *Arabidopsis* overexpressing *AtNDPK2* showed an enhanced tolerance to multiple environmental stresses that elicit ROS accumulation, suggests that AtNDPK2 may positively regulate H_2O_2-mediated MAPK signaling in plants (Moon et al. 2003).

MAPKs can be inactivated by dephosphorylation. *Arabidopsis* phosphotyrosine phosphatase (AtPTP1) inactivates ATMPK4 *in vitro* (Huang et al. 2000). The *AtPTP1* gene is regulated by abiotic stresses such as drought, heat shock, wounding, high salt, and cold temperature. High salt conditions increased the expression level of *AtPTP1,* while cold significantly downregulates the *AtPTP1* gene (Xu et al. 1998). The *Arabidopsis mkp1* mutant is resistant to salinity but hypersensitive to genotoxic stress induced by UV-C and methyl methanesulphonate. *MKP1* encodes a MAPK phosphatase 1 (MKP1). A yeast two-hybrid screen showed that MKP1 could interact with three *Arabidopsis* MAPKs: MPK6, MPK3, and MPK4 and the interaction was strongest with MPK4 (Ulm et al. 2002). These three MAPKs have been implicated in salt stress signaling (Mizoguchi et al. 1996; Ichimura et al. 2000). The activity of MPK6 is regulated by MKP1 *in vivo*. Mutant analysis revealed that either *MKP1* deletion or loss of MKP1 phosphatase activity results in enhanced salt tolerance. This suggests that MKP1 is a negative regulator of salt stress signaling through MAPK, while it functions as positive regulator in genotoxic stress tolerance (Ulm et al. 2002). Microarray analysis showed an increased mRNA level of a putative Na^+/H^+-exchanger (*AT4G23700*) gene in the *mkp1* mutant under salt stress (Ulm et al. 2002), which suggests that AT4G23700 may be upregulated by a MAPK cascade that is under the negative control of MPK1. It is not known whether increased salt tolerance of the *mpk1* mutant is due to increased expression of the putative Na^+/H^+-exchanger. Overexpression of *SOS1*, a plasma membrane Na^+/H^+-exchanger (Shi et al. 2003) and *AtNHX1*, a vacuolar Na^+/H^+-exchanger (Apse et al. 1999; Zhang and Blumwald 2001; Zhang et al. 2001) resulted in enhanced salt stress tolerance. The *AT4G23700* gene is located on chromosome 4 and hence is different from *SOS1* (located on chromosome 2) and *AtNHX1* (located on chromosome 5) (Ulm et al. 2002). This suggests that a salt stress-responsive MAPK cascade in *Arabidopsis* may involve ANP1, MKK1, MPK3, 4, and/or 6, and their negative regulator MKP1(Fig. 2).

9.4 ABA-mediated salt stress signaling

ABA plays an important role in many aspects of plant growth and development from germination to seed development, and also plays a pivotal role in abiotic stress resistance. Salt stress induces ABA accumulation and the amount of the increase depends upon the tissue type. In maize, salt stress increased ABA accumulation up to 10-fold in roots but only 1-fold in leaf tissues. Salt stress induced ABA accumulation appears to be due to both ionic and osmotic stresses in roots, while that in the leaf is mainly due to osmotic stress (Jia et al. 2002). Turgor loss caused by osmotic stress leads to ABA synthesis and accumulation, which in turn regulates part of the cellular response to osmotic stress under salinity. ABA regulates cell water balance through stomatal regulation and genes involved in osmolyte biosynthesis, while it imparts dehydration tolerance through LEA-like genes (Hasegawa et al. 2000; Shinozaki and Yamaguchi-Shinozaki 2000; Zhu 2002). ABA signaling for stomatal closure and gene expression is transduced through Ca^{2+} (Leung and Giraudat 1998; Schroeder et al. 2001). The importance of ABA mediated stomatal regulation in salt tolerance was revealed by the analysis of *OSM1* locus of *Arabidopsis*. Root growth of the *Arabidopsis* T-DNA insertion mutant, *osm1* (for osmotic stress–sensitive mutant), was hypersensitive to NaCl or mannitol stress. Molecular cloning revealed that *OSM1* encodes a protein similar to SNARE type mammalian syntaxins (Zhu et al. 2002). SNARE proteins are required for fusion vesicle trafficking, control membrane Ca^{2+} and Cl^{-}channel activity and guard cell volumes (Schroeder et al. 2001). Consistent with this, ABA-mediated guard cell function is impaired in the *osm1* mutant. *OSM1* is strongly expressed in roots and leaf guard cells. The *osm1* mutant showed enhanced wilting and decreased survival when salt or drought stress was imposed on soil grown plants. Thus, OSM1 plays a critical role in root growth and in ABA regulation of stomatal responses under osmotic stresses (Zhu et al. 2002).

Osmotic stress responsive genes and ion transporters are regulated by ABA under salt stress. ABA induces several LEA-like stress responsive proteins, which are known as RD (responsive to dehydration), ERD (early responsive to dehydration), KIN (cold inducible), and RAB (responsive to ABA). Transient expression studies in isolated protoplasts showed that IP_3 and cADPR gated calcium channels are involved in ABA induced Ca^{2+} signatures. The expression of the stress responsive genes *RD29A* and *KIN2* is activated by ABA signaling through Ca^{2+} (Wu et al. 1997). ABA induces *AtPLC1* expression. Transgenic plants expressing *AtPLC1* in antisense and sense orientation showed that ABA induced expression of *RD22*, *RD29A* and *KIN2* requires AtPLC1 but it is not sufficient for maximal induction of stress responsive genes (Sanchez and Chua 2001). The *RD29A::LUC* reporter genetic screen facilitated isolation of abiotic stress and ABA signaling mutants in *Arabidopsis* (Ishitani et al. 1997). Two of these mutants, *los5* and *los6*, were impaired in the expression of stress responsive genes, such as *RD29A, COR15A, COR47, RD22*, and *P5CS*, under salt and osmotic stresses. Salt induced *RD29A::LUC* expression was restored to the wild type level by exogenous application of ABA. These mutants were also defective in osmotic stress induced ABA biosynthesis. Molecular cloning revealed that *LOS5* encodes a molybdenum cofactor sulfurase, which is allelic to *ABA3* (Xiong et al. 2001a), while *LOS6* encodes zeaxanthin epoxidase, which is allelic to *ABA1* (Xiong et al. 2002a). These results demonstrate that stress responsive gene expression under salinity is mediated by ABA. Salt stress and ABA upregulate a vacuolar Na^{+}/H^{+} antiporter, *AtNHX1*, which was reduced in ABA deficient mutants (*aba2-1* and *aba3-1*), but not in *salt overly sensitive mutants* (*sos1*, *sos2* or *sos3*) mutants. The *abi1-1* but not in *abi2-1* mutation decreased ABA and salt-induced *AtNHX1* expression. *AtNHX1* contains putative ABRE elements between –736 to –728 from the initiation codon. These results suggest that transcriptional upregulation of *AtNHX1* under salt stress is partially dependent on ABA biosynthesis and ABA signaling through the 2C type protein phosphatase ABI1 (Fig. 2; Shi and Zhu 2002).

ABA-deficient *los5/aba3* and *los6/aba1* mutants are more tolerant to salt stress at germination but at the vegetative stage *los5/aba3* is hypersensitive to salt stress (Xiong et al. 2001a & 2002a). *Arabidopsis* T-DNA insertion mutant *sañ5 (SALOBREÑO)* is tolerant to osmotic stress (NaCl, KCl, and mannitol) and ABA during germination. This mutation is allelic to *abi4* (Quesada et al. 2000). The quantitative trait loci (QTL) for the most effective ABA response at germination were mapped very close to the QTL for salt tolerance in germination (Mano and Takeda 1997). The QTLs for salt tolerance at germination were different from those of QTLs controlling salt tolerance at the seedling stage, indicating that salt tolerance at germination and at the seedling stage are regulated by different mechanisms (Mano and Takeda 1997; Quesada et al. 2002). Salt sensitivity in germinating seeds is mainly due to inhibition of germination by salt stress-induced ABA.

9.5 The SOS signaling pathway of ion homeostasis

Cellular ion homeostasis under salinity is achieved by the following strategies: 1) Exclusion of Na^+ from the cell by plasma membrane-bound Na^+/H^+ antiporters or by limiting the Na^+ entry, 2) Utilization of Na^+ for osmotic adjustment by compartmentation of Na^+ into the vacuole through tonoplast Na^+/H^+ antiporters, and 3) Na^+ secretion. Thus, regulation of ion transport systems is fundamental to plant salt tolerance. Genetic analysis of *s*alt *o*verly *s*ensitive (*sos*) mutants of *Arabidopsis*, led to the identification of the SOS pathway, which regulates cellular ion homeostasis and salt tolerance (Fig. 1; Zhu 2002).

The *Arabidopsis sos3* mutant is hypersensitive to salt stress. Molecular cloning revealed that the *SOS3* encodes a Ca^{2+} binding protein homologous to the regulatory subunit of yeast calcineurin and animal neuronal calcium sensors. It has an N-myristoylation motif and three calcium binding EF hands. SOS3 senses salt stress-induced increases in cytosolic Ca^{2+} concentration in plants (Liu and Zhu 1998; Ishitani et al. 2000). Myristoylated SOS3 is recruited to the plasma membrane (Quintero et al. 2002). Mutations that disrupt either myristoylation (G2A) or calcium binding (*sos3-1*) cause salt stress hypersensitivity to *Arabidopsis* plants. Since myristoylation of SOS3 is essential for salt tolerance, it is likely that membrane recruitment of SOS3 is essential for its function. Membrane localization of SOS3 may help in the regulation of its target ion transporters (Ishitani et al. 2000).

Identification of additional *SOS* loci (*SOS2* and *SOS1*) revealed that the SOS pathway regulates cellular ion homeostasis under salt stress. *Arabidopsis sos1* and *sos2* mutants are also hypersensitive to salt stress and *sos1*, *sos2*, and *sos3* mutations do not show an additive effect, implying that they are in the same pathway of salt stress response. SOS2 is a ser/thr protein kinase with an N-terminal kinase catalytic domain and a C-terminal regulatory domain. The SOS2 C-terminal regulatory domain consists of the SOS3-binding, autoinhibitory FISL motif (Liu et al. 2000). Binding of SOS3 activates the SOS2 protein kinase (Halfter et al. 2000). Deletion of the FISL motif from SOS2 leads to constitutive activation of the kinase (Guo et al. 2001). Molecular genetic analysis of the *sos1* mutant led to the identification of a target for the SOS3-SOS2 kinase complex. *SOS1* encodes a plasma membrane Na^+/H^+ antiporter (Shi et al. 2000). The *sos1* mutant accumulates high levels of Na^+ in tissues under salt stress, and isolated plasma membrane vesicles from *sos1* mutants showed significantly less Na^+/H^+ exchange activity than the wild type, suggesting that the SOS1 Na^+/H^+ antiporter is located on the plasma membrane (Qiu et al. 2002). The *sos3* and *sos2* mutants accumulate higher levels of Na^+ than wild type plants. Isolated plasma membranes vesicles from these mutants also showed significantly less Na^+/H^+ exchange activity, and this could be restored to the wild type levels by the addition of activated SOS2. The SOS3-SOS2 kinase complex activates SOS1 by phosphorylation (Quintero et al. 2002). *SOS1* complemented yeast mutants defective in Na^+ transporters. Co-expression of *SOS2* and *SOS3* significantly increased *SOS1*-dependent Na^+ tolerance of the yeast mutant (Quintero et al. 2002). These results show that SOS1 is a Na^+/H^+ antiporter involved in Na^+ efflux, which is activated by the SOS3-SOS2 kinase complex (Fig.1 1; Qiu et al. 2002; Quintero et al. 2002). Constitutive expression of a CaMV 35S promoter driven active form of SOS2 could rescue *sos2* and *sos3* mutants under salt stress (Xiong et al. 2002b).

The expression of *SOS1* is stronger in cells bordering the xylem. Under salt stress (100 mM NaCl), a higher concentration of Na^+ accumulates in shoots of *sos1* mutants than in those of the wild type. These results suggest that SOS1 might retrieve Na^+ from the xylem, thereby preventing excess Na^+ accumulation in the shoot (Shi et al. 2002a). Transgenic *Arabidopsis* plants overexpressing *SOS1* showed improved salt tolerance and accumulated less Na^+ in the xylem transpirational stream as well as in the shoot compared to the wild type plants. This demonstrated that Na^+ efflux from the root cells and long distance Na^+ transport within the plant under salt stress are regulated by SOS1 (Shi et al. 2003), which in turn is regulated by the SOS3-SOS2 kinase complex. In addition to the activation of Na^+/H^+ antiporter activity of SOS1, SOS3-SOS2 kinase complex also is involved in salt stress induced upregulation of *SOS1* expression (Fig. 1; Shi et al. 2000). In the *sos3* mutant salt, stress could not induce *SOS1* expression, while the *sos2* mutant is impaired in *SOS1* expression only in roots, but not in shoots. Interestingly, *SOS1* overexpressing transgenic *Arabidopsis* showed a significantly higher steady state level of *SOS1* mRNA under salt stress than that grown under normal conditions. Since *SOS1* was overexpressed under the control of the CaMV 35S promoter, its higher mRNA abundance under salt stress might be due to an increase in *SOS1* transcript stability (Shi et al. 2003).

In addition to positive control of Na^+ exclusion from the cytosol, the SOS pathway may also negatively regulate Na^+ influx systems. Expression of plant high affinity K^+ transporters, *AtHKT1*, *EcHKT1*, and *EcHKT2*, in *Xenopus laevis* oocytes showed that they could mediate Na^+ uptake. Transgenic wheat plants expressing the wheat *HKT1* in antisense orientation under control of a ubiquitin promoter showed significant downregulation of the native *HKT1* transcript. These lines showed significantly less ^{22}Na uptake and enhanced growth under salinity when compared with the control (Laurie et al. 2002). These results suggest that HKT1 mediates sodium uptake under salinity, and salt tolerance can be improved by downregulation of *HKT1* expression. Consistent with this observation, a suppressor genetic screen for the *sos3* mutation revealed that functional disruption of *AtHKT1* could suppress the salt-sensitive phenotype of *sos3*. In addition, the *athkt1* mutation alleviates the K^+-deficient phenotype of the *sos3* mutant (Rus et al. 2001a), which suggests that the K^+-deficient phenotype of the *sos3* mutant might be due to an excess of cytoplasmic Na^+, as *sos3* impairs the Na^+ efflux mediated by SOS1. These results suggest that ATHKT1 might function as low affinity Na^+ transporter that is involved in Na^+ influx under salinity. Significant amounts of Na^+ enter plant roots through voltage independent channels, which are probably regulated by Ca^{2+} concentrations (Tyerman and Skerret 1999). We do not know whether activity of these channels and their gene expression are also regulated by the calcium dependent SOS3-SOS2 kinase complex. Thus, the SOS3-SOS2 kinase complex positively regulates Na^+ efflux by activating SOS1 and upregulating the *SOS1* transcript level, and may negatively regulate Na^+ influx by downregulating low affinity Na^+ transporter (*HKT1*) genes to restore cellular ion homeostasis under salt stress in plants (Fig. 1; Zhu 2002)

9.6 Osmotic stress management

Plant survival depends on maintaining a positive turgor, which is indispensable for expansion growth of cells and stomatal opening. A decrease in water availability under soil salinity causes osmotic stress, which leads to decreased turgor. Osmotic adjustment is one of the vital cellular tolerance process to osmotic stress, conserved in both halophytic and glycophytic plants. Osmotic stress may induce ion (Na^+ & K^+) uptake and compartmentalization into the vacuole, and synthesis of organic compatible solutes such as proline, betaine, polyols, and soluble sugars. Use of ions for osmotic adjustment may be energetically more favorable than organic osmolyte biosynthesis under stress, as ion uptake and sequestration into the vacuole may cost only 3-4 moles of ATP compared with the 30-50 moles of ATP needed for synthesis of one mole of organic osmolytes (Raven 1985).

9.6.1 Sodium sequestration into the vacuole

Cytoplasmic ion homeostasis by exclusion of excess Na^+ from the cytoplasm may necessitate the plant to synthesize compatible osmolytes to reduce the osmotic potential, which is required for water uptake under salt stress. Hence, compartmentation of Na^+ in the vacuole is an important strategy for plants, to maintain a lower Na^+ concentration at the sites of biochemical reactions in the cytosol, and yet maintain a lower overall osmotic potential. Active transport of solutes across biological membranes utilizes the electrochemical gradient generated by P-type H^+-ATPases (plasma membrane H^+-ATPases), V-type H^+-ATPases (vacuolar H^+-ATPase) and H^+-pyrophosphatase (vacuolar H^+-PPase). The sodium efflux plasma membrane Na^+/H^+ antiporters use a proton electrochemical gradient generated by the plasma membrane H^+-ATPase, which is upregulated under salinity. A salt-tolerant mutant of rice showed higher induction of the plasma membrane H^+-ATPase gene *OSA3* in roots than that of the wild type (Zhang et al. 1999). Influx of Na^+ into the vacuole occurs through Na^+/H^+ antiporters, which use the proton gradient generated by V-type H^+-ATPase and H^+-PPase (Apse et al. 1999). Thus, Na^+ sequestration into the vacuole depends upon the expression and activity of Na^+/H^+ antiporters as well as V-type H^+-ATPase and H^+-PPase. Salinity upregulates the expression of a V-type H^+-ATPase gene (Golldack and Dietz 2001) and a vacuolar Na^+/H^+ antiporter gene (Gaxiola et al. 1999; Shi and Zhu 2002). To investigate the role of tonoplast H^+-PPase in salinity tolerance, the *AVP1* gene (vacuolar H^+-pyrophosphatase) was overexpressed in *Arabidopsis*. The transgenics showed increased sequestration of Na^+ into the vacuole, maintained higher relative water content in leaves and were more tolerant to salt and drought stress than the wild type was (Gaxiola et al. 2001).

In *Arabidopsis*, the *AtNHX1* gene encodes a tonoplast Na^+/H^+ antiporter. Expression of *AtNHX1* in the yeast *nhx1* mutant suppressed some of the mutant phenotypes. Salinity induces *NHX1* expression in *Arabidopsis* (Gaxiola et al. 1999; Shi and Zhu 2002) and rice (Fukuda et al. 1999). Transgenic *Arabidopsis* plants that overexpress *AtNHX1* showed significantly higher salt tolerance than wild type plants (Apse et al. 1999). Similarly, transgenic tomato and canola (*Brassica napus*) plants overexpressing *AtNHX1* accumulated high concentrations of sodium in leaves but not in fruits/seeds. These transgenics were shown to be highly tolerant to salt stress at the same time they maintained the quality of fruit in tomato and oil in canola (Zhang and Blumwald 2001; Zhang et al. 2001). These studies confirm that sequestration of Na^+ into the vacuole is an important trait of salt tolerance in plants.

9.6.2 K^+ Uptake

Plants maintain a high cytosolic K^+/Na^+ ratio under optimal conditions. Salt stress induced decrease in the K^+/Na^+ ratio is inimical to cellular biochemical processes. In addition to this, K^+ provides necessary osmotic potential for water uptake by plant cells (Keller and Van Volkenburgh 1996; Claussen et al. 1997). Thus, K^+ uptake is pivotal for cell turgor and maintenance of biochemical processes under salinity. In plants, Na^+ competes with K^+ for uptake under saline conditions. The *Mesembryanthemum crystallinum* K^+ transporter genes, *McHAK1* and *McHAK2*, are upregulated under K^+ starvation and NaCl stress in both roots and leaves (Su et al. 2002). Low K^+ concentration in the growth medium inhibits the growth of *sos* mutants. The *sos3* mutant could be rescued by increasing Ca^{2+} in a low K^+ medium (Zhu et al. 1998). Hence, expression of transport systems specific for K^+ uptake might help in maintaining ionic balance.

Overexpression of *AtHAL3a* (a regulator of K^+ transport) in yeast and *Arabidopsis* conferred increased salt tolerance (Espinosa-Ruiz et al. 1999), as did transgenic melon plants expressing the *HAL1* gene (Bordás et al. 1997). To investigate the role of HAL1 *in vivo*, tomato plants were engineered to overexpress the yeast *HAL1* gene. This transgenic plant showed increased K^+ accumulation under NaCl stress (Rus et al. 2001b). Transgenics showed better salt tolerance than the control plants (Gisbert et al. 2000; Rus et al. 2001b), suggesting that K^+ accumulation is an important trait of salt tolerance. Further, the *Arabidopsis sos4* mutant defective in the pyridoxal kinase gene showed hypersensitive-root growth under NaCl and KCl stresses and accumulated more Na^+ but less K^+. Pyridoxal-5-phosphate and its derivatives act as ligands for P2X receptor ion channels in animals. ATP is required for K^+ channel activity and a cyclic nucleotide-binding site is required for K^+ channel (KAT1) function. Thus regulation of K^+ and Na^+ channels or transporters by pyridoxal-5-phosphate and its derivatives may be important in plant salt tolerance (Table 1; Shi et al. 2002b)

9.6.3 Osmoprotectant biosynthesis

Organic compatible solutes/osmoprotectants protect plants from stress by (1) osmotic adjustment which helps in turgor maintenance (2) detoxification of reactive oxygen species and (3) stabilization of the quaternary structure of proteins (Yancey et al. 1982; Bohnert and Jensen 1996). Genes involved in osmoprotectant biosynthesis are upregulated under salt and drought stresses (Zhu 2002; Xiong et al. 2001a). Enhanced tolerance to salt stress was observed in transgenic plants engineered to over-accumulate mannitol (Tarczynski et al. 1993; Karakas et al. 1997; Sheveleva et al. 1997; Shen et al. 1997), glycine betaine (Holmstrom et al. 2000; Hayashi et al. 1997; Sakamoto et al. 1998; Kishitani et al. 2000; Prasad et al. 2000), and proline (Kishor et al. 1995; Zhu et al. 1998; Nanjo et al. 1999; Hong et al. 2000). Transgenic rice plants expressing a peroxisomal betaine aldehyde dehydrogenase of barley accumulated fewer Na^+ and Cl^- ions and more K^+ ions (Kishitani et al. 2000).

Further evidence for the involvement of osmoprotectants in salt tolerance came from analysis of the *Arabidopsis* mutant, *t365*, in which the S-adenosyl-L-methionine phosphoethanolamine N-methyltransferase (*PEAMT*) gene is silenced (Table 1). The PEAMT protein catalyzes all three methylation steps required to convert phosphoethanolamine to phosphocholine, which is a precursor of choline biosynthesis. Some plants synthesize the osmoprotectant glycinebetaine from choline. The *t365* mutants produced significantly less choline and showed hypersensitivity to salinity in addition to temperature-sensitive male sterility (Mou et al. 2002), which supports the importance of osmoprotectant in salt tolerance. The ectopic expression studies showed that osmoprotectants increase salt stress tolerance mainly by protection of membranes and proteins against reactive oxygen species (ROS) rather than by increasing osmotic adjustment. ABA regulates the *P5CS* gene involved in proline biosynthesis under osmotic stress (Xiong et al. 2001a). A signaling cascade similar to that of the yeast MAPK HOG1 pathway may also regulate osmolyte biosynthesis.

9.7 Stress damage control and repair

9.7.1 Salt stress induced proteins

In higher plants, osmotic stress induces several proteins in vegetative tissues, which are related to late-embryogenesis-abundant (LEA) proteins. The correlation between LEA protein accumulation in vegetative tissues and stress tolerance in various plant species indicates its protective role under dehydration stress (reviewed by Ingram & Bartels 1996). Engineered rice plants overexpressing a barley *LEA* gene, *HVA1*, under control of the rice actin 1 promoter showed better stress tolerance under 200 mM NaCl and drought stress than did the wild type (Xu et al. 1996). *Arabidopsis* LEA-like stress proteins are encoded by *COR* genes (*RD29A, COR47, COR15, KIN1, KIN2*) which are induced by cold, dehydration (due to water deficit or high salt), or ABA. Promoter analysis of the *COR* genes showed that many of them contain dehydration responsive elements (DRE) or C-Repeat (CRT), as well as ABA-responsive elements or ABREs. Transcription factors that regulate the LEA-like genes include CBFs (C-repeat Binding Proteins, also known as Dehydration Responsive Element Binding Proteins, DREBs) and bZIP proteins. The expression of *COR* genes is regulated by both ABA dependent and independent pathways (Ishitani et al. 1997; Shinozaki and Yamaguchi-Shinozaki 2000). Constitutive overexpression of *CBF3* or stress induced expression of *CBF3* driven by the *RD29A* promoter resulted in enhanced expression of *COR* genes under cold, dehydration, and salt stresses in transgenic *Arabidopsis* and also conferred higher osmotic stress tolerance (Kasuga et al. 1999). *CBF3*-overexpression in *Arabidopsis* also resulted in elevated accumulation of proline and total soluble sugars, including sucrose, raffinose, glucose, and fructose. The increase in proline levels was due to increased expression of the key proline biosynthetic enzyme Δ^1-pyrroline-5-carboxylate synthase (Gilmour et al. 2000). Thus, LEA-like proteins appear to protect plants under salt stress. Osmotic or salt stress-induced calcium signals may activate the LEA-like genes through DREB2 transcription factors, while salt stress induced ABA accumulation appears to induce the genes through ABA responsive element binding factors (Xiong et al. 2002b; Zhu 2002).

The *Alfin1* gene of *Medicago sativa* encodes a member of the zinc-finger family transcription factors, and its expression is correlated with NaCl tolerance (Winicov and Bastola 1999; Winicov 2000). *In vitro*, Alfin1 binds to the promoter of *MsPRP2*, which encodes a salt stress inducible root-specific cell wall protein. The *Alfin1* gene appears to be conserved in alfalfa, rice, and *Arabidopsis*. The role of *Alfin1* in salt stress tolerance was examined in transgenic alfalfa expressing *Alfin1* driven by the *CaMV 35S* promoter in the sense and antisense orientations. Although overexpression lines did not show any growth defect, the antisense transgenic plants grew poorly in soil in a normal environment, demonstrating that *Alfin1* expression is essential for normal plant development. *Alfin1* overexpression enhanced the root growth significantly both under normal and saline conditions, while the antisense plants showed poor root growth (Winicov and Bastola 1999; Winicov 2000). The tobacco-stress-induced-gene 1 (*Tsi1*) encodes a DNA-binding protein with an EREBP/AP2 DNA binding motif, which is involved in defense- and drought-responsive gene expression. *Tsi1* gene expression was rapidly induced by salt stress but not by drought or ABA. Overexpression of *TSI1* in tobacco enhanced retention of chlorophyll content when the leaves were floated on 400 mM NaCl solution for 48 or 72 hr (Park et al. 2001). Further studies are needed to assess the role of Alfin1 and TSI1 in salt stress tolerance, as it is not clear at present whether these proteins and their targets are involved in ion/osmotic homeostasis or in detoxification.

9.8 Oxidative stress management

Reactive oxygen species (ROS) namely, superoxide radicals ($O_2^{\cdot -}$), hydrogen peroxide (H_2O_2), and hydroxyl radicals ($OH^{\cdot}$) are produced in aerobic cellular processes such as mitochondrial and chloroplast electron transport, or oxidation of glycolate (photorespiration), xanthine, and glucose. Due to metabolic disturbance under stress conditions, ROS production increases under abiotic stresses including salinity (Smirnoff 1993; Gomez et al. 1999; Hernandez et al. 2001). The ROS causes oxidative damage to membrane lipids, proteins and nucleic acids. Hence, ROS detoxification forms an important defense against abiotic stresses. The antioxidants employed by plants are ascorbate, glutathione, α-tocopherol, and carotenoids. Detoxifying enzymes include superoxide dismutase (SOD), catalase, and enzymes of the ascorbate- glutathione cycle. The *Arabidopsis* salt tolerant mutant *pst1* (for *photoautotrophic salt tolerance1*) is more tolerant to oxidative stress than is the wild type (Table 1). The *pst1* mutant did not differ in proline accumulation or monovalent cation (sodium, potassium) accumulation when compared to the wild type. Under salt stress, the *pst1* mutant showed significantly higher activity of superoxide dismutase and ascorbate peroxidase than that of wild type *Arabidopsis* (Tsugane et al. 1999). Overexpressing the tobacco *NtGST/GPX* gene (encoding an enzyme with both glutathione S-transferase and glutathione peroxidase activity) in transgenic tobacco plants improved salt and chilling stress tolerance due to enhanced ROS scavenging and prevention of membrane damage (Roxas et al. 1997; Roxas et al. 2000). Transgenic tobacco plants expressing the constitutively active MAPKKK, ANP1, show an activated MAPK cascade that activates the glutathione S-transferase 6 (*GST6*) gene promoter. These transgenic plants were also tolerant to salt and other abiotic stresses (Kovtun et al. 2000). Components of MAPK cascades are activated by ROS and salinity as discussed earlier. Thus, it appears ROS management under salt stress through the induction of genes encoding antioxidant enzymes may be controlled by a MAPK signaling cascade (Fig. 2).

9.9 Growth regulation

Maintenance of root growth at low water potential is an adaptive trait of osmotic stress tolerance. In maize roots, salt stress increased ABA accumulation up to 10-fold (Jia et al. 2002). Root elongation at low water potential might be achieved by an increase in the activity of the putative wall loosening enzyme xyloglucan endotransglycosylase (Wu et al. 1994) and proline accumulation (Ober and Sharp 1994), which are regulated by ABA.

Root elongation at low water potential was impaired in the *vp5* mutant, or by a chemical inhibitor of ABA biosynthesis (fluridone) in maize. However, this could be restored by treatment of roots with a chemical that inhibits ethylene biosynthesis or action. Moreover, treatment of seedlings with fluridone resulted in an increase in the rate of ethylene production. These data suggested that ABA-mediated root cell elongation under osmotic stress might be due to its inhibition of ethylene biosynthesis (Spollen et al. 2000).

In *Arabidopsis*, a null allele of the Gα gene impaired cell division (Ullah et al. 2001) and ABA inhibition of stomatal closure (Wang et al. 2001). A loss-of-function allele of another ABA signaling locus encoding a SNARE protein, *osm1*, also showed impaired root growth under salt and osmotic stress (Zhu et al. 2002). This suggests that ABA may regulate cell division under osmotic stress. Consistent with this, the transcripts of a cyclin-dependent kinase (*AtCDC2a*) and two mitotic cyclin (*AtCycB1* and *AtCycA2*) genes were diminished initially but induced subsequently in the shoot apex during salt stress adaptation (Burssens et al. 2000). SIMK is activated and translocated into the nucleus in suspension-cultured alfalfa cells under salt stress (Baluska et al. 2000). In the root elongation zone, epidermal cells contained much higher SIMK protein than in cortex cells. SIMK showed a cell cycle phase-dependent localization, being predominantly nuclear in interphase but associating with the cell plate and the newly formed cell wall in telophase and early G1 phase (Baluska et al. 2000). It is not clear whether cell division/elongation is regulated through MAPK signaling under salt stress.

In the root tips of *Arabidopsis, AtCDC2a*, *AtCycA2* and *AtCycB1* expression were diminished concomitant with inhibition of root growth under salt stress (Burssens et al. 2000). The activity of CDC2a is negatively regulated by a cyclin-dependent protein kinase inhibitor, ICK1. The expression of *ICK1* is upregulated by ABA in *Arabidopsis* (Wang et al. 1998). The knowledge of tissue- and plant species-specific regulation of cell division/elongation by ABA under salt stress is still in its infancy.

9.10 Conclusions and perspectives

Although a salt stress sensor is yet to be identified, some of the components of salt stress signaling and plant salt tolerance are known today. Genetic evidence demonstrated that a salt stress induced calcium signal is transduced at least in part through the SOS3-SOS2 kinase complex, which activate SOS1, a plasma membrane Na^+/H^+ antiporter. In addition, the SOS3-SOS2 kinase complex positively regulates the *SOS1* transcript level. Correlative evidence implicates the involvement of a putative osmosensory histidine kinase (AtHK1) and a MAPK cascade in osmoprotectant biosynthesis under salt stress. Responses to ion toxicity, osmotic stress and oxidative stress may be integrated by signaling pathways including MAPK and its negative regulator MAPK phosphatase. Except for the SOS pathway, salt stress signaling pathways are not yet understood in terms of their components and targets. Moreover, the evidence summarized here is mainly derived from studying *Arabidopsis*, a glycophytic plant and hence further analysis of salt tolerance mechanisms in halophytic plants is also warranted. Characterization of chloride and sulfate transporters and their regulation under salt stress is also the need of the hour. Salt tolerance varies with plant development, and it is imperative to understand the tissue and developmental specificity of salt stress tolerance. Availability of whole genome sequences in *Arabidopsis* and rice, as well as the use of microarrays to analyze the transcriptome response will facilitate the identification of genes involved in salt tolerance, which can be validated by RNA interference and/or T-DNA/transposon/EMS mutational studies. Continued genetic and biochemical dissection of salt tolerance in the near future may provide us a clear picture of salt tolerance in plants, which will help to engineer agronomically useful salt tolerant crop varieties.

Acknowledgements

Our work has been supported by grants from United States Department of Agriculture – National Research Initiative, Binational Agricultural Research and Development Fund, Southwest Consortium on Plant Genetics and Water Resources, National Science foundation, and National Institutes of Health. We are thankful to Prof. André Jagendorf, Department of Plant Biology, Cornell University, for his critical reading of the manuscript and suggestions.

References

Allen GJ, Muir SR, Sanders D (1995) Release of Ca^{2+} from individual plant vacuoles by $InsP_3$ and cyclic ADP-ribose. Science 268:735-737

Apse MP, Aharon GS, Snedden WS, Blumwald E (1999) Salt tolerance conferred by overexpression of a vacuolar Na^+/H^+ antiport in *Arabidopsis*. Science 285:1256-1258

Baluska F, Ovecka M, Hirt H (2000) Salt stress induces changes in amounts and localization of the mitogen-activated protein kinase SIMK in alfalfa roots. Protoplasma 212:262-267

Bohnert HJ, Jensen RG (1996) Strategies for engineering water stress tolerance in plants. Trends in Biotechnol 14:89–97

Bordas M, Montesinos C, Dabauza M, Salvador A, Roig V, Serrano R, Moreno V (1997) Transfer of the yeast salt tolerance gene *HAL1* to *Cucumis melo* L. cultivars and *in vitro* evaluation of salt tolerance. Transgenic Research 6:41-50

Boyer JS (1982) Plant productivity and environment. Science 218:443-448

Burssens S, Himanen K, Van B, Beeckman T, Van M, Inze D, Verbruggen N (2000) Expression of cell cycle regulatory genes and morphological alterations in response to salt stress in *Arabidopsis thaliana.* Planta 211: 632-640

Cardinale H, Meskiene I, Ouaked F, Hirt H (2002) Convergence and divergence of stress-induced mitogen-activated protein kinase signaling pathways at the level of two distinct mitogen-activated protein kinase kinases. Plant Cell 14:703-711

Chang C, Stewart RC (1998) The two-component system. Plant Physiol 117:723-731

Charrier B, Champion A, Henry Y, Kreis M (2002) Expression profiling of the whole *Arabidopsis* shaggy-like kinase multigene family by real-time reverse transcriptase-polymerase chain reaction. Plant Physiol 130:577-590

Claussen M, Luthen H, Blatt M, Bottger M (1997) Auxin induced growth and its linkage to potassium channels. Planta 201:227-234

Desikan R, Mackerness SA-H, Hancock JT, Neill SJ (2001) Regulation of the *Arabidopsis* transcriptome by oxidative stress. Plant Physiol 127:159-172

DeWald DB, Torabinejad J, Jones CA, Shope JC, Cangelosi AR, Thompson JE, Prestwich GD, Hama H (2001) Rapid accumulation of phosphatidylinositol 4,5-bisphosphate and inositol 1,4,5-trisphosphate correlates with calcium mobilization in salt-stressed *Arabidopsis*. Plant Physiol. 126:759-769

Espinosa-Ruiz A, Belles JM, Serrano R, Culianez-Macla V (1999) *Arabidopsis thaliana AtHAL3*: a flavoprotein related to salt and osmotic tolerance and plant growth. Plant J 20:529-39

Fukuda A, Nakamura A, Tanaka Y (1999) Molecular cloning and expression of the Na^+/H^+ exchanger gene in *Oryza sativa.* Biochim Biophys Acta 1446:149-155

Garcia A, Rizzo CA, Ud-Din J, Bartos SL, Senadhira D, Flowers TJ, Yeo AR (1997) Sodium and potassium transport to the xylem are inherited independently in rice and the mechanism of sodium:potassium selectivity differs from rice and wheat. Plant Cell Environ 20:1167-1174

Gaxiola RA, Li J, Undurraga S, Dang V, Allen GJ, Alper SL, Fink GR (2001) Drought- and salt-tolerant plants result from overexpression of the *AVP1* H^+-pump. Proc Natl Acad Sci USA 98:11444-11449

Gaxiola RA, Rao R, Sherman A, Grisafi P, Alper SL, Fink GR (1999) The *Arabidopsis thaliana* proton transporters, AtNhx1 and Avp1, can function in cation detoxification in yeast. Proc Natl Acad Sci USA 96:1480-1485

Gilmour SJ, Sebolt AM, Salazar MP, Everard JD, Thomashow MF (2000) Overexpression of the *Arabidopsis CBF3* transcriptional activator mimics multiple biochemical changes associated with cold acclimation. Plant Physiol 124:1854-1865

Gisbert C, Rus AM, Bolarín MC, López-Coronado JM, Arrillaga I, Montesinos C, Caro M, Serrano R, Moreno V (2000) The Yeast *HAL1* gene improves salt tolerance of transgenic tomato. Plant Physiol 123:393-402

Golldack D, Dietz KJ (2001) Salt-induced expression of the vacuolar H^+-ATPase in the common ice plant is developmentally controlled and tissue specific. Plant Physiol 125:1643-1654

Gomez JM, Hernandez JA, Jimenez A, del Rio LA, Sevilla F (1999) Differential response of antioxidative enzymes of chloroplast and mitochondria to long term NaCl stress of pea plants. Free Radic Res 31:S11-S18.

Gorham J, Bridges J, Dubcovsky J, Dvorak J, Hollington PA, Luo MC, Khan JA (1997) Genetic analysis and physiology of a trait for enhanced K^+/Na^+ discrimination in wheat. New Phytol 137:109-116

Guo Y, Halfter U, Ishitani M, Zhu J-K (2001) Molecular characterization of functional domains in the protein kinase SOS2 that is required for plant salt tolerance. Plant Cell 13:1383-1400

Halfter U, Ishitani M, Zhu J-K (2000) The *Arabidopsis* SOS2 protein kinase physically interacts with and is activated by the calcium-binding protein SOS3. Proc Natl Acad Sci USA 97:3735-3740

Harmon AC, Gribskov M, Harper JF (2000) CDPKs-A kinase for every Ca^{2+} signal? Trends Plant Sci 5:154-159

Hasegawa PM, Bressan RA, Zhu J-K, Bohnert HJ (2000) Plant cellular and molecular responses to high salinity. Annu Rev Plant Mol Plant Physiol 51:463-499

Hayashi HA, Mustardy L, Deshnium P, Ida M, Murata N (1997) Transformation of *Arabidopsis thaliana* with the *codA* gene for choline oxidase: accumulation of glycine betaine and enhanced tolerance to salt and cold stress. Plant J 12:133-142

Hernández JA, Ferrer MA, Jiménez A, Barceló AR, Sevilla F (2001) Antioxidant systems and O_2^-/H_2O_2 production in the apoplast of pea leaves. Its relation with salt-induced necrotic lesions in minor veins. Plant Physiol 127:817 – 831

Hirayama T, Ohto C, Mizoguchi T, Shinozaki K (1995) A gene encoding a phosphatidylinositol-specific phospholipase C is induced by dehydration and salt stress in *Arabidopsis thaliana*. Proc Natl Acad Sci USA 92:3903-3907

Hirschi KD (1999) Expression of *Arabidopsis CAX1* in tobacco: altered calcium homeostasis and increased stress sensitivity. Plant Cell 11:2113-2122

Holmstrom KO, Somersalo S, Mandal A, Palva TE, Welin B (2000) Improved tolerance to salinity and low temperature in transgenic tobacco producing glycine betaine. J Exp Bot 51:177-185

Hong Z, Lakkineni K, Zhang Z, Verma DPS (2000) Removal of feedback inhibition of Δ^1-pyrroline-5-carboxylate synthetase results in increased proline accumulation and protection of plants from osmotic stress. Plant Physiol 122:1129-1136

Hoyos ME, Zhang S (2000) Calcium-independent activation of salicylic acid-induced protein kinase and a 40-kilodalton protein kinase by hyperosmotic stress. Plant Physiol 122:1355-1364

Huang Y, Li H, Gupta R, Morris PC, Luan S, Kieber JJ (2000) ATMPK4, an *Arabidopsis* homolog of mitogen-activated protein kinase, is activated *in vitro* by AtMEK1 through threonine phosphorylation. Plant Physiol 122:1301–1310

Ichimura K, Mizoguchi T, Irie K, Morris P, Giraudat J, Matsumoto K, Shinozaki K (1998) Isolation of ATMEKK1 (a MAP kinase kinase kinase)-interacting proteins and analysis of a MAP kinase cascade in *Arabidopsis*. Biochem Biophys Res Commun 253:532-543

Ichimura K, Mizoguchi T, Yoshida R, Yuasa T, Shinozaki K (2000) Various Abiotic stresses rapidly activate *Arabidopsis* MAP kinases ATMPK4 and ATMPK6. Plant J 24:655-665

Ingram J, Bartels D (1996) The molecular basis of dehydration tolerance in plants. Annu Rev Plant Physiol Plant Mol Biol 47:377-403

Ishitani M, Liu J, Halfter U, Kim C-S, Shi W, and Zhu, J-K (2000) SOS3 function in plant salt tolerance requires N-myristoylation and calcium-binding. Plant Cell 12:1667-1677

Ishitani M, Xiong L, Stevenson B, Zhu J-K (1997) Genetic analysis of osmotic and cold stress signal transduction in *Arabidopsis thaliana*: Interactions and convergence of abscisic acid-dependent and abscisic acid-independent pathways. Plant Cell 9:1935-1949

Iwata K, Tazawa M, Itoh T (2001) Turgor pressure regulation and the orientation of cortical microtubules in *Spirogyra* cells. Plant Cell Physiol 42:594-598

Jacobsen T, Adams RM (1958) Salt and silt in ancient Mesopotamian agriculture. Science 128:1251-1258

Jia W, Wang Y, Zhang S, Zhang J (2002) Salt-stress-induced ABA accumulation is more sensitively triggered in roots than in shoots. J Exp Bot 53:2201-2206

Karakas B, Ozias-Akins P, Stushnoff C, Suefferheld M, Rieger M (1997) Salinity and drought tolerance of mannitol-accumulating transgenic tobacco. Plant Cell Environ 20: 609–616

Kasuga M, Liu Q, Miura S, Yamaguchi-Shinozaki K, Shinozaki K (1999) Improving plant drought, salt, and freezing tolerance by gene transfer of a single stress-inducible transcription factor. Nature Biotech 17:287–291

Keller CP, Volkenburgh EV (1996) Osmoregulation by oat coleoptile protoplasts (Effect of Auxin). Plant Physiol 110:1007-1016

Kiegerl S, Cardinale F, Siligan C, Gross A, Baudouin E, Liwosz A, Eklof S, Till S, Bogre L, Hirt H, Meskiene I (2000) SIMKK, a mitogen-activated protein kinase (MAPK) kinase, is a specific activator of the salt stress induced MAPK, SIMK. Plant cell 12:2247-2258

Kiegle E, Moore CA, Haseloff J, Tester MA, Knight MR (2000) Cell-type-specific calcium responses to drought, salt and cold in the *Arabidopsis* root. Plant J 23:267-278

Kim SA, Kwak JM, Jae SK, Wang MH, Nam HG (2001) Overexpression of the *AtGluR2* gene encoding an *Arabidopsis* homolog of mammalian glutamate receptors impairs calcium utilization and sensitivity to ionic stress in transgenic plants. Plant Cell Physiol 42:74-84

Kishitani S, Takanami T, Suzuki M, Oikawa M, Yokoi S, Ishitani M, Alvarez-Nakase AM, Takabe T, Takabe T (2000) Compatibility of glycinebetaine in rice plants: evaluation using transgenic rice plants with a gene for peroxisomal betaine aldehyde dehydrogenase from barley. Plant Cell Environ 23:107-114

Kishor PBK, Hong Z, Miao GH, Hu CAA, Verma DPS (1995) Overexpression of [delta]-pyrroline-5-carboxylate synthetase increases proline production and confers osmotolerance in transgenic plants. Plant Physiol 108:1387-1394

Knight H (2000) Calcium signaling during abiotic stress in plants. International Rev Cytol 195:269-325

Knight H, Trewavas AJ, Knight MR (1997) Calcium signaling in *Arabidopsis thaliana* responding to drought and salinity. Plant J 12:1067–1078

Kovtun Y, Chiu W-L, Tena G, Sheen J (2000) Functional analysis of oxidative stress-activated mitogen-activated protein kinase cascade in plants. Proc Natl Acad Sci USA 97:2940-2945.

Koyama ML, Levesley A, Koebner RMD, Flowers TJ, Yeo AR (2001) Quantitative trait loci for component physiological traits determining salt tolerance in rice. Plant Physiol 125:406-422

Laurie S, Feeney KA, Maathuis FJM, Heard PJ, Brown SJ, Leigh RA (2002) A role for HKT1 in sodium uptake by wheat roots. Plant J 32:139-149

Leung J, Giraudat J (1998) Abscisic acid signal transduction. Annu Rev Plant Physiol Plant Mol Biol 49:199-222

Li J, Lee Y-R, Assmann SM (1998) Guard cells possess a calcium-dependent protein kinase that phosphorylates the KAT1 potassium channel. Plant Physiol 116: 785-795

Lino B, Baizabal-Aguirre VM, de la Vara LEG (1998) The plasma-membrane H^+-ATPase from beet root is inhibited by a calcium-dependent phosphorylation. Planta 204:352-359

Liu J, Ishitani M, Halfter U, Kim C-S, Zhu J-K (2000) The *Arabidopsis thaliana SOS2* gene encodes a protein kinase that is required for salt tolerance. Proc Natl Acad Sci USA 97:3730-3734

Liu J, Zhu J-K (1998) A calcium sensor homolog required for plant salt tolerance. Science 280:1943-1945

Liu W, Fairbairn DJ, Reid RJ, Schachtman DP (2001) Characterization of two *HKT1* homologues from *Eucalyptus camaldulensis* that display intrinsic osmosensing capability. Plant Physiol 127:283-294

Lynch J, Polito VS, Läuchli A (1989) Salinity stress increases cytoplasmic Ca activity in maize root protoplasts. Plant Physiol 90:1271-1274

Mano Y, Takeda K (1997) Mapping quantitative trait loci for salt tolerance at germination and the seedling stage in barley (*Hordeum vulgare* L.). Euphytica 94:263-272

Marschner H (1995) Mineral nutrition of higher plants, 2nd edition, Academic Press, London

Mikolajczyk M, Awotunde OS, Muszyska G, Klessig DF, Dobrowolska G (2000) Osmotic stress induces rapid activation of a salicylic acid–induced protein kinase and a homolog of protein kinase ASK1 in tobacco cells. Plant Cell 12:165–178

Miyata S, Urao T, Yamaguchi-Shinozaki K, Shinozaki K (1998). Characterization of genes for two-component phosphorelay mediators with a single HPt domain in *Arabidopsis thaliana*. FEBS Lett 437:11-14.

Mizoguchi T, Irie K, Hirayama T, Hayashida N, Yamaguchi-Shinozaki K, Matsumoto K, Shinozaki K (1996) A gene encoding a mitogen-activated protein kinase kinase kinase is induced simultaneously with genes for a mitogen-activaed protein kinase and an S6 ribosomal protein kinase by touch, cold, and water stress in *Arabidopsis thaliana*. Proc Natl Acad Sci USA 93:765-769

Moon H, Lee B, Choi G, Shin D, Prasad DT, Lee O, Kwak S-S, Kim DH, Nam J, Bahk J, Hong JC, Lee SY, Cho MJ, Lim CO, Yun D-J (2003) NDP kinase 2 interacts with two oxidative stress-activated MAPKs to regulate cellular redox state and enhances multiple stress tolerance in transgenic plants Proc Natl Acad Sci USA 100:358-363

Mou Z, Wang X, Fu Z, Dai Y, Han C, Ouyang J, Bao F, Hu Y, Li J (2002) Silencing of phosphoethanolamine *N*- Methyltransferase results in temperature-sensitive male sterility and salt hypersensitivity in *Arabidopsis*. Plant Cell 14:2031–2043

Munnik T, Ligterink W, Meskiene I, Calderini O, Beyerly J, Musgrave A, Hirt H (1999) Distinct osmosensing protein kinase pathways are involved in signaling moderate and severe hyper-osmotic stress. Plant J 20:381-388

Nanjo T, Kobayashi TM, Yoshida Y, Kakubari Y, Yamaguchi-Shinozaki K, Shinozaki K (1999) Antisense suppression of proline degradation improves tolerance to freezing and salinity in *Arabidopsis thaliana*. FEBS Lett 461:205-210

Navazio L, Bewell MA, Siddiqua A, Dickinson GD, Galione A, Sanders D (2000) Calcium release from the endosplasmic reticulum of higher plants elicited by the NADP metabolite nicotinic acid adenine dinucleotide phosphate. Proc Natl Acad Sci USA 97:8693-8698

Ober ES, Sharp RE (1994) Proline accumulation in maize (Zea mays L.) primary roots at low water potentials. I. Requirement for increased levels of abscisic acid. Plant Physiol 105:981-987

Ohnishi J, Flugge U-I, Heldt HW, Kanai R (1990) Involvement of Na^+ in active uptake of pyruvate in mesophyll chloroplasts of some C_4 plants: Na^+/Pyruvate cotransport. Plant Physiol 94: 950-959

Park JM, Park CJ, Lee SB, Ham BK, Shin R, Paek KH (2001) Overexpression of the tobacco *Tsi1* gene encoding an EREBP/AP2–type transcription factor enhances resistance against pathogen attack and osmotic stress in tobacco. Plant Cell 13:1035-1046

Patharkar OR, Cushman JC (2000) A stress-induced calcium-dependent protein kinase from *Mesembryanthemum crystallinum* phosphorylates a two-component pseudo-response regulator. Plant J 24:679-691

Pei ZM, Murata Y, Benning G, Thomine S, Klusener B, Allen GJ, Grill E, Schroeder JI (2000) Calcium channels activated by hydrogen peroxide mediate abscisic acid signaling in guard cells. Nature 406:731-734

Pei ZM, Ward JM, Harper JF, Schroeder JI (1996) A novel chloride channel in *Vicia faba* guard cell vacuoles activated by the serine/threonine kinase, CDPK. EMBO J 15:6564–6574

Piao HL, Pih KT, Lim JH, Kang SG, Jin JB, Kim SH, IH (1999) An *Arabidopsis GSK3/shaggy*-like gene that complements yeast salt stress-sensitive mutants is induced by NaCl and abscisic acid. Plant Physiol 119:1527-1534

Pical C, Westergren T, Dove SK, Larsson C, Sommarin M (1999) Salinity and hyperosmotic stress induce rapid increases in phosphatidylinositol 4,5-bisphosphate, diacylglycerol pyrophosphate, and phosphatidylcholine in *Arabidopsis thaliana* cells. J Biol Chem 274:38232-38240

Plieth C, Hansen UP, Knight H, Knight MR (1999) Temperature sensing by plants: the primary characteristics of signal perception and calcium response. Plant J 18:491–497

Prasad KVSK, Sharmila P, Kumar PA, Saradhi PP (2000) Transformation of *Brassica juncea* (L.) Czern with bacterial *codA* gene enhances its tolerance to salt stress. Molecular Breed 6:489-499

Qiu QS, Guo Y, Dietrich MA, Schumaker KS, Zhu J-K (2002) Regulation of SOS1, a plasma membrane Na+/H+ exchanger in *Arabidopsis thaliana*, by SOS2 and SOS3. Proc Natl Acad Sci USA 99:8436-8441

Quesada V, Garcia-Martinez S, Piqueras P, Ponce MR, Micol JL (2002) Genetic architecture of NaCl tolerance in *Arabidopsis*. Plant Physiol 130: 951-963

Quesada V, Ponce MR, Micol JL (2000) Genetic analysis of salt-tolerant mutants in *Arabidopsis thaliana*. Genetics 154:421-436

Quintero FJ, Garciadeblas B, Rodriguez-Navarro A (1996) The *SAL1* gene of *Arabidopsis*, encoding an enzyme with 3'(2'),5'-bisphosphate nucleotide and inositol polyphosphate 1-phosphatase activities, increases salt tolerance in yeast. Plant Cell 8:529-537

Quintero FJ, Ohta M, Shi H, Zhu, J-K, Pardo JM (2002) Reconstitution in yeast of the *Arabidopsis* SOS signaling pathway for Na^+ homeostasis. Proc Natl Acad Sci USA 99:9061-9066

Raven JA (1985) Regulation of pH and generation of osmolarity in vascular plants: a cost-benefit analysis in relation to efficiency of use of energy, nitrogen and water. New Phytol 101:25-77

Roxas VP, Lodhi SA, Garrett DK, Mahan JR, Allen RD (2000) Stress tolerance in transgenic tobacco seedlings that overexpress glutathione S-transferase/glutathione peroxidase. Plant Cell Physiol 41:1229-1234

Roxas VP, Smith Jr RK, Allen ER, Allen RD (1997) Overexpression of glutathione S-transferase/glutathione peroxidase enhances the growth of transgenic tobacco seedlings during stress. Nature Biotech 15:988-991

Rubio F, Gassmann W, Schroeder JI (1995) Sodium driven potassium uptake by the plant potassium transporter HKT1 and mutations conferring salt tolerance. Science 270:1660-1663.

Rus A, Yokoi S, Sharkhuu A, Reddy M, Lee B-H, Matsumoto TK, Koiwa H, Zhu J-K, Bressan RA, Hasegawa PM (2001a) AtHKT1 is a salt tolerance determinant that controls $Na^{(+)}$ entry into plant roots. Proc Natl Acad Sci USA 98:14150-14155

Rus AM, Estañ MT, Gisbert C, Garcia-Sogo B, Serrano R, Caro M, Moreno V, Bolarín MC (2001b) Expressing the yeast *HAL1* gene in tomato increases fruit yield and enhances K^+ /Na^+ selectivity under salt stress. Plant Cell Environ 24:875-880

Saijo Y, Hata S, Kyozuka J, Shimamoto K, Izui K (2000) Over-expression of a single Ca^{2+} dependent protein kinase confers both cold and salt/drought tolerance on rice plants. Plant J 23:319-327

Sakamoto A, Alia H, Murata N (1998) Metabolic engineering of rice leading to biosynthesis of glycinebetaine and tolerance to salt and cold. Plant Mol Biol 38:1011-1019

Sanchez J-P, Chua N-H (2001) *Arabidopsis* PLC1 is required for secondary responses to abscisic acid signals. Plant Cell 13:1143-1154

Sanders D, Brownlee C, Harper JF (1999) Communicating with calcium. Plant Cell 11:691-706

Schroeder JI, Allen GJ, Hugouvieux V, Kwak JM, Waner D (2001) Guard cell signal transduction. Annu Rev Plant Physiol Plant Mol Biol 52:627-658

Sheen J (1996) Ca^{2+}-dependent protein kinases and stress signal transduction in plants. Science 274:1900-1902

Shen B, Jensen RG, Bohnert HJ (1997) Increased resistance to oxidative stress in transgenic plants by targeting mannitol biosynthesis to chloroplasts. Plant Physiol 113:1177-1183

Shen QX, Zhang PH, Ho T-HD (1996) Modular nature of abscisic acid (ABA) response complexes: composite promoter units that are necessary and sufficient for ABA induction of gene expression in barley. Plant Cell 8:1107-1119

Sheveleva E, Chmara W, Bohnert HJ, Jensen RG (1997) Increased salt and drought tolerance by D-ononitol production in transgenic *Nicotiana tabacum* L. Plant Physiol 115:1211-1219

Shi H, Ishitani M, Kim C-S, Zhu J-K (2000) The *Arabidopsis thaliana* salt tolerance gene *SOS1* encodes a putative Na^+/H^+ antiporter. Proc Natl Acad Sci USA 97:6896-6901

Shi H, Lee B-H, Wu S-J, Zhu J-K (2003) Overexpression of a plasma membrane Na^+/H^+ antiporter improves salt tolerance in *Arabidopsis*. Nature Biotech 21:81-85

Shi H, Quintero, FJ, Pardo JM, Zhu J-K (2002a) The putative plasma membrane Na^+/H^+ antiporter SOS1 controls long-distance Na^+ transport in plants. Plant Cell 14:465-477

Shi H, Xiong L, Stevenson B, Lu T, Zhu J-K (2002b) The *Arabidopsis* salt overly sensitive 4 mutants uncover a critical role for vitamin B6 in plant salt tolerance. Plant Cell 14:575-588

Shi H, Zhu J-K (2002) Regulation of expression of the vacuolar Na^+/H^+ antiporter gene AtNHX1 by salt stress and ABA. Plant Mol Biol 50:543-550

Shinozaki K, Yamaguchi-Shinozaki K (2000) Molecular response to dehydration and low temperature: Differences and cross-talk between two stress signaling pathways. Curr Opin Plant Biol 3:217-223

Skriver K, Olsen FL, Rogers JC, Mundy J (1991) *Cis*-acting elements responsive to gibberellin and its antagonist abscisic acid. Proc Natl Acad Sci USA 88:7266-7270

Smirnoff N (1993) The role of active oxygen in the response of plants to water deficit and desiccation. New Phytol 125:27-58

Spollen WG, LeNoble ME, Samuels TD, Bernstein N, Sharp RE (2000) Abscisic acid accumulation maintains maize primary root elongation at low water potentials by restricting ethylene production. Plant Physiol 122:967-976

Su H, Golldack D, Zhao C, Bohnert HJ (2002) The expression of HAK-type K^+ transporters is regulated in response to salinity stress in common ice plant. Plant Physiol 129:1482-1493

Suzuki T, Imamura A, Ueguchi C, Mizuno T (1998) Histidine-containing phosphotransfer (HPt) signal transducers implicated in His-to-Asp phosphorelay in *Arabidopsis*. Plant Cell Physiol 39:1258-1268

Suzuki T, Sakurai K, Ueguchi C, Mizuno T (2001) Two types of putative nuclear factors that physically interact with Histidine-containing phosphotransfer (Hpt) domains, signaling mediators in His-to-Asp phosphorelay, in *Arabidopsis thaliana*. Plant Cell Physiol 42:37-45

Takahashi S, Katagiri T, Hirayama T, Yamaguchi-Shinozaki K, Shinozaki K (2001) Hyperosmotic stress induces a rapid and transient increase in inositol 1,4,5-trisphosphate independent of abscisic acid in *Arabidopsis* cell culture. Plant Cell Physiol 42:214-222

Tarczynski MC, Jensen RG, Bohnert HJ (1993) Stress protection of transgenic tobacco by production of the osmolyte mannitol. Science 259: 508-510

Tsugane K, Kobayashi K, Niwa Y, Ohba Y, Wada K, Kobayashi H (1999) A recessive *Arabidopsis* mutant that grows photoautotrophically under salt stress shows enhanced active oxygen detoxification. Plant Cell 11:1195-1206

Tyerman SD, Skerrett IM (1999) Root ion channels and salinity. Sci Hort 78:175-235

Ulm R, Ichimura K, Mizoguchi T, Peck SC, Zhu T, Wang X, Shinozaki K, Paszkowski J (2002) Distinct regulation of salinity and genotoxic stress responses by *Arabidopsis* MAP kinase phosphatase 1. EMBO J 21:6483-6493

Ullah H, Chen JG, Young J, Im K-H, Sussman M, Jones A (2001) Modulation of cell proliferation by heterotrimeric G protein in *Arabidopsis*. Science 292 2066-2069

Uozumi N, Kim EJ, Rubio F, Yamaguchi T, Muto S, Tsuboi A, Bakker EP, Nakamura T, Schroeder JI (2000) The *Arabidopsis HKT1* gene homolog mediates inward Na^+ currents in *Xenopus laevis* oocytes and Na^+ uptake in *Saccharomyces cerevisiae*. Plant Physiol 122:1249-1259

Urao T, Katagiri T, Mizoguchi T, Yamaguchi-Shinozaki K, Hayashida N, Shinozaki K (1994) Two genes that encode Ca^{2+} dependent protein kinases are induced by drought and high-salt stress in *Arabidopsis thaliana*. Mol Gen Genet 244:331-340

Urao T, Yakubov B, Satoh R, Yamaguchi-Shinozaki K, Seki M, Hirayama T, Shinozaki K (1999) A transmembrane hybrid-type histidine kinase in *Arabidopsis* functions as an osmosensor. Plant Cell 11:1743-1754

Viswanathan C, Zhu J-K (2002) Molecular genetic analysis of cold-regulated gene transcription. Phil Trans Royal Soc London. Biol Sci 357: 877-886

Wang QY, Nick P (2001) Cold acclimation can induce microtubular cold stability in a manner distinct from abscisic acid. Plant Cell Physiol 42:999-1005

Wang H, Qi Q, Schorr P, Cutler AJ, Crosby W, Fowke LC (1998) ICK1, a cyclin dependent protein kinase inhibitor from *Arabidopsis* thaliana interacts with both Cdc2a and CycD3, and its expression is induced by abscisic acid. Plant J 15:501-510

Wang X-Q, Ullah H, Jones A, Assmann S (2001) G protein regulation of ion channels and abscisic acid signaling in *Arabidopsis* guard cells. Science 292:2070-2072

Winicov I (2000) *Alfin1* transcription factor overexpression enhances plant root growth under normal and saline conditions and improves salt tolerance in alfalfa. Planta 210:416-422

Winicov I, Bastola DR (1999) Transgenic overexpression of the transcription factor *Alfin1* enhances expression of the endogenous, *MSPRP2* gene in alfalfa and improves salinity tolerance of the plant. Plant Physiol 120:473-480

Wu Y, Spollen WG, Sharp RE, Hetherington PR, Fry SC (1994) Root growth maintenance at low water potentials. Increased activity of xyloglucan endotransglycosylase and its possible regulation by abscisic acid. Plant Physiol 106: 607-615

Wu Y, Kuzma J, Marechal E, Graeff R, Lee HC, Foster R, Chua N-H (1997) Abscisic acid signaling through cyclic ADP-ribose in plants. Science 278:2126–2130

Wurgler-Murphy SM, Saito H (1997) Two-component signal transducers and MAPK cascades. Trends Biochem Sci 22:172-176

Xiong L, Ishitani M, Lee H, Zhu J-K (2001a) The *Arabidopsis LOS5/ABA3* locus encodes a molybdenum cofactor sulfurase and modulates cold stress- and osmotic stress- responsive gene expression. Plant Cell 13:2063-2083

Xiong L, Lee B-H, Ishitani M, Lee H, Zhang C, Zhu J-K (2001b) *FIERY1* encoding an inositol polyphosphate 1-phosphatase is a negative regulator of abscisic acid and stress signaling in *Arabidopsis*. Genes Dev 15:1971-1984

Xiong L, Lee H, Ishitani M, Zhu J-K (2002a) Regulation of osmotic stress responsive gene expression by LOS6/ABA1 locus in *Arabidopsis*. J Biol Chem 277:8588-8596

Xiong L, Schumaker KS, Zhu J-K (2002b) Cell signaling for cold, drought, and salt stresses. Plant Cell 14:S165-183

Xu D, Duan X, Wang B, Hong B, Ho TD, Wu R (1996) Expression of a late embryogenesis abundant protein gene, *HVA1*, from barley confers tolerance to water deficit and salt stress in transgenic rice. Plant Physiol 110:249-257

Xu Q, Fu H-H, Gupta R, Luan S (1998) Molecular characterization of a tyrosine-specific protein phosphatase encoded by a stress-responsive gene in *Arabidopsis*. Plant Cell 10:849–857

Yadav R, Flowers TJ, Yeo AR (1996) The involvement of the transpirational bypass flow in sodium uptake by high- and low-sodium-transporting lines of rice developed through intravarietal selection. Plant Cell Environ 22:329-336

Yancey PH, Clark ME, Hand SC, Bowles RD, Somero GN (1982) Living with water stress: evolution of osmolyte system. Science 217:1214-1222

Yeo AR, Flowers SA, Rao G, Welfare K, Senanayake N, Flowers TJ (1999) Silicon reduces sodium uptake in rice (*Oryza sativa* L.) in saline conditions and this is accounted for by a reduction in the transpirational bypass flow. Plant Cell Environ 22:559-565

Yuasa T, Ichimura K, Mizoguchi T, Shinozaki K (2001) Oxidative stress activates ATMPK6, an *Arabidopsis* homologue of MAP Kinase. Plant Cell Physiol 42:1012-1016

Zhang HX, Blumwald E (2001) Transgenic salt-tolerant tomato plants accumulate salt in foliage but not in fruit Nature Biotech 19:765-768

Zhang HX, Hodson JN, Williams JP, Blumwald E (2001) Engineering salt-tolerant Brassica plants: Characterization of yield and seed oil quality in transgenic plants with increased vacuolar sodium accumulation. Proc Natl Acad Sci USA 98:12832-12836

Zhang J-S, Xie C, Li ZY, Chen SY (1999) Expression of the plasma membrane H^+ATPase gene in response to salt stress in a rice salt-tolerant mutant and its original variety. Theor Appl Genet 99:1006-1011

Zhu B, Su J, Chang MC, Verma DPS, Fan YL, Wu R (1998) Overexpression of a pyrroline-5-carboxylate synthetase gene and analysis of tolerance to water and salt stress in transgenic rice. Plant Sci 139:41-48

Zhu J, Gong Z, Zhang C, Song C-P, Damsz B, Inan G, Koiwa H, Zhu J-K, Hasegawa PM, Bressan RA (2002) OSM1/SYP61: a syntaxin protein in *Arabidopsis* controls abscisic acid–mediated and non-abscisic acid–mediated responses to abiotic stress. Plant Cell 14:3009-3028

Zhu J-K (2001) Plant salt tolerance. Trends Plant Sci 6:66-71

Zhu J-K (2002) Salt and drought stress signal transduction in plants. Annu Rev Plant Biol 53:247-273

Zhu J-K, Liu J, Xiong L (1998) Genetic analysis of salt tolerance in *Arabidopsis thaliana*: evidence of a critical role for potassium nutrition. Plant Cell 10:1181-1192

10 Transcriptome analysis in abiotic stress conditions in higher plants

Motoaki Seki, Ayako Kamei, Masakazu Satou, Tetsuya Sakurai, Miki Fujita, Youko Oono, Kazuko Yamaguchi-Shinozaki and Kazuo Shinozaki

Abstract

Drought, high salinity, and low temperature are major environmental factors that limit plant productivity. Plants respond and adapt to these stresses in order to survive. Recent molecular and genetic studies have revealed the presence of many signaling components are involved in the signaling pathways of these stresses. Furthermore, gene expression profiling using cDNA microarrays or gene chips has identified many genes that are regulated by drought-, cold-, or high-salinity stresses. In this review, we highlight recent progress on the transcriptome analysis in drought-, cold-, or high-salinity stress conditions.

10.1 Introduction

Plant growth is greatly affected by environmental abiotic stresses, such as drought, high salinity and low temperature. Plants respond and adapt to these stresses in order to survive. These stresses induce various biochemical and physiological responses in plants. Several genes that respond to drought, high-salinity, or cold stress at the transcriptional level have been studied (Hasegawa et al. 2000; Shinozaki and Yamaguchi-Shinozaki 2000; Thomashow 1999; Zhu 2002). The products of the stress-inducible genes have been classified into two groups: those that directly protect against environmental stresses and those that regulate gene expression and signal transduction in the stress response. The first group includes proteins that likely function by protecting cells from dehydration, such as the enzymes required for biosynthesis of various osmoprotectants, late-embryogenesis-abundant (LEA) proteins, antifreeze proteins, chaperones, and detoxification enzymes. The second group of gene products includes transcription factors, protein kinases, and enzymes involved in phosphoinositide metabolism. Stress-inducible genes have been used to improve the stress tolerance of plants by gene transfer (Hasegawa et al. 2000; Shinozaki and Yamaguchi-Shinozaki 2000; Thomashow 1999). It is important to analyze the functions of stress-inducible genes not only to understand the molecular mechanisms of stress tolerance and the responses of higher plants but also to improve the stress tolerance of crops by gene manipulation. Hundreds of genes are thought to be involved in abiotic stress responses (Shinozaki and Yamaguchi-Shinozaki 1999, 2000; Xiong and Zhu 2001,

Topics in Current Genetics, Vol. 4
H. Hirt, K. Shinozaki (Eds.) Plant Responses To Abiotic Stress

2002; Xiong et al. 2002; Zhu 2002). In this review, we highlight recent progress on the gene expression in these stress responses.

10.2 Cis- and trans-acting factors involved in regulation of gene expression by drought, high-salinity and cold stress

A number of genes that are induced by osmotic stress have been identified (Shinozaki and Yamaguchi-Shinozaki 1999, 2000; Thomashow 1999; Xiong and Zhu 2001, 2002; Xiong et al. 2002; Zhu 2002). Although the signaling pathways responsible for the activation of these genes are largely unknown, transcriptional activation of some of the stress-responsive genes is understood to a great extent, owing to studies on *RD29A/COR78/LTI78* gene. The promoter of this gene contains both ABA-responsive element (ABRE) and DRE/CRT (Yamaguchi-Shinozaki and Shinozaki 1994). ABRE and DRE/CRT are *cis*-acting elements that function in ABA-dependent and ABA-independent gene expression in response to stress, respectively. Transcription factors belonging to the ERF/AP2 family that bind to DRE/CRT were isolated and termed *DREB1A/CBF3*, *DREB1B/CBF1* and *DREB1C/CBF2* (Liu et al. 1998; Stockinger et al. 1997). These transcription factor genes are induced early and transiently by cold stress, and they, in turn, activate the expression of target genes. Similar transcription factors *DREB2A* and *DREB2B* are induced by dehydration stress to express various genes involved in drought stress tolerance (Liu et al. 1998). Sakuma et al. (2002) precisely analyzed the DNA-binding specificity of *DREB1A/CBF3* and *DREB2* and demonstrated that the core sequence of DRE is the 6-bp A/GCCGAC sequence. The ability of *DREB1/CBF* to activate the DRE/CRT class of stress-responsive genes was further demonstrated by the observation that overexpression or enhanced inducible expression of *DREB1/CBF* could activate the target genes. Overexpression of *DREB1/CBF* also increased the tolerance of the transgenic plants to freezing, drought and high-salinity stresses (Jaglo-Ottosen et al. 1998; Kasuga et al. 1999; Liu et al. 1998; Shinozaki and Yamaguchi-Shinozaki 2000), suggesting that the DREB1/CBF system is important for the development of stress tolerance in plants. The DREB1/CBF pathway is a major transcription system regulating ABA-independent gene expression in response to drought and cold stresses (Shinozaki and Yamaguchi-Shinozaki 2000). Taji et al (2002) showed that galactinol synthase (*AtGolS*) gene is a target gene of *DREB1A/CBF3*. Transgenic *Arabidopsis* plants overexpressing the *AtGolS2* gene accumulated galactinol and raffinose, showed a reduced transpiration rate, and were more tolerant to drought-stress than were control plants. Kim et al. (2002) reported that cold-induced gene expression through DRE/CRT is greatly enhanced by a signal generated by light and that the primary photoreceptor involved in light signaling is phytochrome B.

Several basic leucine zipper (bZIP) transcription factors that can bind to ABRE and activate the expression of ABRE-driven reporter genes also have been isolated: *AREB1/ABF2*, *AREB2/ABF4*, *AREB3*, *ABF1*, and *ABF3* (Choi et al. 2000; Uno et al. 2000). *AREB1/ABF2* and *AREB2/ABF4* genes need ABA for full activation, since the activities of these transcription factors were reduced in the ABA-deficient mutant *aba2* and ABA-insensitive mutant *abi1-1*, but were enhanced in the ABA-hypersensitive *era1* mutant, probably due to ABA-dependent phosphorylation of the proteins (Uno et al. 2000). Recently, Kang et al. (2002) reported that constitutive overexpression of *ABF3* or *AREB2/ABF4* in *Arabidopsis* resulted in ABA hypersensitivity, reduced transpiration rate and enhanced drought tolerance. Changes in phenotypes for loss-of-function mutants have not yet been reported for any *DREB/CBF* or *AREB/ABF* genes. This may be due to functional redundancy between the family members, and hence it may be necessary to combine loss-of-function mutants for two or more members to see the phenotype.

The induction of the drought-inducible genes such as *RD22* is mediated by ABA and requires protein biosynthesis for its ABA-dependent expression (Abe et al. 1997; Shinozaki and Yamaguchi-Shinozaki 2000). A MYC transcription factor, *RD22BP1 (AtMYC2)*, and a MYB transcription factor, *ATMYB2*, were shown to bind *cis*-elements in the *RD22* promoter and cooperatively activate *RD22* (Abe et al. 1997, 2003).

A number of drought- and/or ABA-inducible genes encoding various transcription factors have been reported (Zhu 2002). Among them, ATHB6 containing the homeodomain functions as a negative regulator downstream of ABI1 in the ABA signal transduction pathway (Himmelbach et al. 2002).

10.2.1 Application of cDNA microarray analysis to expression profiling under abiotic stress conditions

Recently, microarray technology has become a powerful tool for the systematic analysis of expression profiles of large numbers of genes (Eisen and Brown 1999; Richmond and Somerville 2000; Seki et al. 2001b). This DNA chip-based technology arrays cDNA sequences or oligonucleotides on a glass slide at a density >1000 genes/cm^2. These arrayed sequences are hybridized simultaneously to a two-color fluorescently labeled cDNA probe pair prepared from RNA samples of different cell or tissue types, allowing direct and large-scale comparative analysis of gene expression. Several groups reported the application of the microarray technology to the analysis of expression profiles in response to drought, cold and high-salinity stresses (Chen et al. 2002; Fowler and Thomashow 2002; Kawasaki et al. 2001; Seki et al. 2001a, 2002b, 2002c). In this review article, first, we summarize the recent progress on the transcriptome analysis under abiotic stress conditions using our RIKEN *Arabidopsis* full-length (RAFL) cDNA microarray.

10.3 Collection and functional annotation of RIKEN *Arabidopsis* full-length (RAFL) cDNAs

We have constructed *Arabidopsis* full-length cDNA libraries from plants grown under different conditions as reported previously (Seki et al. 1998, 2001b, 2002a) by the biotinylated CAP trapper method using trehalose-thermoactivated reverse transcriptase (Carninci et al. 1996, 1997, 1998). Until now, we have constructed 19 full-length cDNA libraries from *Arabidopsis* plants grown under various stress, hormone and light conditions, from plants at various developmental stages, and from various plant tissues (Seki et al. 2002a). We performed single-pass sequencing of the cDNA clones from the 3'-end. We have obtained 155,144 3'-end expressed sequence tags (ESTs) as of February, 2002 (Seki et al. 2002a). The 155,144 3'-ESTs were clustered and then mapped onto the *Arabidopsis* genome. Finally, 14,668 non-redundant RAFL cDNA clones were identified and mapped on the *Arabidopsis* genome (Seki et al. 2002a). The information of the 14,668 RAFL cDNA clones (the "RAFL cDNA" genes) is available at http://www.gsc.riken.go.jp/Plant/index.html (Seki et al. 2002a). Assuming that the total number of *Arabidopsis* genes is about 26,000, the RAFL clones isolated should account for about 60% of all *Arabidopsis* genes.

From the 5'-end sequences of mRNAs, the promoter sequences can be obtained by comparison with the *Arabidopsis* genomic sequences. We also obtained 5'-ESTs of 14,034 RAFL cDNA clones and constructed a promoter database (Seki et al., 2002a) using the Plant cis-acting regulatory DNA elements (PLACE) database (Higo et al., 1999). The *Arabidopsis* promoter database constructed contains the information on the genomic sequences 1000-bp upstream from the 5'-termini of each RAFL cDNA clone, and about 300 *cis*-acting elements known from plants (Seki et al., 2002a). The *Arabidopsis* promoter database constructed is available at http://www.gsc.riken.go.jp/Plant/index.html (Seki et al. 2002a, 2002d). One of the interesting types of the microarray analysis is the identification of novel *cis*-elements that regulate the expression of genes in response to various experimental treatments. By identifying subsets of the genes that have a common expression profile, it might be possible to identify conserved motifs in the promoter regions. We think that our promoter database becomes useful for systematic analysis of *cis*-acting elements in *Arabidopsis* (Seki et al. 2002a, 2002d).

10.3.1 Application of RIKEN *Arabidopsis* full-length (RAFL) cDNA microarray to identify drought-, cold-, or high-salinity-stress-regulated genes

A number of genes have been described that respond to drought, cold, and high-salinity stresses at the transcriptional level as described above. However, many unidentified genes are thought to be involved in drought, cold, and high-salinity stress responses. Therefore, we applied the full-length cDNA microarray containing ca. 1300 *Arabidopsis* full-length cDNAs to identify new drought- or cold-inducible genes (Seki et al. 2001a). Forty-four and nineteen cDNAs for drought- and cold-inducible genes, respectively, were isolated, 30 and 10 of which were novel stress-inducible genes that have not been reported as drought- or cold-inducible genes previously. As described above, we reported that overexpression of the *DREB1A/CBF3* cDNA under the control of the cauliflower mosaic virus (CaMV) 35S promoter or the stress-inducible *rd29A* promoter in transgenic plants gave rise to strong constitutive expression of the stress-inducible DREB1A target genes and increased tolerance to freezing, drought, and salt stresses (Kasuga et al. 1999; Liu et al. 1998). Kasuga et al. (1999) identified 6 DREB1A target genes. However, it is not well understood how overexpression of the *DREB1A* cDNA in transgenic plants increases stress tolerance to freezing, drought, and high-salinity stresses. To study the molecular mechanisms of drought and freezing tolerance, it is important to identify and analyze more genes that are controlled by DREB1A. Therefore, we applied the full-length cDNA microarray containing ca. 1300 *Arabidopsis* full-length cDNAs to identify new target genes of *DREB1A* (Seki et al. 2001a). Twelve stress-inducible genes were identified as target stress-inducible genes of DREB1A, and six of them were novel. All *DREB1A* target genes identified contained DRE- or DRE-related CCGAC core motif sequences in their promoter regions (Seki et al. 2001a). These results show that our full-length cDNA microarray is a useful material with which to analyze the expression pattern of *Arabidopsis* genes under drought and cold stresses, to identify target genes of stress-related transcription factors, and to identify potential *cis*-acting DNA elements by combining the expression data with the genomic sequence data.

Recently, we prepared a new full-length cDNA microarray containing ca. 7000 independent *Arabidopsis* full-length cDNA groups, including drought-inducible genes, *responsive to dehydration* (*rd*) and *early responsive to dehydration* (*erd*) (Taji et al. 1999), as positive controls, the PCR-amplified fragment from lambda control template DNA fragment (Takara) as an external control, and the mouse nicotinic acetylcholine receptor epsilon-subunit (nAChRE) gene and the mouse glucocorticoid receptor homolog gene, which have no substantial homology to any sequences in the *Arabidopsis* database, to assess for nonspecific hybridization as negative controls.(Seki et al. 2002b). We applied the cDNA microarray containing ca. 7000 *Arabidopsis* full-length cDNA groups to identify new drought-, cold-, high-salinity-, or ABA-inducible genes. In this study, we used the PCR-amplified fragment from lambda control template DNA fragment (Takara) as an external control gene to equalize hybridization signals generated from different samples and regarded the genes with expression ratios (dehydration/unstressed, cold/unstressed, or high-salinity/unstressed) greater than five times that of the lambda control template DNA fragment in at least 1 time-course point as dehydration-, cold-, or high-salinity-stress-inducible genes. We identified 299 drought-inducible genes, 54 cold-inducible genes, 213 high-salinity-stress-inducible genes and 245 ABA-inducible genes (Fig. 1)(Seki et al. 2002b, 2002c). Information on each stress-inducible gene is available at http://www.gsc.riken.go.jp /Plant/index.html. Venn diagram analysis indicated the existence of significant crosstalk between drought and high-salinity stress signaling processes (Fig. 1) (Seki et al. 2002b). Many ABA-inducible genes are induced after drought - and

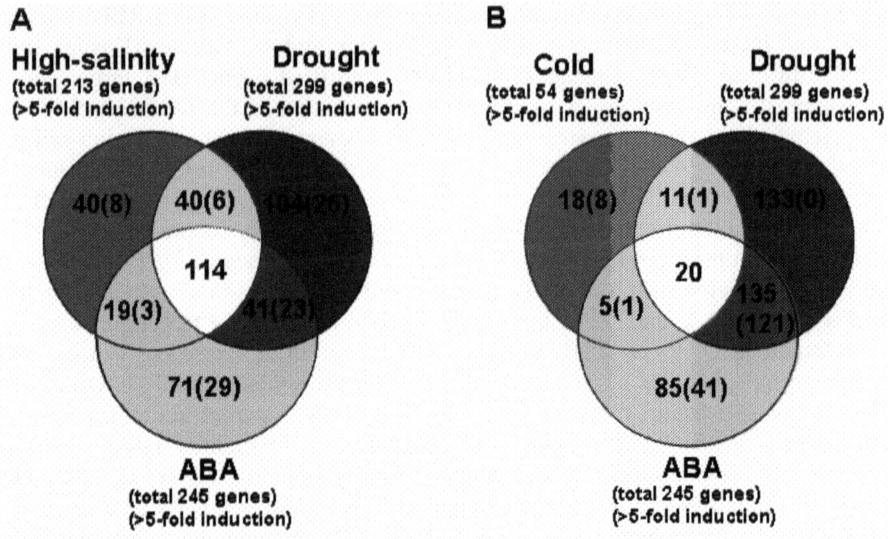

Fig. 1. Classification of the drought-, cold-, high-salinity-stress- or ABA- inducible genes on the basis of their expression pattern: A The ABA-, drought-, or high-salinity-stress-inducible genes identified were grouped into the following 7 groups: (1) ABA-highly-inducible genes, (2) drought-stress-highly-inducible genes, (3) high-salinity-stress-highly-inducible genes, (4) ABA- and drought-stress-highly-inducible genes, (5) ABA- and high-salinity-stress-highly-inducible genes, (6) drought- and high-salinity-stress-highly-inducible genes, and (7) ABA-, drought- and high-salinity-stress-inducible genes. B The ABA-, drought-, or cold-stress-inducible genes identified were grouped into the following 7 groups: (1) ABA-highly-inducible genes, (2) drought-highly-inducible genes, (3) cold-stress-highly-inducible genes, (4) ABA- and drought-stress-highly-inducible genes, (5) ABA- and cold-stress-highly-inducible genes, (6) drought- and cold-stress-highly-inducible genes, and (7) ABA-, drought- and cold-stress-inducible genes. The number of genes whose expression ratio is more than 5-fold for each treatment and less than 5-fold for the other treatments is indicated. Numbers in parentheses represent the number of genes whose expression ratio is more than 5-fold for each treatment and less than 3-fold for the other treatments. The list of the genes is available at http://pfgweb.gsc.riken.go.jp/index.html (Seki et al. 2002c).

high-salinity-stress treatments, which indicates the existence of significant crosstalk between drought and ABA responses (Fig. 1) (Seki et al. 2002c). These results supported our previous model on strong overlap of gene expression in response to drought, high-salinity, and ABA (Shinozaki and Yamaguchi-Shinozaki 2000), and partial overlap of gene expression in response to cold and osmotic stress.

10.4 Stress-inducible genes and functions of their gene products identified by RAFL cDNA microarray

The products of the drought-, high-salinity-, or cold-stress-inducible gene products can be classified into 2 groups (Figs. 2, 3 and 4; Supplemental tables 1, 2, and 3) (Seki et al. 2002b; Shinozaki and Yamaguchi-Shinozaki 1999, 2000). The first group includes functional proteins, or proteins that probably function in stress tolerance. They are late-embryogenesis-abundant (LEA) proteins, heat shock proteins, KIN (cold-inducible) proteins, osmoprotectant-biosynthesis-related proteins, carbohydrate-metabolism-related proteins, water channel proteins, sugar transporters, potassium transporters, detoxification enzymes, proteases, senescence-related genes, protease inhibitors, ferritin, and lipid transfer proteins (Figs. 2, 3, and 4; Supplemental tables 1, 2, and 3) (Seki et al. 2002b). LEA proteins and heat shock proteins have been shown to be involved in protecting macromolecules like enzymes and lipids (Shinozaki and Yamaguchi-Shinozaki 1999). Proline, sugars and raffinose family oligosaccharides (RFO) probably function as osmolytes in protecting cells from dehydration (Cushman and Bohnert 2000; Taji et al. 2002). KIN proteins may have a unique ability to neutralize ice nucleators and inhibit ice recrystallization (Holmberg and Bülow 1998). Water channel proteins and sugar transporters are thought to function in transport of water and sugars through plasma membranes and tonoplast to adjust the osmotic pressure under stress conditions. Potassium transporters may function in transport of K^+, which is an essential cofactor for many enzymes (Hasegawa et al. 2000) or control K^+ uptake and regulate Na^+ uptake, which can be an important determinant of salinity tolerance (Bray 1997). Detoxification enzymes, such as glutathione S-transferase are thought to be involved in protection of cells from active oxygens. Proteases including cysteine proteases, Clp protease, and ubiquitin-conjugating enzyme are thought to be required for protein turnover and recycling of amino acids. Drought stress has been shown to accelerate leaf senescence, which is characterized by many subcellular changes, including an increase in protease activities (Thomas and Stoddart 1980). The protease inhibitors may perform a defensive role against the proteases. Ferritin may have a function in protecting the cells from oxidative damage caused by various stresses by sequestering intracellular iron involved in the generation of various reactive hydroxyl radicals through a Fenton reaction (Bajaj et al. 1999). Lipid transfer proteins and fatty acid-metabolism-related genes may have a function in repair of stress-induced damage in membranes or changes in the lipid composition of membranes, perhaps to regulate the permeability to toxic ions and the fluidity of the membrane (Holmberg and Bülow 1998; Torres-Schumann et al. 1992).

The second group contains regulatory proteins, that is, protein factors involved in further regulation of signal transduction and gene expression that probably function in stress response (Figs. 2, 3, and 4; Supplemental tables 1, 2 and 3) (Seki et al. 2002b; Shinozaki and Yamaguchi-Shinozaki 1999, 2000). They are various transcription factors, protein kinases, protein phosphatases, enzymes involved in

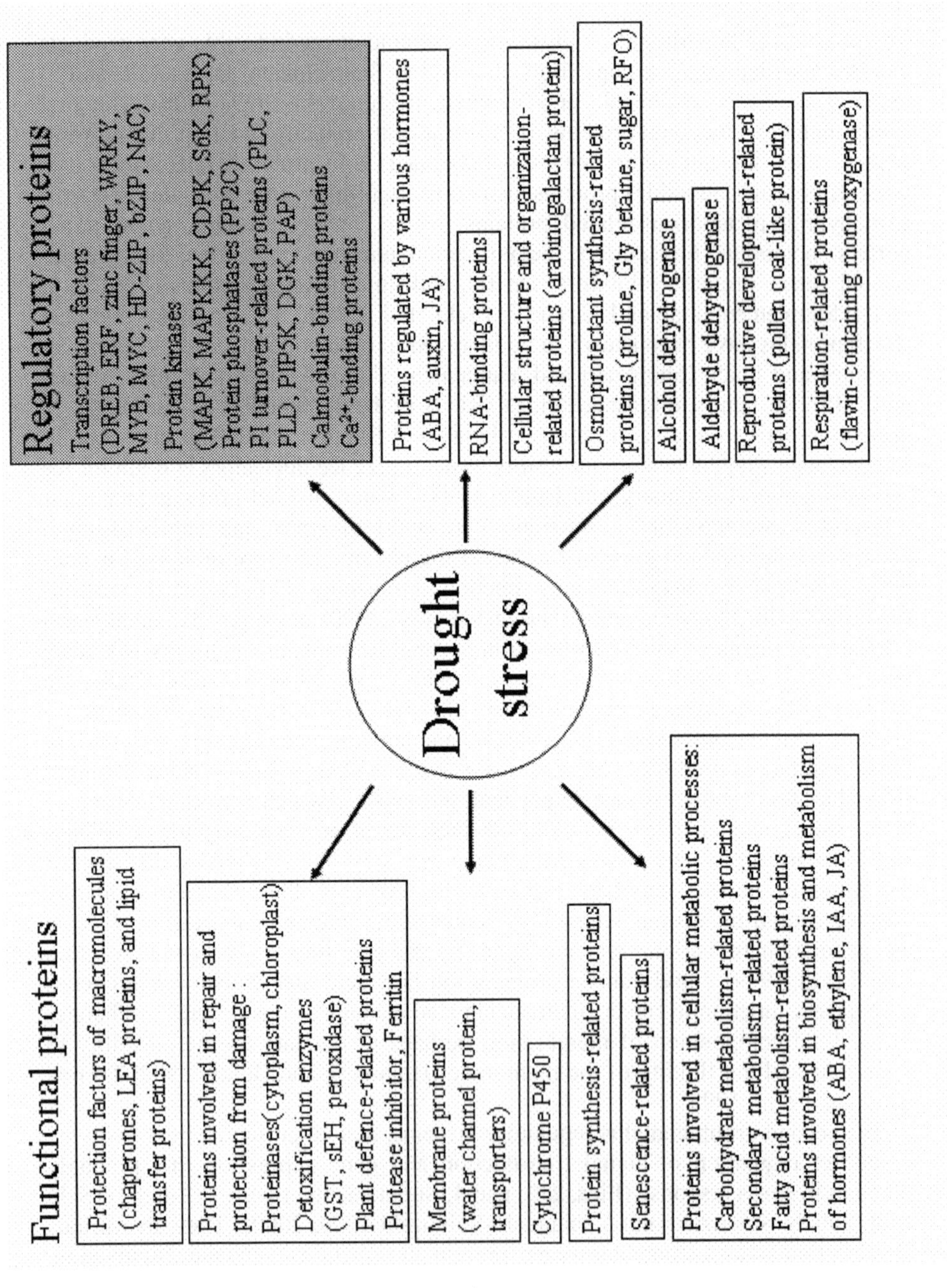
Regulatory proteins
Transcription factors
(DREB, ERF, zinc finger, WRKY, MYB, MYC, HD-ZIP, bZIP, NAC)
Protein kinases
(MAPK, MAPKKK, CDPK, S6K, RPK)
Protein phosphatases (PP2C)
PI turnover-related proteins (PLC, PLD, PIP5K, DGK, PAP)
Calmodulin-binding proteins
Ca2+-binding proteins
Proteins regulated by various hormones (ABA, auxin, JA)
RNA-binding proteins
Cellular structure and organization-related proteins (arabinogalactan protein)
Osmoprotectant synthesis-related proteins (proline, Gly betaine, sugar, RFO)
Alcohol dehydrogenase
Aldehyde dehydrogenase
Reproductive development-related proteins (pollen coat-like protein)
Respiration-related proteins (flavin-containing monooxygenase)
Drought stress
Functional proteins
Protection factors of macromolecules (chaperones, LEA proteins, and lipid transfer proteins)
Proteins involved in repair and protection from damage :
Proteinases(cytoplasm, chloroplast)
Detoxification enzymes
(GST, sEH, peroxidase)
Plant defence-related proteins
Protease inhibitor, Ferritin
Membrane proteins
(water channel protein, transporters)
Cytochrome P450
Protein synthesis-related proteins
Senescence-related proteins
Proteins involved in cellular metabolic processes:
Carbohydrate metabolism-related proteins
Secondary metabolism-related proteins
Fatty acid metabolism-related proteins
Proteins involved in biosynthesis and metabolism of hormones (ABA, ethylene, IAA, JA)

Fig. 2. Drought stress-inducible genes and their possible functions in stress tolerance and response. Gene products are classified into two groups. The first group (Functional proteins) includes proteins that probably function in stress tolerance. They are protection factors such as chaperones, LEA proteins, and lipid transfer proteins, proteins involved in repair and protection from damages, such as proteinases, detoxification enzymes, protease inhibitors, ferritin and plant defense-related proteins, membrane proteins such as water channel protein and transporters, protein synthesis-related proteins, proteins involved in synthesis of osmoprotectant (proline, glycine betaine, sugars and RFO), proteins involved in cellular metabolic processes, such as carbohydrate metabolism, secondary metabolism, fatty acid metabolism, biosynthesis of plant hormones (ABA, ethylene, IAA and JA), proteins regulated by plant hormones (ABA, auxin and JA), RNA-binding proteins, cellular structure and organization-related proteins such as arabinogalactan protein, senescence-related proteins, cytochrome P450, alcohol dehydrogenase, aldehyde dehydrogenase, reproduction development-related proteins such as pollen coat-like protein and respiration-related proteins such as flavin-containing monooxygenase. The second group (Regulatory proteins) contains protein factors involved in further regulation of signal transduction and gene expression that probably function in stress response. They are transcription factors such as DREB family, ERF family, zinc finger family, WRKY family, MYB family, MYC family, HD-ZIP family, bZIP family and NAC family, protein kinases such as MAPK (Mizoguchi et al. 1996), MAPKKK (Mizoguchi et al. 1996), CDPK (Urao et al. 1994), S6K (Mizoguchi et al. 1996) and RPK (Hong et al. 1997), protein phosphatases such as PP2C, PI turnover-related proteins, such as PLC (Hirayama et al. 1995), PLD (Katagiri et al. 2001), PIP5K (Mikami et al. 1998), DGK (Shinozaki and Yamaguchi-Shinozaki 1999), and PAP (Shinozaki and Yamaguchi-Shinozaki 1999), and calmodulin-binding protein and Ca^{2+}-binding protein. The list of the drought stress-inducible genes identified by the cDNA microarray analysis (Seki et al. 2002b) is available at http://pfgweb.gsc.riken.go.jp /index.html (as supplemental table 1).

phospholipid metabolism, and other signaling molecules, such as calmodulin-binding protein (Figs. 2, 3 and 4; Supplemental tables 1, 2, and 3) (Seki et al. 2002b). Among the drought-, cold-, or high-salinity-stress-inducible genes identified, we found ca. 40 (corresponding to ca. 11% of all stress-inducible genes identified) transcription factor genes, suggesting that various transcriptional regulatory mechanisms function in the drought-, cold-, or high-salinity-stress signal transduction pathways (Seki et al. 2002b, 2002c). Among these stress-inducible transcription factors, there are 6 DREB family cDNAs, 2 ethylene-responsive element binding factor (ERF) family cDNAs, 10 zinc finger family cDNAs, 4 WRKY family cDNAs, 3 MYB family cDNAs, 2 basic helix-loop-helix (bHLH) family cDNAs, 4 bZIP family cDNAs, 5 NAC family cDNAs, and 3 homeodomain-leucine zipper (HD-ZIP) transcription factor family cDNAs. These transcription factors probably regulate various stress-inducible genes cooperatively or separately. Among 6 protein kinase genes, we found 2 receptor-like protein kinase genes. These regulatory proteins are thought to function in further regulating various functional genes under stress conditions. Functional analysis of these stress-inducible transcription factors or protein kinase genes should provide more information on signal transduction in responses to drought, cold and high-salinity stresses.

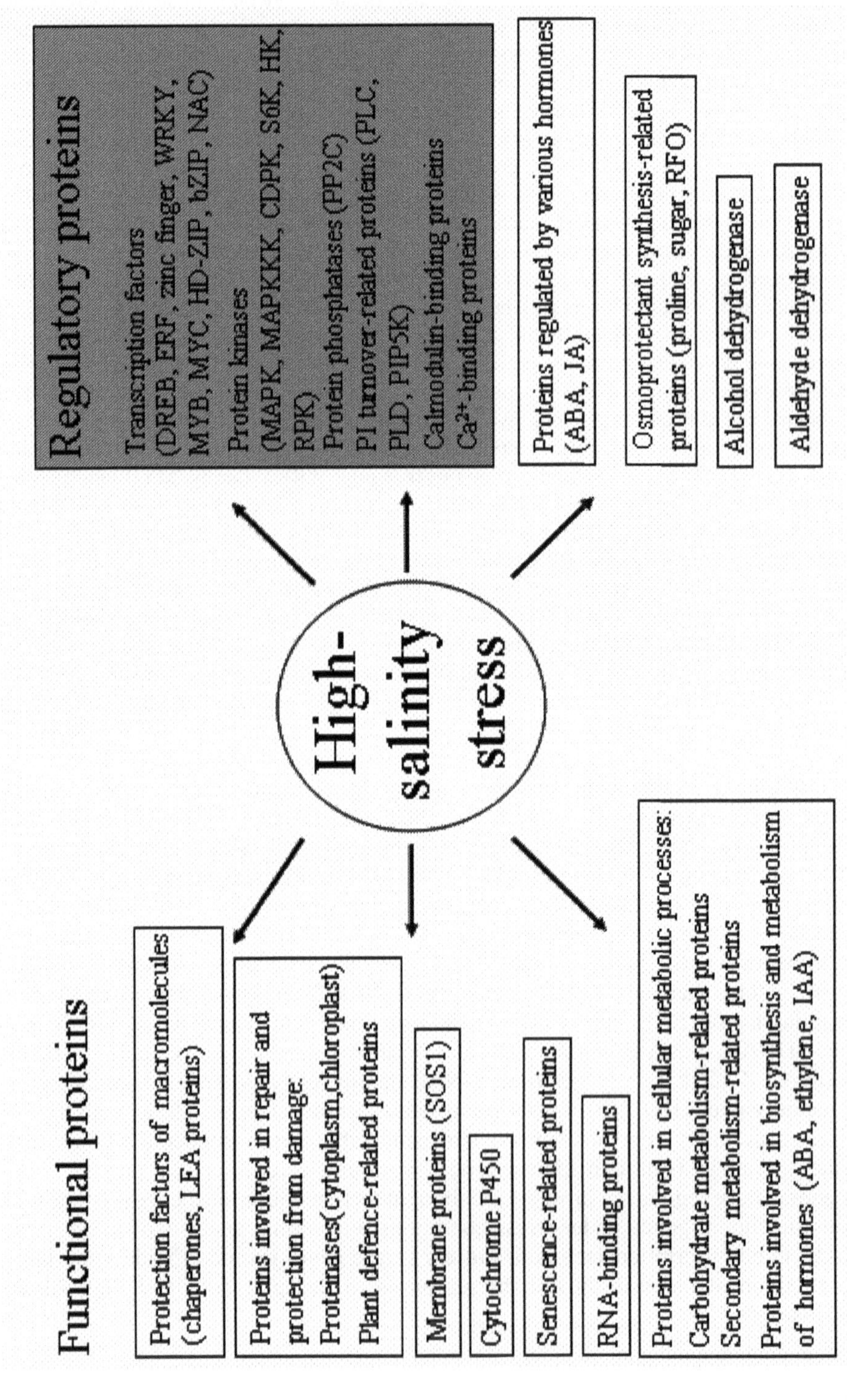

Functional proteins
Protection factors of macromolecules (chaperones, LEA proteins)
Proteins involved in repair and protection from damage:
Proteinases(cytoplasm,chloroplast)
Plant defence-related proteins
Membrane proteins (SOS1)
Cytochrome P450
Senescence-related proteins
RNA-binding proteins
Proteins involved in cellular metabolic processes:
Carbohydrate metabolism-related proteins
Secondary metabolism-related proteins
Proteins involved in biosynthesis and metabolism of hormones (ABA, ethylene, IAA)
High-salinity stress
Regulatory proteins
Transcription factors
(DREB, ERF, zinc finger, WRKY, MYB, MYC, HD-ZIP, bZIP, NAC)
Protein kinases
(MAPK, MAPKKK, CDPK, S6K, HK, RPK)
Protein phosphatases (PP2C)
PI turnover-related proteins (PLC, PLD, PIP5K)
Calmodulin-binding proteins
Ca2+-binding proteins
Proteins regulated by various hormones (ABA, JA)
Osmoprotectant synthesis-related proteins (proline, sugar, RFO)
Alcohol dehydrogenase
Aldehyde dehydrogenase

Fig. 3. High salinity stress-inducible genes and their possible functions in stress tolerance and response. Gene products are classified into two groups. The first group (Functional proteins) includes proteins that probably function in stress tolerance. They are protection factors such as chaperones and LEA proteins, proteins involved in repair and protection from damages, such as proteinases, and plant defense-related proteins, membrane proteins such as SOS1 (Shi et al. 2000), protein synthesis-related proteins, proteins involved in synthesis of osmoprotectant (proline, sugars and RFO), senescence-related proteins, proteins involved in cellular metabolic processes, such as carbohydrate metabolism, secondary metabolism, biosynthesis of plant hormones (ABA, ethylene and IAA), proteins regulated by plant hormones (ABA and JA), RNA-binding proteins, cytochrome P450, alcohol dehydrogenase and aldehyde dehydrogenase. The second group (Regulatory proteins) contains protein factors involved in further regulation of signal transduction and gene expression that probably function in stress response. They are transcription factors such as DREB family, ERF family, zinc finger family, WRKY family, MYB family, MYC family, HD-ZIP family, bZIP family and NAC family, protein kinases such as MAPK (Mizoguchi et al. 1996), MAPKKK (Mizoguchi et al. 1996), CDPK (Urao et al. 1994), S6K (Mizoguchi et al. 1996), HK (Urao et al. 1999) and RPK (Hong et al. 1997), protein phosphatases such as PP2C, PI turnover-related proteins, such as PLC (Hirayama et al. 1995), PLD (Katagiri et al. 2001), and PIP5K (Mikami et al. 1998), calmodulin-binding proteins and Ca^{2+}-binding proteins. The list of the high-salinity-stress-inducible genes identified by the cDNA microarray analysis (Seki et al. 2002b) is available at http://pfgweb.gsc.riken.go.jp/index.html (as supplemental table 2).

Various genes involved in the metabolism of ABA, ethylene, jasmonic acid (JA), and auxin, and JA- or auxin-regulated genes were identified as drought-inducible genes (Fig. 2; Supplemental table 1) (Seki et al. 2002b), suggesting the link between ethylene, JA, and auxin, and drought-stress-signaling pathways. Also, aldehyde dehydrogenase genes, genes related to secondary metabolism, genes involved in various cellular metabolic processes, genes encoding membrane proteins and cytochrome P450 were identified as drought- or high-salinity-stress-inducible genes (Figs. 2 and 3; Supplemental tables 1 and 2) (Seki et al. 2002b). At present, the functions of most of these genes are not fully understood. Furthermore, we found many drought-, cold-, or high-salinity-stress-inducible genes whose functions are unknown.

10.4.1 Cold-inducible genes and stress-downregulated genes identified using RAFL cDNA microarray

Among the cold-inducible genes identified, 9 genes did not contain DRE or DRE-related CCGAC core motif in their promoters. These results suggest the existence of novel *cis*-acting elements involved in cold-inducible gene expression (Seki et al. 2002b).

Analysis of the expression profiles of cold-inducible genes during cold treatment showed the existence of at least 2 groups that show different expression profiles (Seki et al. 2002b). In one group containing the *DREB1A* gene, gene expression was rapid and transient in response to cold treatment, reached a maximum at

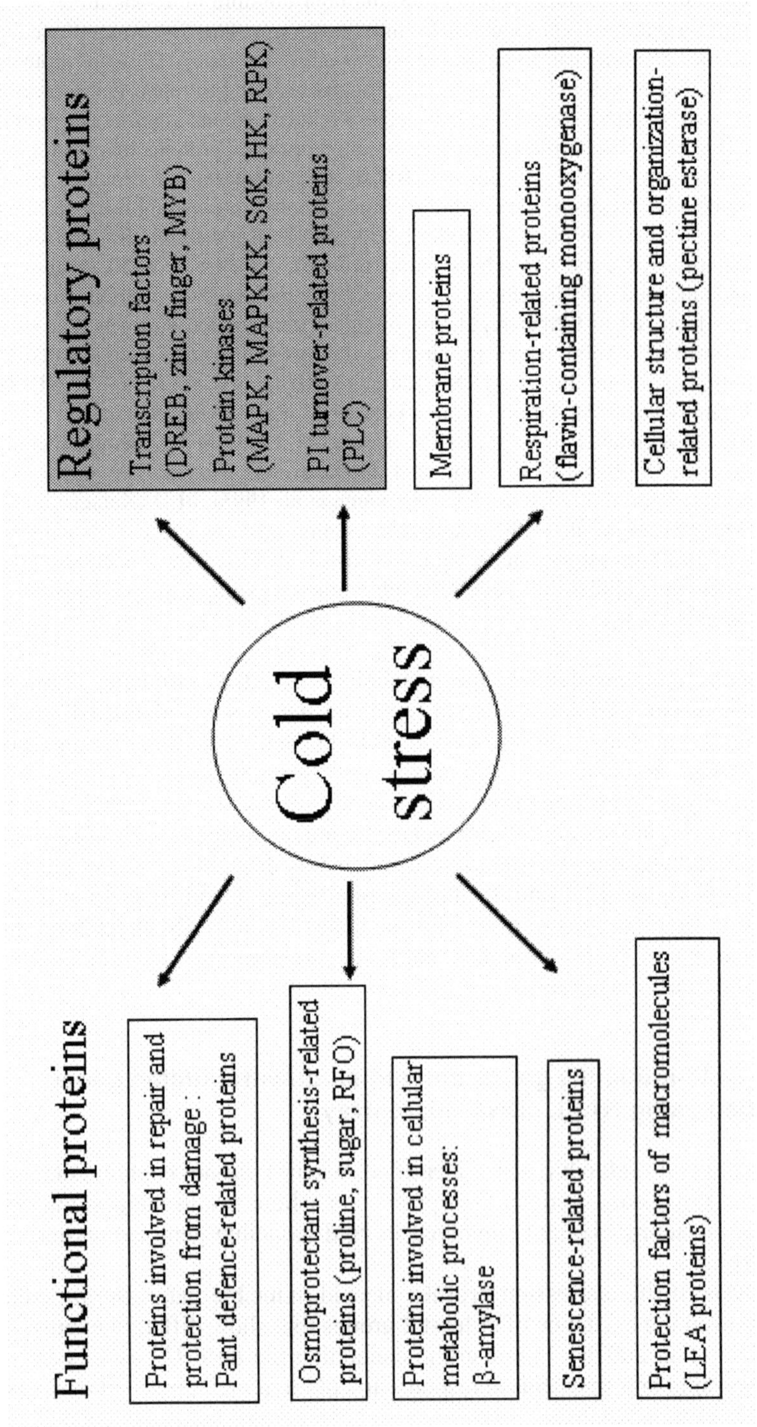
Functional proteins
Proteins involved in repair and protection from damage : Pant defence-related proteins
Osmoprotectant synthesis-related proteins (proline, sugar, RFO)
Proteins involved in cellular metabolic processes: β-amylase
Senescence-related proteins
Protection factors of macromolecules (LEA proteins)
Cold stress
Regulatory proteins
Transcription factors (DREB, zinc finger, MYB)
Protein kinases (MAPK, MAPKKK, S6K, HK, RPK)
PI turnover-related proteins (PLC)
Membrane proteins
Respiration-related proteins (flavin-containing monooxygenase)
Cellular structure and organization-related proteins (pectine esterase)

Fig. 4. Cold stress-inducible genes and their possible functions in stress tolerance and response. Gene products are classified into two groups. The first group (Functional proteins) includes proteins that probably function in stress tolerance. They are protection factors such as LEA proteins, proteins involved in repair and protection from damages, such as plant defense-related proteins, membrane proteins, proteins involved in synthesis of osmoprotectant (proline, sugars and RFO), proteins involved in cellular metabolic processes, such as β-amylase, cellular structure and organization-related proteins such as pectine esterase, senescence-related proteins, and respiration-related proteins such as flavin-containing monooxygenase. The second group (Regulatory proteins) contains protein factors involved in further regulation of signal transduction and gene expression that probably function in stress response. They are transcription factors such as DREB family, zinc finger family, MYB family, protein kinases such as MAPK (Mizoguchi et al. 1996), MAPKKK (Mizoguchi et al. 1996), S6K (Mizoguchi et al. 1996), HK (Urao et al. 1999) and RPK (Hong et al. 1997), PI turnover-related proteins, such as PLC (Hirayama et al. 1995). The list of the cold stress-inducible genes identified by the cDNA microarray analysis (Seki et al. 2002b) is available at http://pfgweb.gsc.riken.go.jp/index.html (as supplemental table 3).

2 hours, and then decreased (Seki et al. 2002b). In the other group containing DREB1A target genes, such as *rd29A*, *erd10, cor15A, rd17, kin2*, and RAFL06-16-B22 genes, their expression increased slowly and gradually after cold treatment within 10 hours (Seki et al. 2002b). The expression of the *DREB1A* gene during cold stress preceded that of the DREB1A target genes. These results support our previous results that *DREB1A* regulates the expression of the DREB1A target genes, such as *rd29A, erd10, cor15A, rd17, kin2*, and RAFL06-16-B22 genes (Kasuga et al. 1999; Seki et al. 2001a).

Analysis of stress-downregulated as well as stress-upregulated genes is important in understanding of molecular responses to abiotic stresses. We identified many drought-, high-salinity-, cold-stress-, or ABA-downregulated genes by microarray analysis (Seki et al. 2002b, 2002c). The list and the expression data on these drought-, cold-, high-salinity-stress-, or ABA-downregulated genes is available at http://www.gsc.riken.go.jp/Plant/index.html. Among the drought-, cold-, high-salinity-stress-, or ABA-downregulated genes, we found many photosynthesis-related genes, such as ribulose 1,5-bisphosphate carboxylase small subunit (*rbcS*), chlorophyll a/b-binding protein (*cab*), and the components of photosystem I and II. These results are consistent with the previous report that water stress inhibits photosynthesis (Tezara et al. 1999).

10.4.2 Application of RAFL cDNA microarray to study the expression profiles under abiotic stress conditions

Simpson et al. (2003) showed that a 14-bp region (CACTAAATTGTCAC; site-1-like sequence) from –599 to –586, and a *myc* recognition element (CATGTG) from –466 to –461 in the promoter region of the *erd1* gene encoding a regulatory subunit of Clp protease (Kiyosue et al., 1993, 1994; Nakashima et al., 1998) are responsible for gene expression during dehydration. To assess the frequency with which the sequence with homology to the core sequence of the site-1 motif, and the myc recognition element (CATGTG), occur together in the promoter regions of dehydration-inducible genes, a homology search for these two sequences within the promoter regions of dehydration-inducible genes was performed. Of the 100 drought-, cold-, high-salinity-stress-, or ABA-inducible genes (Seki et al. 2002b, 2002c) that show the greatest degree of induction by dehydration, 22 contained both the putative core motif of the site-1-like sequence (in either forward or complementary orientation) and the putative myc recognition sequence in their promoter regions (Simpson et al. 2003). Examination of the data revealed that just under 50% of the 22 genes show similar pattern of expression in response to dehydration, high salinity, ABA and cold treatment as that of the *erd1* gene; such that induction by dehydration > high salinity > ABA > cold, and that 21 have low levels of induction in response to cold stress. These results suggest that these sequences may also function as novel *cis*-acting elements in stress-responsive gene expression (Simpson et al. 2003).

The transgenic plants overexpressing AtMYC2 and/or AtMYB2 cDNAs have higher sensitivity to ABA (Abe et al. 2003). Abe et al. (2003) studied the expression profiles in the transgenic plants overexpressing AtMYC2 and/or AtMYB2 cDNAs using the RAFL cDNA microarray. mRNAs prepared from 35S:AtMYC2/AtMYB2 and wild type plants were used for the generation of Cy3-labeled and Cy5-labeled cDNA probes, respectively. Microarray analysis of the transgenic plants revealed that several ABA-inducible genes were upregulated in the 35S:AtMYC2/AtMYB2 transgenic plants. Abe et al. (2003) searched for the MYC recognition sequence (CANNTG) and the MYB recognition sequences (A/TAACCA and C/TAACG/TG) located within the 10- to 600-bp upstream region from each putative TATA box in the promoter regions of the 32 upregulated genes identified. Abe et al. (2003) found that 29 genes have the MYC recognition sequence, 29 genes have the MYB recognition sequence, and 26 genes have both MYC and MYB recognition sequences in their promoter regions. Ds insertion mutant of the *AtMYC2* gene was less sensitive to ABA and showed significantly decreased ABA-induced gene expression of *rd22* and *AtADH1*. These results indicated that both AtMYC2 and AtMYB2 function as transcriptional activators in ABA-inducible gene expression under drought stress conditions in plants.

In rice, Dubouzet et al. (2003) isolated five cDNAs for DREB homologs: *OsDREB1A*, *OsDREB1B*, *OsDREB1C*, *OsDREB1D*, and *OsDREB2A*. Expression of OsDREB1A and OsDREB1B was induced by cold stress, whereas expression of OsDREB2A was induced by dehydration and high-salinity stresses. The OsDREB1A and OsDREB2A proteins specifically bound to DRE and activated the transcription of the GUS reporter gene driven by DRE in rice protoplasts. Overexpression of OsDREB1A in transgenic *Arabidopsis* plants resulted in improved tolerance to drought, high-salinity, and freezing stresses, indicating that OsDREB1A has functional similarity to DREB1A (Dubouzet et al. 2003). Several OsDREB1A target genes were identified by the cDNA microarray and RNA gel blot analyses. Computer analysis showed that the seven OsDREB1A target genes have at least one core GCCGAC sequence as the DRE core motif in their promoter regions. Some of the DREB1A target genes such as *kin1*, *kin2*, and *erd10*, containing ACCGAC but not GCCGAC as the DRE core motifs in their promoter regions, were not overexpressed in the 35S:OsDREB1A plants. These results indicated that the OsDREB1A protein binds more preferentially to GCCGAC than to ACCGAC in the promoter regions, whereas the DREB1A protein binds to both GCCGAC and ACCGAC efficiently (Dubouzet et al. 2003).

Proline (Pro) is one of the most widely distributed osmolytes in water-stressed plants. L-Pro is metabolized to L-Glu via Δ^1-pyrroline-5-carboxylate (P5C) by two enzymes, Pro dehydrogenase (ProDH) and P5C dehydrogenase (Strizhov et al. 1997; Yoshiba et al. 1997). The ProDH gene in *Arabidopsis* is upregulated not only by rehydration after dehydration, but also by L-Pro and hypoosmolarity (Kiyosue et al. 1996; Nakashima et al. 1998). Satoh et al. (2002) analyzed the promoter regions of ProDH to identify *cis*-acting elements involved in L-Pro-induced and hypoosmolarity-induced expression in transgenic tobacco and *Arabidopsis* plants. Satoh et al. (2002) found that a 9-bp sequence, ACTCATCCT, in the ProDH promoter is necessary for the efficient expression of ProDH in response to L-Pro and hypoosmolarity and that the ACTCAT sequence is a core *cis*-acting element. To elucidate whether the promoter region of the other L-Pro-inducible genes have the ACTCAT sequence, Satoh et al. (2002) used the RAFL cDNA microarray and found that 27 L-Pro-inducible genes identified have the ACTCAT sequence in their promoter regions. 21 genes among the 27 genes showed L-Pro-inducible expression based on RNA gel blot analysis. These results suggest that the ACTCAT sequence is conserved in many L-Pro-inducible promoters and plays a key role in L-Pro-inducible gene expression. The microarray analysis also showed that some L-Pro-inducible genes do not have the ACTCAT sequences in their promoter regions, suggesting the existence of other *cis*-acting elements for L-Pro-inducible gene expression.

10.5 Application of *Arabidopsis* GeneChip to study the expression profiles under abiotic stress conditions

Recently, several studies on the expression profiling under abiotic stress conditions using *Arabidopsis* GeneChip provided by Affymetrix Co. (Zhu et al. 2001) have been published. The GeneChip used includes probes for 8,300 *Arabidopsis* genes and forty probes for spiking and negative controls (Zhu et al. 2001). For each gene, there are sixteen probe pairs (probe sets) including perfect match probes and mismatch probes to control for non-specific binding (Zhu et al. 2001). In this session, we also summarize the studies on the expression profiling under abiotic stress conditions using *Arabidopsis* GeneChip.

Recently, Kreps et al. (2002) studied the expression profiles in leaves and roots from *Arabidopsis* subjected to salt (100mM NaCl), hyperosmotic (200 mM mannitol), and cold (4°C) stress treatments. RNA samples were collected separately from leaves and roots after 3- and 27-hour stress treatments. Kreps et al. (2002) identified a total of 2,409 unique stress-regulated genes that displayed a greater than 2-fold change in expression compared with control. The results suggested the majority of changes were each stress-specific. At the 3-hour time point, less than

5% (118 genes) of the changes were observed as shared by all three stress responses, and by 27 hours, the number of shared responses was reduced more than 10-fold (< 0.5%). Roots and leaves displayed very different changes. For example, less than 14% of the cold-specific changes were shared between roots and leaves at both 3 and 27 hours. The gene with the largest induction under all three stress treatments was *rd29A/lti78/cor78*, with induction levels in roots greater than 250-fold for cold, 40-fold for mannitol, and 57-fold for NaCl. Kreps et al. (2002) identified 306 stress-regulated genes among the 453 known circadian controlled genes (Harmer et al. 2000). These results suggested that ca. 68% of the circadian controlled genes are linked to a stress response pathway and supported the hypothesis that an important function of the circadian clock is to "anticipate" predictable stresses such as cold nights.

Chen et al. (2002) used the expression profiles generated from the GeneChip experiments to deduce the functions of genes encoding known and putative *Arabidopsis* transcription factors. The expression levels of the 402 transcription factor genes were monitored in various organs, at different developmental stages, and under various biotic and abiotic stresses. A two-dimensional matrix (genes versus treatments or developmental stages/tissues) describing the changes in the mRNA levels of the 402 transcription factor genes was constructed for these experiments. The data represent 19 independent experiments, with samples derived from different organs such as roots, leaves, inflorescence stems, flowers, and siliques and at different developmental stages (Zhu et al. 2001) and >80 experiments representing 57 independent treatments with cold, salt, osmoticum, wounding, jasmonic acid, and different types of pathogens at different time points. The results showed that the transcription factors potentially controlling downstream gene expression in stress signal transduction pathways were identified by observed activation and repression of the genes after certain stress treatments and that the mRNA levels of a number of previously characterized transcription factor genes were changed significantly in connection with other regulatory pathways, suggesting their multifunctional nature (Chen et al. 2002). Among the 43 transcription factor genes that are induced during senescence, 28 of them also are induced by stress treatment, suggesting that the signaling pathway activated by senescence may overlap substantially with the stress signaling pathways (Chen et al. 2002). The statistical analysis of the promoter regions of the genes responsive to cold stress indicated that two elements, the ABRE-like element and the DRE-like element (Shinozaki and Yamaguchi-Shinozaki 2000) occur at significantly higher frequencies in the promoters from the late cold response cluster than their average frequency in all of the promoters of the genes on the *Arabidopsis* Genechip (Chen et al. 2002). These results suggest that ABRE-like element and DRE-like element are two major elements that are important for the transcriptional regulation of genes in the late cold response cluster.

Hugouvieux et al. (2001) isolated a recessive ABA hypersensitive *Arabidopsis* mutant, *abh1. ABH1* encodes a functional mRNA cap binding protein. DNA chip experiments showed that 18 genes including *RD20, KIN2*, and *COR15b* had significant and 3-fold reduced transcript levels in the *abh1* mutant, and 7 of these genes are ABA-regulated in the wild type. Consistent with these results, *abh1* plants showed ABA-hypersensitive stomatal closing and reduced wilting during drought. Hugouvieux et al. (2001) showed ABA-hypersensitive cytosolic calcium increases in *abh1* guard cells. These results indicate a functional link between mRNA processing and modulation of early ABA signal transduction.

Recently, Fowler and Thomashow (2002) identified 306 cold-regulated genes and 41 *DREB/CBF*-regulated genes using Affymetrix Gene Chip (Fowler and Thomashow 2002). This report expanded on the findings of Seki et al. (2001a) by describing the expression of ca. 8,000 genes after transfer of plants to cold temperature. Several differences between our results (Seki et al. 2001a, 2002b) and those of Fowler and Thomashow exist. This difference may be due to differences in gene annotation, expression profiling methods, ecotypes used and plant growth conditions. As the original Affymetrix annotation table was outdated, the Affymetrix annotation table should be updated with the improved *Arabidopsis* genome annotation data (Ghassemian et al. 2001; Garcia-Hernandez et al. 2002), and then the expression profiling data should be rechecked.

10.6 Abiotic stress-inducible genes identified using microarrays in monocots

Several other similar studies reported gene expression profile analysis under abiotic stress in other plant species, such as rice (Bohnert et al. 2001; Kawasaki et al. 2001) and barley (Ozturk et al. 2002). Kawasaki et al. (2001) analyzed the expression profiles using cDNA microarray including ca. 1700 cDNAs under salt stress conditions in rice and reported similar results that transcripts of protease inhibitor, beta-glucosidase, detoxification enzyme, water channel protein, and protein synthesis-related genes are upregulated after salt stress. Ozturk et al. (2002) analyzed the expression profiles using cDNA microarray including ca. 1500 cDNAs under drought and salt stressed conditions in barley and also reported similar results that transcripts of Δ1-pyrroline-5-carboxylate synthetase (P5CS) and ERD1 homologs in barley are upregulated after drought and salt stress treatments.

10.7 Conclusions and perspectives

The cDNA microarray analysis includes useful material with which to analyze the expression pattern of *Arabidopsis* genes under drought-, cold-, or high-salinity-stresses, to identify target genes of stress-related transcription factors, and to identify potential *cis*-acting DNA elements by combining the expression data with the genomic sequence data. By the expression profiling approach, more than 300 drought-, cold-, or high-salinity-stress-inducible genes and 40 drought-, cold-, or high-salinity-stress-inducible transcription factor genes have been identified, suggesting that various transcriptional regulatory mechanisms function in these stress signal transduction pathways. Functional analysis of these drought-, cold-, or high-salinity-stress-inducible genes should provide more information on the signal transduction in these stress responses.

By genetic approaches and biochemical analyses of signal transduction and stress tolerance of drought, cold, and high-salinity stress, several mutants on the signal transduction and stress tolerance of these stresses have been identified (Browse and Xin 2001; Finkelstein et al. 2002; Knight and Knight 2001; Xiong et al. 2002; Zhu 2002). In a genetic screen using a firefly luciferase reporter gene (*LUC*) under the control of the *RD29A* promoter, Zhu's group isolated several *Arabidopsis* mutants with altered induction of stress-responsive genes under drought, high-salinity, cold and ABA treatments (Ishitani et al. 1997). Compared with wild type *RD29A-LUC* plants, mutants exhibited either a constitutive (*cos*), high (*hos*), or low (*los*) level of *RD29A-LUC* expression in response to various stress or ABA treatments (Ishitani et al. 1997; Xiong and Zhu 2001, 2002; Xiong et al. 2002; Zhu 2002). These mutants might be involved in the activation of the DRE/CRT class genes. The *Arabidopsis salt overly sensitive (sos)* mutants (*sos1, sos2, sos3* and *sos4*) were also identified by genetic screening for seedlings that were hypersensitive to growth inhibition by NaCl stress (Liu and Zhu 1998; Liu et al. 2000; Shi et al. 2000, 2002). The *sos1*, *sos2* and *sos3* mutants are hypersensitive to salt stress, but activation of the DRE/CRT class of genes seems to be unchanged in them. Reverse genetic approaches, such as transgenic analyses, have also become useful for studying the function of the signaling components (Apse and Blumwald 2002; Finkelstein et al. 2002; Gong et al. 2002; Guo et al. 2002; Hasegawa et al. 2000; Iuchi et al. 2001; Xiong et al. 2002).

The availability of the *Arabidopsis* genome sequence will not only greatly facilitate the isolation of mutations identified by the above genetic screen, but also offer many other useful opportunities to study stress signal transduction. Genome-wide expression profiling of the stress-resistant or stress-sensitive mutants, and mutants on the stress signal transduction should help identify more genes that are regulated at the transcriptional level by the signaling components. Moreover, full-length cDNAs (Seki et al. 2002a) are useful resources for transgenic analyses, such as overexpression, antisense suppression, and double-stranded RNA interference (dsRNAi) and biochemical analyses to study the function of the encoded proteins. T-DNA- or transposon-knockout mutants also offer the opportunity to study the function of the genes. Genome-wide protein interaction studies will help to identify the interactions among signaling components and to construct the signal networks 'dissected' with the above genetic analysis. The information generated by focused studies of gene function in *Arabidopsis* will be the springboard for a new wave of strategies to improve the dehydration, high-salinity, and cold tolerance in agriculturally important crops.

Acknowledgements

This work was supported in part by a grant for Genome Research from RIKEN, the Program for Promotion of Basic Research Activities for Innovative Biosciences, the Special Coordination Fund of the Science and Technology Agency, and a Grant-in-Aid from the Ministry of Education, Culture, Sports, Science and Technology of Japan (MECSST) to K.S. It was also supported in part by a Grant-in-Aid for Scientific Research on Priority Areas (C) 'Genome Science' from MECSST to M.S.

References

Abe H, Yamaguchi-Shinozaki K, Urao T, Iwasaki T, Hosokawa D, Shinozaki K (1997) Role of *Arabidopsis* MYC and MYB homologs in drought- and abscisic acid-regulated gene expression. Plant Cell 9:1859-1868

Abe H, Urao T, Ito T, Seki M, Shinozaki K, Yamaguchi-Shinozaki K (2003) *Arabidopsis* AtMYC2 (bHLH) and AtMYB2 (MYB) Function as Transcriptional Activators in Abscisic Acid Signaling. Plant Cell 15:63-78

Apse MP and Blumwald E (2002) Engineering salt tolerance in plants. Curr Opin Biotechnol 13:146-150

Bajaj S, Targolli J, Liu LF, Ho THD, Wu R (1999) Transgenic approaches to increase dehydration-stress tolerance in plants. Mol Breed 5:493-503

Bohnert HJ, Ayoubi P, Borchert C et al. (2001) A genomics approach towards salt stress tolerance. Plant Physiol Biochem 39:295-311

Bray EA (1997) Plant responses to water deficit. Trends Plant Sci 2:48-54

Browse J, Xin Z (2001) Temperature sensing and cold acclimation. Curr Opin Plant Biol 4:241-246

Carninci P, Kvam C, Kitamura A, Ohsumi T, Okazaki Y, Itoh M, Kamiya M, Shibata K, Sasaki N, Izawa M et al. (1996) High-efficiency full-length cDNA cloning by biotinylated CAP trapper. Genomics 37:327-336

Carninci P, Westover A, Nishiyama Y, Ohsumi T, Itoh M, Nagaoka S, Sasaki N, Okazaki Y, Muramatsu M, Schneider C et al. (1997) High efficiency selection of full-length cDNA by improved biotinylated cap trapper. DNA Research 4:61-66

Carninci P, Nishiyama Y, Westover A, Itoh M, Nagaoka S, Sasaki N, Okazaki Y, Muramatsu M, Hayashizaki Y (1998) Thermostabilization and thermoactivation of thermolabile enzymes by trehalose and its application for the synthesis of full length cDNA. Proc Natl Acad Sci USA 95:520-524

Chen W, Provart NJ, Glazebrook J, Katagiri F, Chang HS, Eulgem T, Mauch F, Luan S, Zou G, Whitham SA, Budworth PR, Tao Y, Xie Z, Chen X, Lam S, Kreps JA, Harper JF, Si-Ammour A, Mauch-Mani B, Heinlein M, Kobayashi K, Hohn T, Dangl JL, Wang X, Zhu T (2002) Expression profile matrix of *Arabidopsis* transcription factor genes suggests their putative functions in response to environmental stresses. Plant Cell 14:559-574

Choi H, Hong JH, Ha J, Kang JY, Kim SY (2000) ABFs, a family of ABA-responsive element binding factors. J Biol Chem 275:1723-1730

Cushman JC and Bohnert HJ (2000) Genomic approaches to plant stress tolerance. Curr Opin Plant Biol 3:117-124

Dubouzet JG, Sakuma Y, Ito Y, Kasuga M, Dubouzet EG, Miura S, Seki M, Shinozaki K, Yamaguchi-Shinozaki K (2003) OsDREB genes in rice, *Oryza sativa* L., encode transcription activators that function in drought-, high-salt- and cold-responsive gene expression. Plant J 33:751-763

Eisen MB and Brown PO (1999) DNA arrays for analysis of gene expression. Methods Enzymol 303:179-205

Finkelstein RR, Gampala SSL, Rock CD (2002) Abscisic acid signaling in seeds. Plant Cell 14:S15-S45

Fowler S and Thomashow MF (2002) *Arabidopsis* transcriptome profiling indicates that multiple regulatory pathways are activated during cold acclimation in addition to the CBF cold response pathway. Plant Cell 14:1675-1690

Garcia-Hernandez M, Berardini TZ, Chen G, Crist D, Doyle A, Huala E, Knee E, Lambrecht M, Miller N, Mueller LA, Mundodi S, Reiser L, Rhee SY, Scholl R, Tacklind J, Weems DC, Wu Y, Xu I, Yoo D, Yoon J, Zhang P (2002) TAIR: a resource for integrated *Arabidopsis* data. Funct Integr Genomics 2: 239-253

Ghassemian M, Waner D, Tchieu J, Gribskov M, Schroeder JI (2001) An integrated *Arabidopsis* annotation database for Affymetrix Genechip data analysis, and tools for regulatory motif searches. Trends Plant Sci 6:448-449

Gong D, Zhang C, Chen X, Gong Z, Zhu, JK (2002) Constitutive activation and transgenic evaluation of the function of an *Arabidopsis* PKS protein kinase. J Biol Chem 277:42088-42096

Guo Y, Xiong L, Song CP, Gong D, Halfter U, Zhu JK (2002) A calcium sensor and its interacting protein kinase are global regulators of abscisic acid signaling in *Arabidopsis.* Dev Cell 3: 233-244

Harmer SL, Hogenesch JB, Straume M, Chang HS, Han B, Zhu T, Wang X, Kreps JA, Kay SA (2000) Orchestrated transcription of key pathways in *Arabidopsis* by the circadian clock. Science 290:2110-2113

Hasegawa PM, Bressan RA, Zhu JK, Bohnert HJ (2000) Plant cellular and molecular responses to high salinity. Annu Rev Plant Physiol Plant Mol Biol 51:463-499

Higo K, Ugawa Y, Iwamoto M, Korenaga T (1999) Plant *cis*-acting regulatory DNA elements (PLACE) database. Nucleic Acids Res 27:297-300

Himmelbach A, Hoffmann T, Leube M, Hohener B, Grill E (2002) Homeodomain protein ATHB6 is a target of the protein phosphatase ABI1 and regulates hormone responses in *Arabidopsis*. EMBO J 21: 3029-3038

Hirayama T, Ohto C, Mizoguchi T, Shinozaki K (1995) A gene encoding a phosphatidylinositol-specific phospholipase C is induced by dehydration and salt stress in *Arabidopsis thaliana.* Proc Natl Acad Sci USA 92: 3903-3907

Holmberg N and Bülow L (1998) Improving stress tolerance in plants by gene transfer. Trends Plant Sci 3:61-66

Hong SW, Jon JH, Kwak JM, Nam HG (1997) Identification of a receptor-like protein kinase gene rapidly induced by abscisic acid, dehydration, high salt, and cold treatments in *Arabidopsis thaliana.* Plant Physiol 113:1203-1212

Hugouvieux V, Kwak JM, Schroeder JI (2001) An mRNA cap binding protein, ABH1, modulates early abscisic acid signal transduction in *Arabidopsis*. Cell 106:477-487

Ishitani M, Xiong L, Stevenson B, Zhu JK (1997) Genetic analysis of osmotic and cold stress signal transduction in *Arabidopsis*: interactions and convergence of abscisic acid-dependent and abscisic acid-independent pathways. Plant Cell 9:1935-1949

Iuchi S, Kobayashi M, Taji T, Naramoto M, Seki M, Kato T, Tabata S, Kakubari Y, Yamaguchi-Shinozaki K, Shinozaki K (2001) Regulation of drought tolerance by gene manipulation of 9-*cis*-epoxycarotenoid dioxygenase, a key enzyme in abscisic acid biosynthesis, in *Arabidopsis*. Plant J 27:325-333

Jaglo-Ottosen KR, Gilmour SJ, Zarka DG, Schabenberger O, Thomashow MF (1998) *Arabidopsis* CBF1 overexpression induces *cor* genes and enhances freezing tolerance. Science 280:104-106

Kang JY, Choi HI, Im MY, Kim SY (2002) *Arabidopsis* basic leucine zipper proteins that mediate stress-responsive abscisic acid signaling. Plant Cell 14:343-357

Kasuga M, Liu Q, Miura S, Yamaguchi-Shinozaki K, Shinozaki K (1999) Improving plant drought, salt, and freezing tolerance by gene transfer of a single stress-inducible transcription factor. Nature Biotechnol 17:287-291

Katagiri T, Takahashi S, Shinozaki K (2001) Involvement of a novel *Arabidopsis* phospholipase D, AtPLDdelta, in dehydration-inducible accumulation of phosphatidic acid in stress signaling. Plant J 26:595-605

Kawasaki S, Borchert C, Deyholos M, Wang H, Brazille S, Kawai K, Galbraith D, Bohnert H (2001) Gene expression profiles during the initial phase of salt stress in rice. Plant Cell 13:889-905

Kim HJ, Kim YK, Park JY, Kim J (2002) Light signaling mediated by phytochrome plays an important role in cold-induced gene expression through the C-repeat/dehydration responsive element (C/DRE) in *Arabidopsis thaliana*. Plant J 29:693-704

Kiyosue T, Yamaguch-Shinozaki K, Shinozaki K (1993) Characterization of a cDNA for for a dehydration-inducible gene that encodes a Clp A, B-like protein in *Arabidopsis thaliana* L. Biochem Biophys Res Commun 196:1214-1220

Kiyosue T, Yamaguchi-Shinozaki K, Shinozaki K (1994) Cloning of cDNA for genes that are early responsive to dehydration-stress (ERDs) in *Arabidopsis thaliana* L. identification of three ERDs as HSP cognate genes. Plant Mol. Biol 25:791-798

Kiyosue T, Yoshiba Y, Yamaguchi-Shinozaki K, Shinozaki K (1996) A nuclear gene encoding mitochondrial proline dehydrogenase, an enzyme involved in proline metabolism, is upregulated by proline but downregulated by dehydration in *Arabidopsis*. Plant Cell 8:1323-1335

Knight H and Knight MR (2001) Abiotic stress signalling pathways: specificity and cross-talk. Trends Plant Sci 6: 262-267

Kreps JA, Wu Y, Chang HS, Zhu T, Wang X, Harper JF (2002) Transcriptome changes for Arabidopsis in response to salt, osmotic, and cold stress. Plant Physiol 130:2129-2141

Liu J, Ishitani M, Halfter U, Kim CS, Zhu JK (2000) The *Arabidopsis thaliana SOS2* gene encodes a protein kinase that is required for salt tolerance. Proc Natl Acad Sci USA 97:3730-3734

Liu J and Zhu JK (1998) A calcium sensor homolog required for plant salt tolerance. Science 280:1943-1945

Liu Q, Kasuga M, Sakuma Y, Abe H, Miura S, Yamaguchi-Shinozaki K, Shinozaki K (1998) The transcription factors, DREB1 and DREB2, with an EREBP/AP2 DNA binding domain separate two cellular signal transduction pathways in drought- and low-temperature-responsive gene expression, respectively, in *Arabidopsis*. Plant Cell 10:1391-1406

Mikami K, Katagiri T, Iuchi S, Yamaguchi-Shinozaki K, Shinozaki (1998) A gene encoding phosphatidylinositol-4-phosphate 5-kinase is induced by water stress and abscisic acid in *Arabidopsis thaliana*. Plant J 15:563-568

Mizoguchi T, Irie K, Hirayama T, Hayashida N, Yamaguchi-Shinozaki K, Matsumoto K, Shinozaki K (1996) A gene encoding a mitogen-activated protein kinase kinase kinase is induced simultaneously with genes for a mitogen-activated protein kinase and an S6 ribosomal protein kinase by touch, cold, and water stress in *Arabidopsis thaliana*. Proc Natl Acad Sci USA 93:765-769

Nakashima K, Satoh R, Kiyosue T, Yamaguchi-Shinozaki K, Shinozaki K (1998) A gene encoding proline dehydrogenase is not only induced by proline and hypoosmolarity, but is also developmentally regulated in the reproductive organs of *Arabidopsis*. Plant Physiol 118:1233-1241

Ozturk ZN, Talame V, Deyholos M, Michalowski CB, Galbraith DW, Gozukirmizi N, Tuberosa R, Bohnert HJ (2002) Monitoring large-scale changes in transcript abundance in drought- and salt-stressed barley. Plant Mol Biol 48:551-573

Richmond T and Somerville S (2000) Chasing the dream: plant EST microarrays. Curr Opin Plant Biol 3:108-116

Sakuma Y, Liu Q, Dubouzet JG, Abe H, Shinozaki K, Yamaguchi-Shinozaki K (2002) DNA-binding specificity of the ERF/AP2 domain of *Arabidopsis* DREBs, transcription factors involved in dehydration- and cold-inducible gene expression. Biochem Biophys Res Com 290:998-1009

Satoh R, Nakashima K, Seki M, Shinozaki K, Yamaguchi-Shinozaki K (2002) ACTCAT, a novel *cis*-acting element for proline- and hypoosmolarity-responsive expression of the *ProDH* gene encoding proline dehydrogenase in *Arabidopsis*. Plant Physiol 130:709-719

Seki M, Carninci P, Nishiyama Y, Hayashizaki Y, Shinozaki K (1998) High-efficiency cloning of *Arabidopsis* full-length cDNA by biotinylated CAP trapper. Plant J 15:707-720

Seki M, Narusaka M, Abe H, Kasuga M, Yamaguchi-Shinozaki K, Carninci P, Hayashizaki Y, Shinozaki K (2001a) Monitoring the expression pattern of 1300 *Arabidopsis* genes under drought and cold stresses using a full-length cDNA microarray. Plant Cell 13:61-72

Seki M, Narusaka M, Yamaguchi-Shinozaki K, Carninci P, Kawai J, Hayashizaki Y, Shinozaki K (2001b) *Arabidopsis* encyclopedia using full-length cDNAs and its application. Plant Physiol Biochem 39:211-220

Seki M, Narusaka M, Kamiya A, Ishida J, Satou M, Sakurai T, Nakajima M, Enju A, Akiyama K, Oono Y, Muramatsu M, Hayashizaki Y, Kawai J, Carninci P, Itoh M, Ishii Y, Arakawa T, Shibata K, Shinagawa A, Shinozaki K (2002a) Functional annotation of a full-length *Arabidopsis* cDNA collection. Science 296:141-145

Seki M, Narusaka M, Ishida, J, Nanjo T, Fujita M, Oono Y, Kamiya A, Nakajima M, Enju A, Sakurai T, Satou M, Akiyama K, Taji T, Yamaguchi-Shinozaki K, Carninci P, Kawai J, Hayashizaki Y, Shinozaki K (2002b) Monitoring the expression profiles of 7000 *Arabidopsis* genes under drought, cold, and high-salinity stresses using a full-length cDNA microarray. Plant J 31:279-292

Seki M, Ishida J, Narusaka M, Fujita M, Nanjo T, Umezawa T, Kamiya A, Nakajima M, Enju A, Sakurai T, Satou M, Akiyama K, Yamaguchi-Shinozaki K, Carninci P, Kawai J, Hayashizaki Y, Shinozaki K (2002c) Monitoring the expression pattern of ca. 7000

Arabidopsis genes under ABA treatments using a full-length cDNA microarray. Funct Integr Genom 2:282-291

Seki M, Satou M, Sakurai T, Shinozaki K (2002d) RIKEN *Arabidopsis* full-length (RAFL) cDNA database. Trends Plant Sci 7:562-563

Shi H, Ishitani M, Kim C, Zhu JK (2000) The *Arabidopsis thaliana* salt tolerance gene *SOS1* encodes a putative Na^+/H^+ antiporter. Proc Natl Acad Sci USA 97:6896-6901

Shi H, Xiong L, Stevenson B, Lu T, Zhu JK (2002) The *Arabidopsis salt ovary sensitive 4* mutants uncover a critical role for vitamin B6 in plant salt tolerance. Plant Cell 14:575-588

Shinozaki K and Yamaguchi-Shinozaki K (1999) Molecular responses to drought stress. In *Molecular Responses to Cold, Drought, Heat and Salt Stress in Higher Plants*. Edited by Shinozaki, K. and Yamaguchi-Shinozaki, K. Austin, TX: RG Landes; pp. 11-28

Shinozaki K and Yamaguchi-Shinozaki K (2000) Molecular responses to dehydration and low temperature: Differences and cross-talk between two stress signaling pathways. Curr Opin Plant Biol 3:217-223

Simpson SD, Nakashima K, Narusaka Y, Seki M, Shinozaki K, Yamaguchi-Shinozaki K (2003) Two different novel *cis*-acting elements of *erd1*, a *clpA* homologous *Arabidopsis* gene function in induction by dehydration stress and dark-induced senescence. Plant J 33:259-270

Stockinger EJ, Gilmour SJ, Thomashow MF (1997) *Arabidopsis thaliana CBF1* encodes an AP2 domain-containing transcription activator that binds to the C-repeat/DRE, a *cis*-acting DNA regulatory element that stimulates transcription in response to low temperature and water deficit. Proc Natl Acad Sci USA 94:1035-1040

Strizhov N, Abraham E, Okresz L, Blickling S, Zilberstein A, Schell J, Koncz C, Szabados L (1997) Differential expression of two P5CS genes controlling proline accumulation during salt-stress requires ABA and is regulated by *ABA1*, *ABI1* and *AXR2* in *Arabidopsis*. Plant J 12:557-569

Taji T, Seki M, Yamaguchi-Shinozaki K, Kamada H, Giraudat J, Shinozaki K (1999) Mapping of 25 drought-inducible genes, RD and ERD, in *Arabidopsis thaliana*. Plant Cell Physiol 40:119-123

Taji T, Ohsumi C, Iuchi S, Seki M, Kasuga M, Kobayashi M, Yamaguchi-Shinozaki K, Shinozaki K (2002) Important roles of drought- and cold-inducible genes for galactinol synthase in stress tolerance in *Arabidopsis thaliana*. Plant J 29:417-426

Tezara W, Mitchell VJ, Driscoll SD, Lawlor DW (1999) Water stress inhibits plant photosynthesis by decreasing coupling factor and ATP. Nature 401:914-917

Thomas H and Stoddart JL (1980) Leaf senescence. Annu Rev Plant Physiol 31:83-111

Thomashow MF (1999) Plant cold acclimation: freezing tolerance genes and regulatory mechanisms. Annu Rev Plant Physiol Plant Mol Biol 50: 571-599

Torres-Schumann S, Godoy JA, Pintor-Toro JA (1992) A probable lipid transfer protein gene is induced by NaCl in stems of tomato plants. Plant Mol Biol 18:749-757

Uno Y, Furihata T, Abe H, Yoshida R, Shinozaki K, Yamaguchi-Shinozaki K (2000) *Arabidopsis* basic leucine zipper transcription factors involved in an abscisic acid-dependent signal transduction pathway under drought and high-salinity conditions. Proc Natl Acad Sci USA 97: 11632-11637

Urao T, Katagiri T, Mizoguchi T, Yamaguchi-Shinozaki K, Hayashida N, Shinozaki K (1994) Two genes that encode Ca^{2+}-dependent protein kinases are induced by drought and high-salt stresses in *Arabidopsis thaliana*. Mol Gen Genet 244:331-340

Urao T, Yakubov B, Satoh R, Yamaguchi-Shinozaki K, Seki M, Hirayama T, Shinozaki K (1999) A transmembrane hybrid-type histidine kinase in *Arabidopsis* functions as an osmosensor. Plant Cell 11:1743-1754

Xiong L, Schumaker KS, Zhu JK (2002) Cell signaling during cold, drought, and salt stress. Plant Cell Suppl:S165-183

Xiong L and Zhu JK (2001) Abiotic stress signal transduction in plants: Molecular and genetic perspectives. Physiol Plant 112: 152-166

Xiong L and Zhu JK (2002) Molecular and genetic aspects of plant responses to osmotic stress. Plant Cell Environment 25: 131-139

Yamaguchi-Shinozaki K and Shinozaki K (1994) A novel *cis*-acting element in an *Arabidopsis* gene is involved in responsiveness to drought, low temperature, or high-salt stress. Plant Cell 6:251-264

Yoshiba Y, Kiyosue T, Nakashima K, Yamaguchi-Shinozaki K, Shinozaki K (1997) Regulation of levels of proline as an osmolyte in plants under water stress. Plant Cell Physiol 38:1095-1102

Zhu JK (2002) Salt and drought stress signal transduction in plants. Annu Rev Plant Biol 53: 247-273

Zhu T, Budworth P, Han B, Brown D, Chang HS, Zou G, Wang X (2001) Toward elucidating the global gene expression patterns of developing *Arabidopsis*: Parallel analysis of 8300 genes by a high-density oligonucleotide probe array. Plant Physiol Biochem 39:221-242

Abbreviations

ABA: abscisic acid
aba: ABA-deficient
ABF: ABRE-binding factor
abh: ABA-hypersensitive
abi: ABA-insensitive
ABRE: ABA-responsive element
AREB: ABRE-binding protein
AtGolS: *Arabidopsis* galactinol synthase
bHLH: basic helix-loop-helix
bp: base pairs
bZIP: basic-domain leucine zipper
CAB: chlorophyll a/b-binding protein
CaMV: cauliflower mosaic virus
CBF: C-repeat-binding factor
CDPK: calcium-dependent protein kinase
COR: cold-regulated
cos: constitutive expression of osmotically responsive genes
CRT: C-repeat
DGK: diacylglycerol kinase
DRE: dehydration-responsive element
DREB: DRE-binding protein

dsRNAi: double-stranded RNA interference
era: enhanced response to ABA
ERD: early responsive to dehydration
ERF: ethylene-responsive element binding factor
ESTs: expressed sequence tags
GST: glutathione-S-transferase
HD-ZIP: homeodomain-leucine zipper
HK: histidine kinase
hos: high expression of osmotically responsive genes
IAA: indole-3-acetic acid
JA: jasmonic acid
KIN: cold-inducible
LEA: late embryogenesis abundant
los: low expression of osmotically responsive genes
LTI: low-temperature-induced
LUC: firefly luciferase
MAPK: mitogen-activated protein kinase
MAPKKK: mitogen-activated protein kinase kinase kinase
PAP: phosphatidic acid phosphatase
P5CS: Δ^1-pyrroline-5-carboxylate synthetase
P5C: Δ^1-pyrroline-5-carboxylate
PIP5K: phosphatidylinositol-4-phosphate-5-kinase
PLC: phospholipase C
PLD: phospholipase D
PP2C: protein phosphatase 2C
ProDH: proline dehydrogenase
RAFL: RIKEN *Arabidopsis* full-length
rbcs: ribulose 1,5-bisphosphate carboxylase small subunit
RFO: raffinose family oligosaccharides
RD: responsive to dehydration
RPK: receptor-like protein kinase
sEH: soluble epoxide hydrolase
S6K: ribosomal protein S6 kinase
sos: salt overly sensitive

Index

MIX
Papier aus verantwortungsvollen Quellen
Paper from responsible sources
FSC® C105338

If you have any concerns about our products,
you can contact us on
ProductSafety@springernature.com

In case Publisher is established outside the EU,
the EU authorized representative is:
Springer Nature Customer Service Center GmbH
Europaplatz 3, 69115 Heidelberg, Germany

Printed by Libri Plureos GmbH
in Hamburg, Germany